Survival Skills

Department of Defense

Pentagon
Publishing

TABLE OF CONTENTS

Chapter 6

Water Procurement

Chapter 7

Firecraft

Chapter 8

Food Procurement

Chapter 9

Survival Use of Plants

Chapter 10

Poisonous Plants

Chapter 11

Dangerous Animals

Chapter 12

Field-Expedient Weapons, Tools, and Equipment

Chapter 13

Desert Survival

Chapter 14

Tropical Survival

Chapter 23

Survival in Man-Made Hazards

PREFACE

As a soldier you can be sent to any area of the world. It may be in a temperate, tropical, arctic, or subarctic region. You expect to have all your personal equipment and your unit or other members of your unit with you wherever you go. There is, however, no guarantee it will be so. You could find yourself alone in a remote area-possibly enemy territory— with little or no personal gear. This manual provides information and describes basic techniques to enable you to survive and return alive should you find yourself in such a situation.

If you are a trainer, use the information in this manual as a base on which to build survival training. You know the areas to which your unit is likely to be deployed, the means by which it will travel, and the territory through which or over which it will travel. Read what this manual says about survival in those particular areas. Find out all you can about those areas. Read other books on survival. Develop a survival training program that will enable the people in your unit to meet any survival situation they may face. It can make the difference between life and death.

We would like your comments or recommendations for improving this manual. Please make your comments on a DA Form 2028 (Recommended Changes to Publications and Blank Forms), keying your comments to specific pages and paragraphs and stating your reasons for the recommended change. Mail your comments to:

Commander
U.S. Army John F. Kennedy Special Warfare
Center and School
ATTN: AOJK-DT-DM
Fort Bragg, NC 28307-5000

Unless this publication states otherwise, masculine nouns and pronouns do not refer exclusively to men.

CHAPTER 1

INTRODUCTION

This manual is based entirely on the keyword ***SURVIVAL.*** *The letters in this word can help guide you in your actions in any survival situation. Whenever faced with a survival situation, remember the word* ***SURVIVAL***

SURVIVAL ACTIONS

The following paragraphs expand on the meaning of each letter of the word survival. Study and remember what each letter signifies because you may some day have to make it work for you.

S - Size Up the Situation

If you are in a combat situation, find a place where you can conceal yourself from the enemy. Remember, security takes priority. Use your senses of hearing, smell, and sight to get a feel for the battlefield. What is the enemy doing? Advancing? Holding in place? Retreating? You will have to consider what is developing on the battlefield when you make your survival plan.

Size Up Your Surroundings

Determine the pattern of the area. Get a feel for what is going on around you. Every environment, whether forest, jungle, or desert, has a rhythm or pattern. This rhythm or pattern includes animal and bird noises and movements and insect sounds. It may also include enemy traffic and civilian movements.

Size Up Your Physical Condition

The pressure of the battle you were in or the trauma of being in a survival situation may have caused you to overlook wounds you received. Check your wounds and give yourself first aid. Take care to prevent further bodily harm. For instance, in any climate, drink plenty of water to prevent dehydration. If you are in a cold or wet climate, put on additional clothing to prevent hypothermia.

Size Up Your Equipment

Perhaps in the heat of battle, you lost or damaged some of your equipment. Check to see what equipment you have and what condition it is in.

Now that you have sized up your situation, surroundings, physical condition, and equipment, you are ready to make your survival plan. In doing so, keep in mind your basic physical needs—water, food, and shelter.

U - Use All Your Senses, Undue Haste Makes Waste

You may make a wrong move when you react quickly without thinking or planning. That move may result in your capture or death. Don't move just for the sake of taking action. Consider all aspects of your situation (size up your situation) before you make a decision and a move. If you act in haste, you may forget or lose some of your equipment. In your haste you may also become disoriented so that you don't know which way to go. Plan your moves. Be ready to move out quickly without endangering yourself if the enemy is near you. Use all your senses to evaluate the situation. Note sounds and smells. Be sensitive to temperature changes. Be observant.

R - Remember Where You Are

Spot your location on your map and relate it to the surrounding terrain. This is a basic principle that you must always follow. If there are other persons with you, make sure they also know their location. Always know who in your group, vehicle, or aircraft has a map and compass. If that person is killed, you will have to get the map and compass from him. Pay close attention to where you are and to where you are going. Do not rely on others in the group to keep track of the route. Constantly orient yourself. Always try to determine, as a minimum, how your location relates to—

- The location of enemy units and controlled areas.
- The location of friendly units and controlled areas.
- The location of local water sources (especially important in the desert).
- Areas that will provide good cover and concealment.

This information will allow you to make intelligent decisions when you are in a survival and evasion situation.

V - Vanquish Fear and Panic

The greatest enemies in a combat survival and evasion situation are fear and panic. If uncontrolled, they can destroy your ability to make an intelligent decision. They may cause you to react to your feelings and imagination rather than to your situation. They can drain your energy and thereby cause other negative emotions. Previous survival and evasion training and self-confidence will enable you to vanquish fear and panic.

I - Improvise

In the United States, we have items available for all our needs. Many of these items are cheap to replace when damaged. Our easy come, easy go, easy-to-replace culture makes it unnecessary for us to improvise. This inexperience in improvisation can be an enemy in a survival situation. Learn to improvise. Take a tool designed for a specific purpose and see how many other uses you can make or it.

Learn to use natural objects around you for different needs. An example is using a rock for a hammer. No matter how complete a survival kit you have with you, it will run out or wear out after a while. Your imagination must take over when your kit wears out.

V - Value Living

All of us were born kicking and fighting to live, but we have become used to the soft life. We have become creatures of comfort. We dislike inconveniences and discomforts. What happens when we are faced with a survival situation with its stresses, inconveniences, and discomforts? This is when the will to live-placing a high value on living-is vital. The experience and knowledge you have gained through life and your Army training will have a bearing on your will to live. Stubbornness, a refusal to give in to problems and obstacles that face you, will give you the mental and physical strength to endure.

A - Act Like the Natives

The natives and animals of a region have adapted to their environment. To get a feel of the area, watch now the people go about their daily routine. When and what do they eat? When, where, and how do they get their food? When and where do they go for water? What time do they usually go to bed and get up? These actions are important to you when you are trying to avoid capture.

Animal life in the area can also give you clues on how to survive. Animals also require food, water, and shelter. By watching them, you can find sources of water and food.

> **WARNING**
>
> Animals cannot serve as an absolute guide to what you can eat and drink. Many animals eat plants that are toxic to humans.

Keep in mind that the reaction of animals can reveal your presence to the enemy.

If in a friendly area, one way you can gain rapport with the natives is to show interest in their tools and how they get food and water. By studying the people, you learn to respect them, you often make valuable friends, and, most important, you learn how to adapt to their environment and increase your chances of survival.

L - Live by Your Wits, *But for Now,* Learn Basic Skills

Without training in basic skills for surviving and evading on the battlefield, your chances of living through a combat survival and evasion situation are slight.

Learn these basic skills **now** -not when you are headed for or are in the battle. How you decide to equip yourself before deployment will impact on whether or not you survive. You need to know about the environment to which you are going, and you must practice basic skills geared to that environment. For instance, if you are going to a desert, you need to know how to get water in the desert.

Practice basic survival skills during all training programs and exercises. Survival training reduces fear of the unknown and gives you self-confidence. It teaches you to *live by your wits.*

S — **Size Up the Situation (Surroundings, Physical Condition, Equipment)**

U — **Use All Your Senses, Undue Haste Makes Waste**

R — **Remember Where You Are**

V — **Vanquish Fear and Panic**

I — **Improvise**

V — **Value Living**

A — **Act Like the Natives**

L — **Live by Your Wits, *But for Now,* Learn Basic Skills**

PATTERN FOR SURVIVAL

Develop a survival pattern that lets you beat the enemies of survival. This survival pattern must include food, water, shelter, fire, first aid, and signals placed in order of importance. For example, in a cold environment, you would need a *fire* to get warm; a *shelter* to protect you from the cold, wind, and rain or snow; traps or snares to get *food;* a means to *signal* friendly aircraft; and *first aid* to maintain health. *If injured, first aid has top priority* no matter what climate you are in.

Change your survival pattern to meet your immediate physical needs as the environment changes.

As you read the rest of this manual, keep in mind the keyword SURVIVAL and the need for a survival pattern.

C H A P T E R 2

PSYCHOLOGY OF SURVIVAL

It takes much more than the knowledge and skills to build shelters, get food, make fires, and travel without the aid of standard navigational devices to live successfully through a survival situation. Some people with little or no survival training have managed to survive life-threatening circumstances. Some people with survival training have not used their skills and died. A key ingredient in any survival situation is the mental attitude of the individual(s) involved. Having survival skills is important; having the will to survive is essential. Without a desk to survive, acquired skills serve little purpose and invaluable knowledge goes to waste.

There is a psychology to survival. The soldier in a survival environment faces many stresses that ultimately impact on his mind. These stresses can produce thoughts and emotions that, if poorly understood, can transform a confident, well-trained soldier into an indecisive, ineffective individual with questionable ability to survive. Thus, every soldier must be aware of and be able to recognize those stresses commonly associated with survival. Additionally, it is imperative that soldiers be aware of their reactions to the wide variety of stresses associated with survival. This chapter will identify and explain the nature of stress, the stresses of survival, and those internal reactions soldiers will naturally experience when faced with the stresses of a real-world survival situation. The knowledge you, the soldier, gain from this chapter and other chapters in this manual, will prepare you to come through the toughest times alive.

A LOOK AT STRESS

Before we can understand our psychological reactions in a survival setting, it is helpful to first know a little'bit about stress.

Stress is not a disease that you cure and eliminate. Instead, it is a condition we all experience. Stress can be described as our reaction to pressure. It is the name given to the experience we have as we physically, mentally, emotionally, and spiritually respond to life's tensions.

Need for Stress

We need stress because it has many positive benefits. Stress provides us with challenges; it gives us chances to learn about our values and strengths. Stress can snow our ability to handle pressure without breaking; if tests our adaptability and flexibility; it can stimulate us to do our best. Because we usually do not consider unimportant events stressful, stress can also be an excellent indicator of the significance we attach to an event—in other words, it highlights what is important to us.

We need to have some stress in our lives, but too much of anything can be bad. The goal is to have stress, but not an excess of it. Too much stress can take its toll on people and organizations. Too much stress leads to distress. Distress causes an uncomfortable tension that we trv to escape and, preferably, avoid. Listed below are a few of the common signs of distress you may find in your fellow soldiers or yourself when faced with too much stress:

- Difficulty making decisions.
- Angry outbursts.
- Forgetfulness.
- Low energy level.
- Constant worrying.
- Propensity for mistakes.
- Thoughts about death or suicide.
- Trouble getting along with others.
- Withdrawing from others.
- Hiding from responsibilities.
- Carelessness.

As you can see, stress can be constructive or destructive. It can encourage or discourage, move us along or stop us dead in our tracks, and make life meaningful or seemingly meaningless. Stress can inspire you to operate successfully and perform at your maximum efficiency in a survival situation. It can also cause you to panic and forget all your training. Key to your survival is your ability to manage the inevitable stresses you will encounter. The survivor is the soldier who works with his stresses instead of letting his stresses work on him.

Survival Stressors

Any event can lead to stress and, as everyone has experienced, events don't always come one at a time. Often, stressful events occur simultaneously. These events are not stress, but they produce it and are called "stressors." Stressors are the obvious cause while stress is the response. Once the body recognizes the presence of a stressor, it then begins to act to protect itself.

In response to a stressor, the body prepares either to "fight or flee." This preparation involves an internal SOS sent throughout the body. As the body responds to this SOS, several actions take place. The body releases stored fuels (sugar and fats) to provide quick energy; breathing rate increases to supply more oxygen to the blood; muscle tension increases to prepare for action; blood clotting mechanisms are activated to reduce bleeding from cuts; senses become more acute (hearing becomes more sensitive, eyes become big, smell becomes sharper) so that you are more aware of your surrounding and heart rate and blood pressure rise to provide more blood to the muscles. This protective posture lets a person cope with potential dangers; however, a person cannot maintain such a level of alertness indefinitely.

Stressors are not courteous; one stressor does not leave because another one arrives. Stressors add up. The cumulative effect of minor stressors can be a major distress if they all happen too close together. As the body's resistance to stress wears down and the sources of stress continue (or increase), eventually a state of exhaustion arrives. At this point, the ability to resist stress or use it in a positive way gives out and signs of distress appear. Anticipating stressors and developmg strategies to cope with them are two ingredients in the effective management of stress. It is therefore essential that the soldier in a survival setting be aware of the types of stressors he will encounter. Let's take a look at a few of these.

Injury, Illness, or Death

Injury, illness, and death are real possibilities a survivor has to face. Perhaps nothing is more stressful than being alone in an unfamiliar environment where you could die from hostile action, an accident, or from eating something lethal. Illness and injury can also add to stress by limiting your ability to maneuver, get food and drink, find shelter, and defendyourself. Even if illness and injury don't lead to death, they add to stress through the pain and discomfort they generate. It is only "by controlling the stress associated with the vulnerability to injury, illness, and death that a soldier can have the courage to take the risks associated with survival tasks.

Uncertainly and Lack of Control

Some people have trouble operating in settings where everything is not clear-cut. The only guarantee in a survival situation is that nothing is guaranteed. It can be extremely stressful operating on limited information in a setting where you have limited control ofyour surroundings. This uncertainty and lack of control also add to the stress of being ul, injured, or killed.

Environment

Even under the most ideal circumstances, nature is quite formidable. In survival, a soldier will have to contend with the stressors of weather, terrain, and the variety of creatures inhabiting an area. Heat, cold, rain, winds, mountains, swamps, deserts, insects, dangerous reptiles, and other animals are just a few of the challenges awaiting the soldier working to survive. Depending on how a soldier handles the stress of his environment, his surroundings can be either a source of food and protection or can be a cause of extreme discomfort leading to injury, illness, or death.

Hunger and Thirst

Without food and water a person will weaken and eventually die. Thus, getting and preserving food and water takes on increasing importance as the length or time in a survival setting increases. For a soldier used to havingTiis provisions issued, foraging can be a big source of stress.

Fatigue

Forcing yourself to continue surviving is not easy as you grow more tired. It is possible to become so fatigued that the act of just staying awake is stressful in itself.

Isolation

There are some advantages to facing adversity with others. As soldiers we learn individual skills, but we train to function as part of a team. Although we, as soldiers, complain about higher headquarters, we become used to the information and guidance it provides, especially during times of confusion. Being in contact with others also provides a greater sense of security and a feeling someone is available to help if problems occur. A significant stressor in survival situations is that often a person or team has to rely solely on its own resources.

The survival stressors mentioned in this section are by no means the only ones you may face. Remember, what is stressful to one person may not be stressful to another. Your experiences, training, personal outlook on life, physical and mental conditioning, and level of self-confidence contribute to what you will find stressfufin a survival environment. The object is not to avoid stress, but rather to manage the stressors of survival and make them work for you.

We now have a general knowledge of stress and the stressors common to survival; the next step is to examine our reactions to the stressors we may face.

NATURAL REACTIONS

Man has been able to survive many shifts in his environment throughout the centuries. His ability to adapt physically and mentally to a changing world kept him alive while other species around him gradually diecTofr. The same survival mechanisms that kept our forefathers alive can help keep us alive as well! However, these survival mechanisms that can help us can also work against us if we don't understand and anticipate their presence.

It is not surprising that the average person will have some psychological reactions in a survival situation. We will now examine some of the major internal reactions you and anyone with you might experience with the survival stressors addressed in the earlier paragraphs. Let's begin.

Fear

Fear is our emotional response to dangerous circumstances that we believe have the potential to cause death, injury, or illness. This harm is not just limited to physical damage; the threat to one's emotional and mental well-being can generate fear as well. For the soldier trying to survive, fear can have a positive function if it encourages him to be cautious in situations where recklessness could result m injury. Unfortunately, fear can also immobilize a person. It can cause him to become so frightened that he fails to perform activities essential for survival. Most soldiers will have some degree of fear when placed in unfamiliar surroundings under adverse conditions. There is no shame in this! Each soldier must train himself not to be overcome by his fears. Ideally, through realistic training, we can acquire the knowledge and skills needed to increase our confidence and thereby manage our fears.

Anxiety

Associated with fear is anxiety. Because it is natural for us to be afraid, it is also natural for us to experience anxiety. Anxiety can be an uneasy, apprehensive feeling we get when faced with dangerous situations (physical, mental, and emotional). When used in a healthy way, anxiety urges us to act to end, or at least master, the dangers that threaten our existence. If we were never anxious, there would be little motivation to make changes in our lives. The soldier in a survival setting reduces his anxiety by performing those tasks that will ensure his coming through the ordeal alive. As he reduces his anxiety, the soldier is also bringing under control the source of that anxiety-his fears. In this form, anxiety is good; however, anxiety can also have a devastating impact. Anxiety can overwhelm a soldier to the point where he becomes easily confused and has difficulty thinking. Once this happens, it becomes more and more difficult for him to make good judgments and sound decisions. To survive, the soldier must learn techniques to calm his anxieties and keep them in the range where they help, not hurt.

Anger and Frustration

Frustration arises when a person is continually thwarted in his attempts to reach a goal. The goal of survival is to stay alive until you can reach help or until help can reach you. To achieve this goal, the soldier must complete some tasks with minimal resources. It is inevitable, in trying to do these tasks, that something will go wrong; that something will happen beyond the soldier's control; and that with one's life at stake, every mistake is magnified in terms of its importance. Thus, sooner or later, soldiers will have to cope with frustration when a few of their plans run into trouble. One outgrowth of this frustration is anger. There are many events in a survival situation that can frustrate or anger a soldier. Getting lost, damaged or forgotten equipment, the weather, inhospitable terrain, enemy patrols, and physical limitations are just a few sources of frustration and anger. Frustration and anger encourage impulsive reactions, irrational behavior, poorly thought-out decisions, and, in some instances, an "I quit" attitude (people sometimes avoid doing something they can't master). If the soldier can harness and properly channel the emotional intensity associated with anger and frustration, he can productively act as he answers the challenges of survival. If the soldier does not properly focus his angry feelings, he can waste much energy in activities that do little to further either his chances of survival or the chances of those around him.

Depression

It would be a rare person indeed who would not get sad, at least momentarily, when faced with the privations of survival. As this sadness deepens, we label the feeling "depression." Depression is closely linked with frustration and anger. The frustrated person becomes more and more angry as he fails to reach his goals. If the anger does not help the person to succeed, then the frustration level goes even higher. A destructive cycle between anger and frustration continues until the person becomes worn down-physically, emotionally, and mentally. When a person reaches this point, he starts to give up, and his focus shifts from "What can I do" to "There is nothing I can do." Depression is an expression of this hopeless, helpless feeling. There is nothing wrong with being sad as you temporarily think about your loved ones and remember what life is like back in "civilization" or 'the world." Such thoughts, in fact, can give you the desire to try harder and live one more day.T)n the other hand, if you allow yourself to sink into a depressed state, then it can sap all your energy and, more important, your will to survive. It is imperative that each soldier resist succumbing to depression.

Loneliness and Boredom

Man is a social animal. This means we, as human beings, enjoy the company of others. Very few people want to be alone *all the time!* As you are aware, there is a distinct chance of isolation in a survival setting. This is not bad. Loneliness and boredom can bring to the surface qualities you thought only others had. The extent of your imagination and creativity may surprise you. When required to do so, you may discover some hidden talents and abilities. Most of all, you may tap into a reservoir of inner strength and fortitude you never knew you had. Conversely, loneliness and boredom can be another source of depression. As a soldier surviving alone, or with others, you must find ways to keep your mind productively occupied. Additionally, you must develop a degree of self-sufficiency. You must have faith in your capability to "go it alone."

Guilt

The circumstances leading to your being in a survival setting are sometimes dramatic and tragic. It may be the result of an accident or military mission where there was a loss of life. Perhaps you were the only, or one of a few, survivors. While naturally relieved to be alive, you simultaneously may be mourning the deaths of others who were less fortunate. It is not uncommon for survivors to feel guilty about being spared from death while others were not. This feeling, when used in a positive way, has encouraged people to try harder to survive with the belief they were allowed to live for some greater purpose in life. Sometimes, survivors tried to stay alive so that they could carry on the work of those killed. Whatever reason you give yourself, do not let guilt feelings prevent you from living. The living who abandon their chance to survive accomplish nothing. Such an act would be the greatest tragedy.

PREPARING YOURSELF

Your mission as a soldier in a survival situation is to stay alive. As you can see, you are going to experience an assortment of thoughts and emotions. These can work for you, or they can work to your downfall. Fear, anxiety, anger, frustration, guilt, depression, and loneliness are all possible reactions to the many stresses common to survival. These reactions, when controlled in a healthy way, help to increase a soldier's likelihood of surviving. They prompt the soldier to pay more attention in training, to fight back when scared, to take actions that ensure sustenance and security, to keep faith with his fellow soldiers, and to strive against large odds. When the survivor cannot control these reactions in a healthy way, they can bring him to a standstill. Instead of rallying his internal resources, the soldier listens to his internal fears. This soldier experiences psychological defeat long before he physically succumbs. Remember, survival is natural to everyone; beingunexpecfedly thrust into the life and death struggle of survival is notfDon't De afraid of your "natural reactions to this unnatural situation." Prepare yourself to rule over these reactions so they serve your ultimate interest—staying alive with the honor and dignity associated with being an American soldier.

It involves preparation to ensure that your reactions in a survival setting are productive, not destructive. The challenge of survival has produced countless examples of heroism, courage, and self-sacrifice. These are the qualities it can bring out in you if you nave prepared yourself. Below are a few tips to help prepare yourself psychologically for survival. Through studying this manual and attending survival training you can develop the *survival attitude.*

Know Yourself

Through training, family, and friends take the time to discover who you are on the inside. Strengthen your stronger qualities and develop the areas that you know are necessary to survive.

Anticipate Fears

Don't pretend that you will have no fears. Begin thinking about what would frighten you the most if forced to survive alone. Train in those areas of concern to you. The goal is not to eliminate the fear, but to build confidence in your ability to function despite your fears.

Be Realistic

Don't be afraid to make an honest appraisal of situations. See circumstances as they are, not as you want them to be. Keep your hopes and expectations within the estimate of the situation. When you go into a survival setting with unrealistic expectations, you may be laying the groundwork for bitter disappointment. Follow the adage, "Hope for the best, prepare for the worst." It is much easier to adjust to pleasant surprises about one's unexpected good fortunes than to be upset by one's unexpected harsh circumstances.

Adopt a Positive Attitude

Learn to seethe potential good in everything. Looking for the good not only boosts morale, it also is excellent for exercising your imagination and creativity.

Remind Yourself What Is at Stake

Remember, failure to prepare yourself psychologically to cope with survival leads to reactions such as depression, carelessness, inattention, loss of confidence, poor decision-making, and giving up before the body gives in. At stake is your life and the lives of others who are depending on you to do your share.

Train

Through military training and life experiences, begin today to prepare yourself to cope with the rigors of survival. Demonstrating your skills in training will give you the confidence to call upon them should the need arise. Remember, the more realistic the training, the less overwhelming an actual survival setting will be.

Learn Stress Management Techniques

People under stress have a potential to panic if they are not well-trained and not prepared psychologically to face whatever the circumstances may be. While we often cannot control the survival circumstances in which we find ourselves, it is within our ability to control our response to those circumstances. Learning stress management techniques can enhance significantly your capability to remain calm and focused as you work to keep yourself and others alive. A few good techniques to develop include relaxation skills, time management skills, asserfiveness skills, and cognitive restructuring skills (the ability to control how you view a situation).

Remember, "the will to survive" can also be considered to be "the refusal to give up."

CHAPTER 3

SURVIVAL PLANNING AND SURVIVAL KITS

Survival planning is nothing more than realizing something could happen that would put you in a survival situation and, with that in mind, taking steps to increase your chances of survival. Thus, survival planning means preparation.

Preparation means having survival items and knowing how to use them People who live insnow regions prepare their vehicles for poor road conditions. They put snow tires on their vehicles, add extra weight in the back for traction, and they carry a shovel, salt, and a blanket. Another example of preparation is finding the emergeny exits on an aircraft when you board it for a flight. Preparation could also mean knowing your intended route of travel and familiarizing yourself with the area. Finally, emergency planning is essential.

IMPORTANCE OF PLANNING

Detailed prior planning is essential in potential survival situations. Including survival considerations in mission planning will enhance your chances of survival if an emergency occurs. For example, if your job requires that you work in a small, enclosed area that limits what you can carry on your person, plan where you can put your rucksack or your load-bearing equipment. Put it where it will not prevent you from getting out of the area quickly, yet where it is readily accessible.

One important aspect of prior planning is preventive medicine. Ensuring that you have no dental problems and that your immunizations are current will help you avoid potential dental or health problems. A dental problem in a survival situation will reduce your ability to cope with other problems that you face. Failure to keep your shots current may mean your body is not immune to diseases that are prevalent in the area.

Preparing and carrying a survival kit is as important as the considerations mentioned above. AH Army aircraft normally have survival kits on board for the type area(s) over which they will fly. There are kits for overwater survival, for hot climate survival, and an aviator survival vest (see Appendix A for a description of these survival kits and their contents). If you are not an aviator, you will probably not have access to the survival vests or survival kits. However, if you know what these kits contain, it will help you to plan and to prepare your own survival kit.

Even the smallest survival kit, if properly prepared, is invaluable when faced with a survival problem. Before making your survival kit, however, consider your unit's mission, the operational environment, and the equipment and vehicles assigned to your unit.

SURVIVAL KITS

The environment is the key to the types of items you will need in your survival kit. How much equipment you put in your kit depends on now you will carry the kit. A kit carried on your body will have to be smaller than one carried in a vehicle. Always layer your survival kit, keeping the most important items on your body. For example, your map and compass should always be on your body. Carry less important items on your load-bearing equipment. Place bulky items in the rucksack.

In preparing your survival kit, select items you can use for more than one purpose. If you have two items that will serve the same function, pick the one you can use for another function. Do not duplicate items, as this increases your kit's size and weight.

Your survival kit need not be elaborate. You need only functional items that will meet your needs and a case to hold the items. For the case, you might want to use a Band-Aid box, a first aid case, an ammunition pouch, or another suitable case. This case should be-

- Water repellent or waterproof.
- Easy to carry or attach to your body.
- Suitable to accept varisized components.
- Durable.

In your survival kit, you should have—

- First aid items.
- Water purification tablets or drops.
- Fire starting equipment.
- Signaling items.
- Food procurement items.
- Shelter items.

Some examples of these items are-

- Lighter, metal match, waterproof matches.
- Snare wire.
- Signaling mirror.
- Wrist compass.
- Fish and snare line.
- Fishhooks.
- Candle.
- Small hand lens.
- Oxytetracycline tablets (diarrhea or infection).
- Water purification tablets.
- Solar blanket.
- Surgical blades.
- Butterfly sutures.
- Condoms for water storage.
- Chap Stick.
- Needle and thread.
- Knife.

Include a weapon only if the situation so dictates. Read about and practice the survival techniques in this manual. Consider your unit's mission and the environment in which your unit will operate. Then prepare your survival kit.

CHAPTER 4

BASIC SURVIVAL MEDICINE

Foremost among the many problems that can compromise a survivor's ability to return to safety are medical problems resulting from parachute descent and landing, extreme climates, ground combat, evasion, and illnesses contracted in captivity.

Many evaders and survivors have reported difficulty in treating injuries and illness due to the lack of training and medical supplies. For some, this led to capture or surrender.

Survivors have related feeling of apathy and helplessness because they could not treat themselves in this environment. The ability to treat themselves increased their morale and cohesion and aided in their survival and eventual return to friendly forces.

One man with a fair amount of basic medical knowledge can make a difference in the lives of many. Without qualified medical personnel available, it is you who must know what to do to stay alive.

REQUIREMENTS FOR MAINTENANCE OF HEALTH

To survive, you need water and food. You must also have and apply high personal hygiene standards.

Water

Your body loses water through normal body processes (sweating, urinating, and defecating). During average daily exertion when the atmospheric temperature is 20 degrees Celsius (C) (68 degrees Fahrenheit), the average adult loses and therefore requires 2 to 3 liters of water daily. Other factors, such as heat exposure, cold exposure, intense activity, high altitude, burns, or illness, can cause your body to lose more water. You must replace this water.

Dehydration results from inadequate replacement of lost body fluids. It decreases your efficiency and, if injured, increases your susceptibility to severe shock. Consider the following results of body fluid loss:

- A 5 percent loss of body fluids results in thirst, irritability, nausea, and weakness.
- A 10 percent loss results in dizziness, headache, inability to walk, and a tingling sensation in the limbs.
- A 15 percent loss results in dim vision, painful urination, swollen tongue, deafness, and a numb feeling in the skin.
- A loss greater than 15 percent of body fluids may result in death.

The most common signs and symptoms of dehydration are—

- Dark urine with a very strong odor.
- Low urine output.
- Dark, sunken eyes.
- Fatigue.
- Emotional instability.

- Loss of skin elasticity.
- Delayed capillary refill in fingernail beds.
- Trench line down center of tongue.
- Thirst. Last on the list because you are already 2 percent dehydrated by the time you crave fluids.

You replace the water as you lose it. Trying to make up a deficit is difficult in a survival situation, and thirst is not a sign of how much water you need.

Most people cannot comfortably drink more than 1 liter of water at a time. So, even when not thirsty, drink small amounts of water at regular intervals each hour to prevent dehydration.

If you are under physical and mental stress or subject to severe conditions, increase your water intake. Drink enough liquids to maintain a urine output or at least 0.5 liter every 24 hours.

In any situation where food intake is low, drink 6 to 8 liters of water per day. In an extreme climate, especially an arid one, the average person can lose 2.5 to 3.5 liters of water ***per now.*** In this type of climate, you should drink 14 to 30 liters of water per day.

With the loss of water there is also a loss of electrolytes (body salts). The average diet can usually keep up with these losses but in an extreme situation or illness, additional sources need to be provided. A mixture of 0.25 teaspoon of salt to 1 liter of water will provide a concentration that thehody tissues can readily absorb.

Of all the physical problems encountered in a survival situation, the loss of water is the most preventable. The following are basic guidelines for the prevention of dehydration:

- *Always drink water when eating.* Water is used and consumed as a part of the digestion process and can lead to dehydration.
- *Acclimatize.* The body performs more efficiently in extreme conditions when acclimatized.
- *Conserve sweat not water.* Limit sweat-producing activities but drink water.
- *Ration water.* Until you find a suitable source, ration your water sensibly. A daily intake of 500 cubic centimeter (0.5 liter) of a sugar-water mixture (2 teaspoons per liter) will suffice to prevent severe dehydration for at least a week, provided you keep water losses to a minimum by limiting activity and heat gain or loss.

You can estimate fluid loss by several means. A standard field dressing holds about 0.25 liter (one-fourth canteen) of blood. A soaked T-shirt holds 0.5 to 0.75 liter.

You can also use the pulse and breathing rate to estimate fluid loss. Use the following as a guide:

- With a 0.75 liter loss the wrist pulse rate will be under 100 beats per minute and the breathing rate 12 to 20 breaths per minute.
- With a 0.75 to 1.5 liter loss the pulse rate will be 100 to 120 beats per minute and 20 to 30 breaths per minute.
- With a 1.5 to 2 liter loss the pulse rate will be 120 to 140 beats per minute and 30 to 40 breaths per minute. Vital signs above these rates require more advanced care.

Food

Although you can live several weeks without food, you need an adequate amount to stay healthy. Without food your mental and physical capabilities will deteriorate rapidly, and you will become weak. Food replenishes the substances that your body burns and provides energy. It provides vitamins, minerals, salts, and other elements essential to good health. Possibly more important, it helps morale.

The two basic sources of food are plants and animals (including fish). In varying degrees both provide the calories, carbohydrates, fats, and proteins needed for normal daily body functions.

Calories are a measure of heat and potential energy. The average person needs 2,000 calories per day to function at a minimum level. An adequate amount of carbohydrates, fats, and proteins without an adequate caloric intake will lead to starvation and cannibalism of the body's own tissue for energy.

Plant Foods

These foods provide carbohydrates—the main source of energy. Many plants provide enough protein to keep the body at normal efficiency. Although plants may not provide a balanced diet, they will sustain you even in the arctic, where meat's heat-producing qualities are normally essential. Many plant foods such as nuts and seeds will give you enough protein and oils for normal efficiency. Roots, green vegetables, and plant rood containing natural sugar will provide calories and carbohydrates that give the body natural energy.

The food value of plants becomes more and more important if you are eluding the enemy or if you are in an area where wildlife is scarce. For instance—

- You can dry plants by wind, air, sun, or fire. This retards spoilage so that you can store or carry the plant food with you to use when needed.
- You can obtain plants more easily and more quietly than meat. This is extremely important when the enemy is near.

Animal Foods

Meat is more nourishing than plant food. In fact, it may even be more readily available in some places. However, to get meat, you need to know the habits of, and how to capture, the various wildlife.

To satisfy your immediate food needs, first seek the more abundant and more easily obtained wildlife, such as insects, crustaceans, mollusks, fish, and reptiles. These can satisfy your immediate hunger while you are preparing traps and snares for larger game.

Personal Hygiene

In any situation, cleanliness is an important factor in preventing infection and disease. It becomes even more important in a survival situation. Poor hygiene can reduce your chances of survival.

A daily shower with hot water and soap is ideal, but you can stay clean without this luxury. Use a cloth and soapy water to wash yourself. Pay special attention to the feet, armpits, crotch, hands, and hair as these are prime areas for infestation and infection. If water is scarce, take an "air' bath. Remove as much of your clothing as practical and expose your body to the sun and air for at least 1 hour. Be careful not to sunburn.

If you don't have soap, use ashes or sand, or make soap from animal fat and wood ashes, if your situation allows. To make soap—

- Extract grease from animal fat by cutting the fat into small pieces and cooking them in a pot.
- Add enough water to the pot to keep the fat from sticking as it cooks.
- Cook the fat slowly, stirring frequently.
- After the fat is rendered, pour the grease into a container to harden.
- Place ashes in a container with a spout near the bottom.
- Pour water over the ashes and collect the liquid that drips out of the spout in a separate container. This liquid is the potash or lye. Another way to get the lye is to pour the slurry (the mixture of ashes and water) through a straining cloth.
- In a cooking pot, mix two parts grease to one part potash.
- Place this mixture over a fire and boil it until it thickens.

After the mixture—the soap—cools, you can use it in the semiliquid state directly from the pot. You can also pour it into a pan, allow it to harden, and cut it intolbars for later use.

Keep Your Hands Clean

Germs on your hands can infect food and wounds. Wash your hands after handling any material that is likely to carry germs, after visiting the latrine, after caring for the sick, and before handlmg any food, food utensils, or drinking water. Keep your fingernails closely trimmed and clean, and keep your fingers out of your mouth.

Keep Your Hair Clean

Your hair can become a haven for bacteria or fleas, lice, and other parasites. Keeping your hair clean, combed, and trimmed helps you avoid this danger.

Keep Your Clothing Clean

Keep your clothing and bedding as clean as possible to reduce the chance of skin infection as well as to decrease the danger of parasitic infestation. Clean your outer clothing whenever it becomes soiled. Wear clean underclothing and socks each day. If water is scarce, "air" clean your clothing by shaking, airing, and sunning it for 2 hours. If you are using a sleeping bag, turn it inside out after each use, fluff it, and air it.

Keep Your Teeth Clean

Thoroughly clean your mouth and teeth with a toothbrush at least once each day. If you don't have a toothbrush, make a chewing stick. Find a twig about 20 centimeters long and 1 centimeter wide. Chew one end of the stick to separate the fibers. Now brush your teeth thoroughly. Another way is to wrap a clean strip of cloth around your fingers and rub your teeth with it to wipe away rood particles. You can also brush your teeth with small amounts of sand, baking soda, salt, or soap. Then rinse your mouth with water, salt water, or willow bark tea. Also, flossing your teeth with string or fiber helps oral hygiene.

If you have cavities, you can make temporary fillings by placing candle wax, tobacco, aspirin, hot pepper, tooth paste or powder, or portions of a ginger root into the cavity. Make sure you clean the cavity by rinsing or picking the particles out of the cavity before placing a filling in the cavity.

Take Care of Your Feet

To prevent serious foot problems, break in your shoes before wearing them on any mission. Wash and massage your feet daily. Trim your toenails straight across. Wear an insole and the proper size of dry socks. Powder and check your feet daily for blisters.

If you get a small blister, do not open it. An intact blister is safe from infection. Apply a padding material around the blister to relieve pressure and reduce friction. If the blister bursts, treat it as an open wound. Clean and dress it daily and pad around it. Leave large blisters intact. To avoid having the blister burst or tear under pressure and cause a painful and open sore, do the following:

- Obtain a sewing-type needle and a clean or sterilized thread.
- Run the needle and thread through the blister after cleaning the blister.
- Detach the needle and leave both ends of the thread hanging out of the blister. The thread will absorb the liquid inside. This reduces the size of the hole and ensures that the hole does not close up.
- Pad around the blister.

Get Sufficient Rest

You need a certain amount of rest to keep going. Plan for regular rest periods of at least 10 minutes per hour during your daily activities. Learn to make yourself comfortable under less than ideal conditions. A change from mental to physical activity or vice versa can be refreshing when time or situation does not permit total relaxation.

Keep Camp Site Clean

Do not soil the ground in the camp site area with urine or feces. Use latrines, if available. When latrines are not available, dig "cat holes" and cover the waste. Collect drinking water upstream from the camp site. Purify all water.

MEDICAL EMERGENCIES

Medical problems and emergencies you may be faced with include breathing problems, severe bleeding, and shock.

Breathing Problems

Any one of the following can cause airway obstruction, resulting in stopped breathing

- Foreign matter in mouth of throat that obstructs the opening to the trachea.
- Face or neck injuries.
- Inflammation and swelling of mouth and throat caused by inhaling smoke, flames, and irritating vapors or by an allergic reaction.
- "Kink" in the throat (caused by the neck bent forward so that the chin rests upon the chest) may block the passage of air.
- Tongue blocks passage of air to the lungs upon unconsciousness. When an individual is unconscious, the muscles of the lower jaw and tongue relax as the neck drops forward, causing the lower jaw to sag and the tongue to drop back and block the passage of air.

Severe Bleeding

Severe bleeding from any major blood vessel in the body is extremely dangerous. The loss of 1 liter of blood will produce moderate symptoms of shock. The loss of 2 liters will produce a severe state of shock that places the body in extreme danger. The loss of 3 liters is usually fatal.

Shock

Shock (acute stress reaction) is not a disease in itself. It is a clinical condition characterized by symptoms that arise when cardiac output is insufficient to fill the arteries with blood under enough pressure to provide an adequate blood supply to the organs and tissues.

LIFESAVING STEPS

Control panic, both your own and the victim's. Reassure him and try to keep him quiet.

Perform a rapid physical exam. Look for the cause of the injury and follow the ABCs of first aid, starting with the airway and breathing, but be discerning. A person may die from arterial bleedmg more quickly than from an airway obstruction in some cases.

Open Airway and Maintain

You can open an airway and maintain it by using the following steps.

Step 1, Check if the victim has a partial or complete airway obstruction. If he can cough or speak, allow him to clear the obstruction naturally. Stand by, reassure the victim, and be ready to clear his airway and perform mouth-to-mouth resuscitation should he become unconscious. If his airway is completely obstructed, administer abdominal thrusts until the obstruction is cleared.

Step *2.* Using a finger, quickly sweep the victim's mouth clear of any foreign objects, broken teeth, dentures, sand.

Step 3. Using the jaw thrust method, grasp the angles of the victim's lower jaw and lift with both hands, one on each side, moving the jaw forward. For stability, rest your elbows on the surface on which the victim is lying. If his lips are closed, gently open the lower lip with your thumb (Figure 4-1).

Step *4.* With the victim's airway open, pinch his nose closed with your thumb and forefinger and blow two complete breaths into his lungs. Allow the lungs to deflate after the second inflation and perform the following:

- *Look* for his chest to rise and fall.
- *Listen* for escaping air during exhalation.
- *Feel* for flow of air on your cheek.

Step 5. If the forced breaths do not stimulate spontaneous breathing, maintain the victim's breathing by performing mouth-to-mouth resuscitation.

Step *6.* There is danger of the victim vomiting during mouth-to-mouth resuscitation. Check the victim's mouth periodically for vomit and clear as needed.

Note: Cardiopulmonary resuscitation (CPR) may be necessary after cleaning the airway, but only after major bleeding is under control. See FM 21-20, the American Heart Association manual, the Red Cross manual, or most other first aid books for detailed instructions on CPR.

Control Bleeding

In a survival situation, you must control serious bleeding immediately because replacement fluids normally are not available and the victim can die within a matter of minutes. External bleeding falls into the following classifications (according to its source):

- *Arterial.* Blood vessels called arteries carry blood away from the heart and through the body. A cut artery issues *bright red* blood from the wound in *distinct spurts* or pulses that correspond to the rhythm of the heartbeat. Because the blood in the arteries is under high pressure, an individual can lose a large volume of blood in a short period when damage to an artery of significant size occurs. Therefore, arterial bleeding is the most serious type of bleeding. If not controlled promptly, it can be fatal.
- *Venous.* Venous blood is blood that is returning to the heart through blood vessels called veins. A *steady flow of dark red, maroon, or bluish*

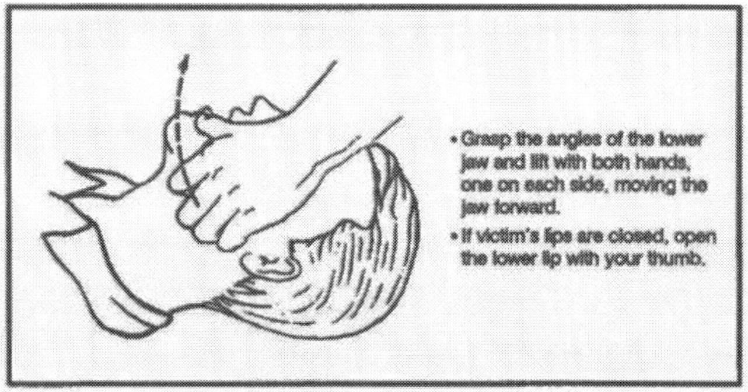

Figure 4-1. Jaw thrust method.

blood characterizes bleeding from a vein. You can usually control venous bleeding more easily than arterial bleeding.

- *Capillary.* The capillaries are the extremely small vessels that connect the arteries with the veins. Capillary bleeding most commonly occurs in minor cuts and scrapes. This type of bleeding is not difficult to control.

You can control external bleeding by direct pressure, indirect (pressure points) pressure, elevation, digital ligation, or tourniquet.

Direct Pressure

The most effective way to control external bleeding is by applying pressure directly over the wound. This pressure must not only be firm enough to stop the bleeding, but it must also be maintained long enough to "seal off" the damaged surface.

If bleeding continues after having applied direct pressure for 30 minutes, apply a pressure dressing. This dressing consists of a thick dressing of gauze or other suitable material applied directly over the wound and held in place with a tightly wrapped bandage (Figure 4-2). It should be tighter than an ordinary compression bandage but not so tight that it impairs circulation to the rest of the limb. Once you apply the dressing, *do not remove it,* even when the dressing becomes blood soaked.

Leave the pressure dressing in place for 1 or 2 days, after which you can remove and replace it with a smaller dressing.

In the long-term survival environment, make fresh, daily dressing changes and inspect for signs of infection.

Elevation

Raising an injured extremity as high as possible above the heart's level slows blood loss by aiding the return of olood to the heart and lowering the blood pressure at the wound. However, elevation alone *will not* control bleeding entirely; you must also apply direct pressure over the wound. When treating a snakebite, however, keep the extremity lower than the heart.

Pressure Points

A pressure point is a location where the main artery to the wound lies near the surface of the skin or where the artery passes directly over a bony prominence (Figure 4-3). You can use digital pressure on a pressure point to slow arterial bleeding until the application of a pressure dressing. Pressure point control is not as effective for controlling bleeding as direct pressure exerted on the wound. It is rare when a single major compressible artery supplies a damaged vessel.

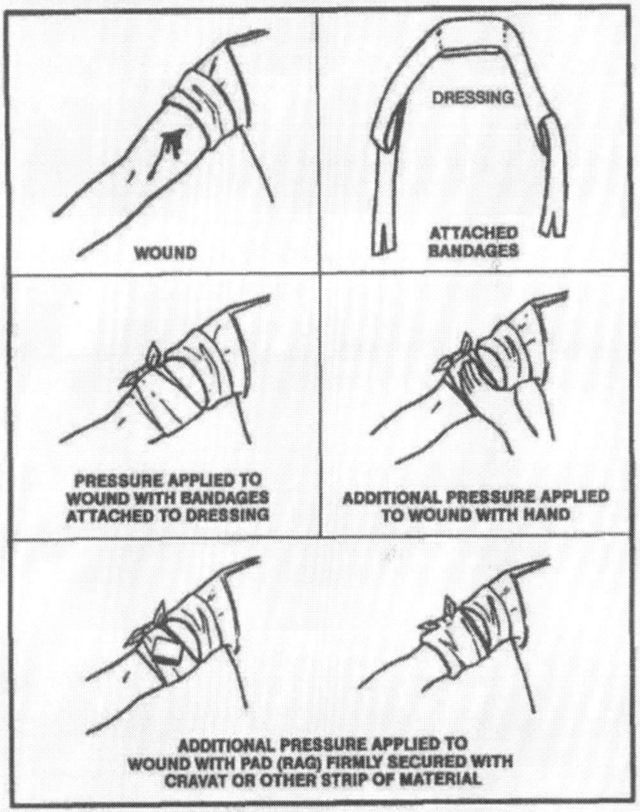

Figure 4-2. Application of a pressure dressing.

If you cannot remember the exact location of the pressure points, follow this rule: Apply pressure at the end of the joint just above the injured area. On hands, feet, and head, this will be the wrist, ankle, and neck, respectively.

> **WARNING**
>
> Use caution when applying pressure to the neck. Too much pressure for too long may cause unconsciousness or death. Never place a tourniquet around the neck.

Maintain pressure points by placing a round stick in the joint, bending the joint over the stick, and then keeping it tightly bent by lashing. By using this method to maintain pressure, it frees your hands to work in other areas.

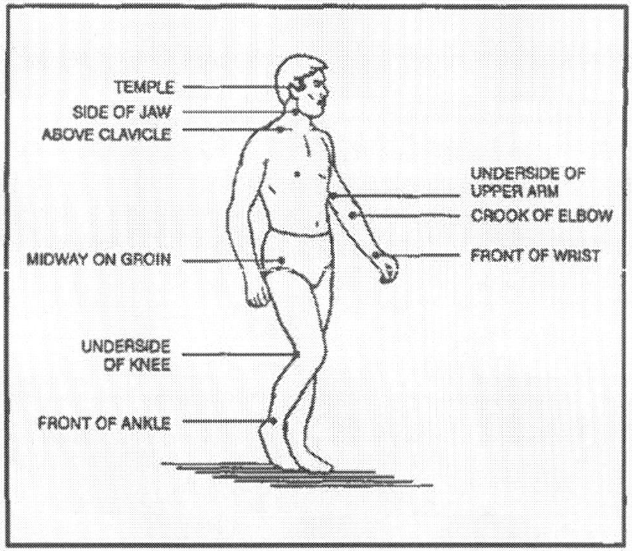

Figure 4-3. Pressure points.

Digital Ligation

You can stop major bleeding immediately or slow it down by applying pressure with a finger or two on the bleeding end of the vein or artery. Maintain the pressure until the bleeding stops or slows down enough to apply a pressure bandage, elevation, and so forth.

Tourniquet

Use a tourniquet only when direct pressure over the bleeding point and all other methods did not control the bleeding. If you leave a tourniquet in place too long, the damage to the tissues can progress to gangrene, with a loss of the limb later. An improperly applied tourniquet can also cause permanent damage to nerves and other tissues at the site of the constriction.

If you must use a tourniquet, place it around the extremity, between the wound and the heart, 5 to 10 centimeters above the wound site (Figure 44). Never place it directly over the wound or a fracture. Use a stick as a handle to tighten the tourniquet and tighten it only enough to stop blood flow. When you have tightened the tourniquet, bind the free end of the stick to the limb to prevent unwinding.

After you secure the tourniquet, clean and bandage the wound. A lone survivor **does not** remove or release an applied tourniquet. In a buddy system, however, the buddy can release the tourniquet pressure every 10 to 15 minutes for 1 or 2 minutes to let blood flow to the rest of the extremity to prevent limb loss.

Prevent and Treat Shock

Anticipate shock in all injured personnel. Treat all injured persons as follows, regardless of what symptoms appear (Figure 4-5):

- If the victim is consdous, place him on a level surface with the lower extremities elevated 15 to 20 centimeters.
- If the victim is unconscious, place him on his side or abdomen with his head turned to one side to prevent choking on vomit, blood, or other fluids.
- If you are unsure of the best position, place the victim perfectly flat. Once the victim is in a shock position, do not move him.
- Maintain body heat by insulating the victim from the surroundings and, in some instances, applying external heat.

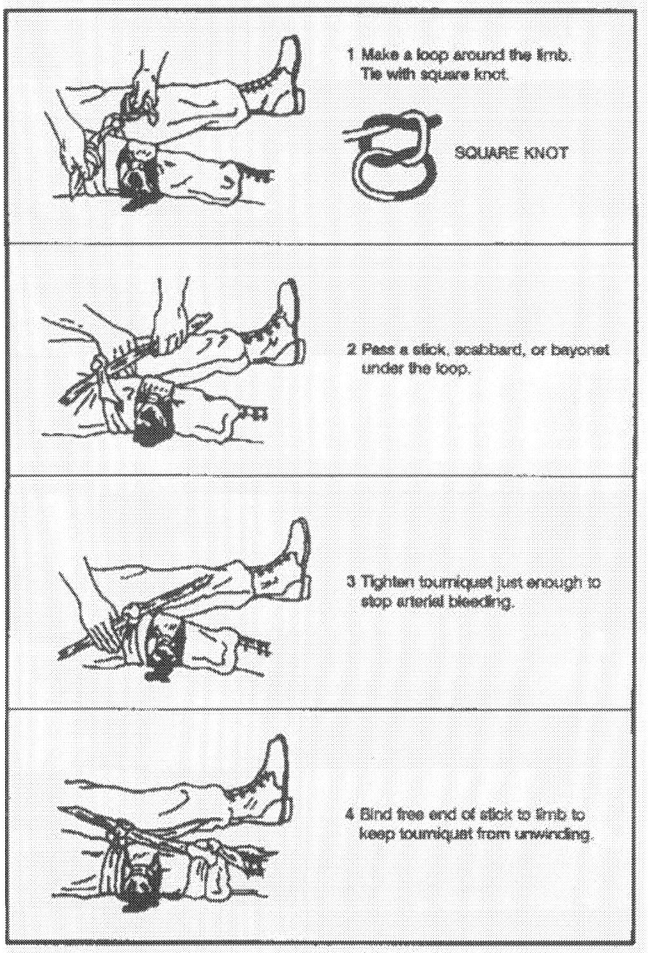

Figure 4-4. Application of tourniquet.

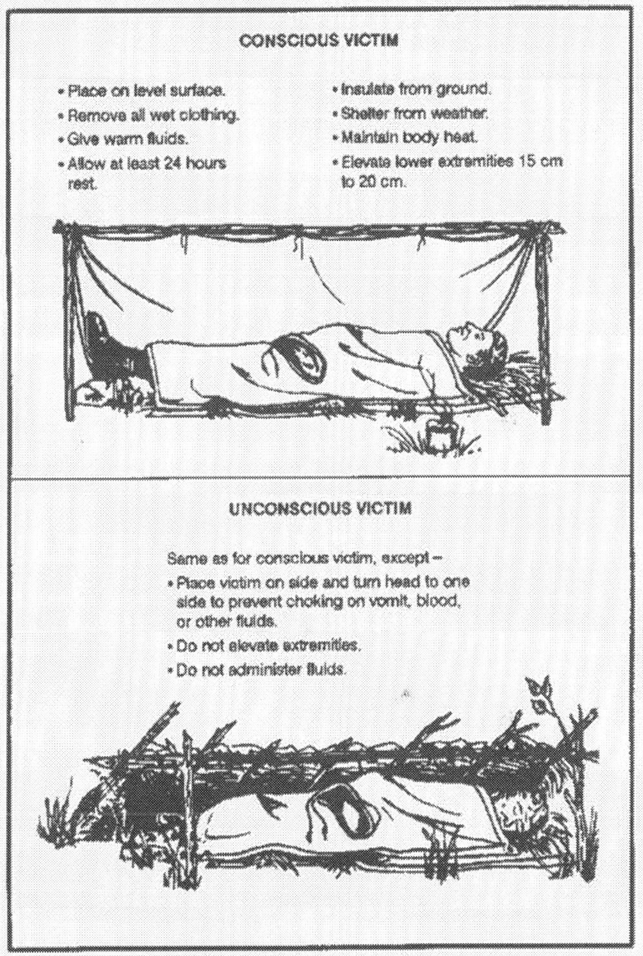

Figure 4-5. Treatment for shock.

- If wet, remove all the victim's wet clothing as soon as possible and replace with dry clothing.
- Improvise a shelter to insulate the victim from the weather.
- Use warm liquids or foods, a prewarmed sleeping bag, another person, warmed water in canteens, hot rocks wrapped in clothing, or fires on either side of the victim to provide external warmth.
- If the victim is conscious, slowly administer small doses of a warm salt or sugar solution, if available.
- If the victim is unconscious or has abdominal wounds, do not give fluids by mouth.
- Have the victim rest for at least 24 hours.
- If you are a lone survivor, lie in a depression in the ground, behind a tree, or any other place out of the weather, with your head lower than your feet.
- If you are with a buddy, reassess your patient constantly.

BONE AND JOINT INJURY

You could face bone and joint injuries that include fractures, dislocations, and sprains.

Fractures

There are basically two types of fractures: open and closed. With an open (or compound) fracture, the bone protrudes through the skin and complicates the actual fracture with an open wound. After setting the fracture, treat the wound as any other open wound.

The closed fracture has no open wounds. Follow the guidelines for immobilization, and set and splint the fracture.

The signs and symptoms of a fracture are pain, tenderness, discoloration, swelling deformity, loss of function, and grating (a sound or feeling that occurs when broken bone ends rub together).

The dangers with a fracture are the severing or the compression of a nerve orblood vessel at the site of fracture. For this reason minimum manipulation should be done, and only very cautiously. If you notice the area below the break becoming numb, swollen, cool to the touch, or turning pale, and the victim shows signs of shock, a major vessel may have been severed. You must control this internal bleeding. Rest the victim for shock, and replace lost fluids.

Often you must maintain traction during the splinting and healing process. You can effectively pull smaller bones such as the arm or lower leg by hand. You can create traction by wedging a hand or foot in the V-notch of a tree and pushing against the tree with the other extremity. You can then splint the break.

Very strong muscles hold a broken thighbone (femur) in place making it difficult to maintain traction during healing. You can make an improvised traction splint using natural material (Figure 4-6) as follows:

- Get two forked branches or saplings at least 5 centimeters in diameter. Measure one from the patient s armpit to 20 to 30 centimeters past his unbroken leg. Measure the other from the groin to 20 to 30 centimeters past the unbroken leg. Ensure that both extend an equal distance beyond the end of the leg.

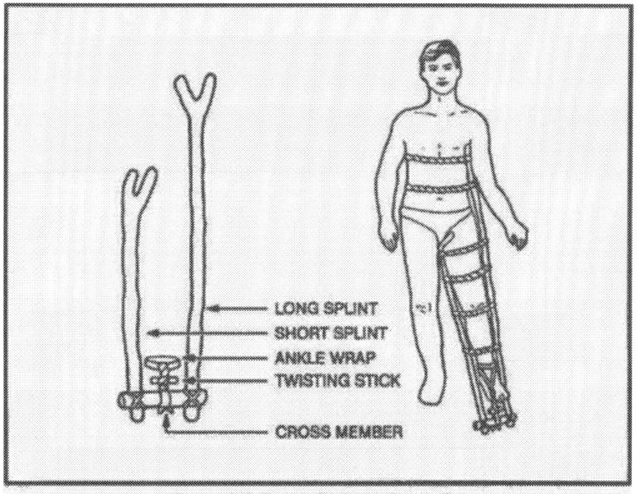

Figure 4-6. Improvised traction splint.

- Pad the two splints. Notch the ends without forks and lash a 20- to 30-centimeter cross member made from a 5-centimeter diameter branch between them.
- Using available material (vines, cloth, rawhide), tie the splint around the upper portion of the body and down the length of the broken leg. Follow the splinting guidelines.
- With available material, fashion a wrap that will extend around the ankle, with the two free ends tied to the cross member.
- Place a 10- by 2.5-centimeter stick in the middle of the free ends of the ankle wrap between the cross member and the foot. Using the stick, twist the material to make the traction easier.
- Continue twisting until the broken leg is as long or slightly longer than the unbroken leg.
- Lash the stick to maintain traction.

Note: Over time you may lose traction because the material weakened. Check the traction periodically. If you must change or repair the splint, maintain the traction manually for a short time.

Dislocations

Dislocations are the separations of bone joints causing the bones to go out of proper alignment. These misalignments can be extremely painful and can cause an impairment of nerve or circulatory function below the area affected. You must place these joints back into alignment as quickly as possible.

Signs and symptoms of dislocations are joint pain, tenderness, swelling, discoloration, limited range of motion, and deformity of the joint. You treat dislocations by reduction, immobilization, and rehabilitation.

Reduction or "setting" is placing the bones back into their proper alignment. You can use several methods, but manual traction or the use of weights to pull the bones are the safest and easiest. Once performed, reduction decreases the victim's pain and allows for normal function and circulation. Without an X ray, you can judge proper alignment by the look and feel of the joint and by comparing it to the joint on the opposite side.

Immobilization is nothing more than splinting the dislocation after reduction. You can use any field-expedient material for a splint or you can splint an extremity to the body. The basic guidelines for splinting are—

- Splint above and below the fracture site.
- Pad splints to reduce discomfort.
- Check circulation below the fracture after making each tie on the splint.

To rehabilitate the dislocation, remove the splints after 7 to 14 days. Gradually use the injured joint until fully healed.

Sprains

The accidental overstretching of a tendon or ligament causes sprains. The signs and symptoms are pain, swelling, tenderness, and discoloration (black and blue).

When treating sprains, think RICE—

R - Rest injured area.
I - Ice for 24 hours, then heat after that.
C - Compression-wrapping and/or splinting to help stabilize. If possible, leave the boot on a sprained ankle unless circulation is compromised.
E - Elevation of the affected area.

BITES AND STINGS

Insects and related pests are hazards in a survival situation. They not only cause irritations, but they are often carriers of diseases that cause severe allergic reactions in some individuals. In many parts of the world you will be exposed to serious, even fatal, diseases not encountered in the United States.

Ticks can carry and transmit diseases, such as Rocky Mountain spotted fever common in many parts of the United States. Ticks also transmit the Lyme disease.

Mosquitoes may carry malaria, dengue, and many other diseases.

Flies can spread disease from contact with infectious sources. They are causes of sleeping sickness, typhoid, cholera, and dysentery.

Fleas can transmit plague.

lice can transmit typhus and relapsing fever.

The best way to avoid the complications of insect bites and stings is to keep immunizations (including booster shots) up-to-date, avoid insect-infested areas, use netting and insect repellent, and wear all clothing properly.

If you get bitten or stung, do not scratch the bite or sting, it might become infected, inspect your body at least once a day to ensure there are no insects attached to you. If you find ticks attached to your body, cover them with a substance, such as Vaseline, heavy oil, or tree sap, that will cut off their air supply. Without air, the tick releases its hold, and you can remove it. Take care to remove the whole tick. Use tweezers if you have them. Grasp the tick where the mouth parts are attached to the skin. Do not squeeze the tick's body. Wash your hands after touching the tick. Clean the tick wound daily until healed.

Treatment

It is impossible to list the treatment of all the different types of bites and stings. Threat bites and stings as follows:

- If antibiotics are available for your use, become familiar with them before deployment and use them.
- Predeployment immunizations can prevent most of the common diseases carried by mosquitoes and some carried by flies.
- The common fly-borne diseases are usually treatable with penicillins or erythromycin.
- Most tick-, flea-, louse-, and mite-bome diseases are treatable with tetracycline.
- Most antibiotics come in 250 milligram (mg) or 500 mg tablets. If you cannot remember the exact dose rate to treat a disease, 2 tablets, 4 times a day for 10 to 14 days will usually kill any bacteria.

Bee and Wasp Stings

If stung by a bee, immediately remove the stinger and venom sac, if attached, by scraping with a fingernail or a knife blade. Do not squeeze or grasp the stinger or venom sac, as squeezing will force more venom into the wound. Wash the sting site thoroughly with soap and water to lessen the chance of a secondary infection.

If you know or suspect that you are allergic to insect stings, always carry an insect sting kit with you.

Relieve the itching and discomfort caused by insect bites by applying—

- Cold compresses.
- A cooling paste of mud and ashes.
- Sap from dandelions.
- Coconut meat.
- Crushed cloves of garlic.
- Onion.

Spider Bites and Scorpion Stings

The black widow spider is identified by a red hourglass on its abdomen. Only the female bites, and it has a neurotoxic venom. The initial pain is not severe, but severe local pain rapidly develops. The pain gradually spreads over the entire body and settles m the abdomen and legs. Abdominal cramps and progressive nausea, vomiting, and a rash may occur. Weakness, tremors, sweating, and salivation may occur. Anaphylactic reactions can occur. Symptoms begin to regress after several hours and are usually gone in a few days. Threat for shock. Be ready toperform CPR. Clean and dress the bite area to reduce the risk of infection. An antivenin is available.

The funnelweb spider is a large brown or gray spider found in Australia. The symptoms and the treatment for its bite are as for the black widow spider.

The brown house spider or brown recluse spider is a small, light brown spider identified by a dark brown violin on its back. There is no pain, or so little pain, that usually a victim is not aware of the bite. Within a few hours a painful red area with a mottled cyanotic center appears. Necrosis does not occur in all bites, but usually in 3 to 4 days, a star-shaped, firm area of deep purple discoloration appears at the bite site. The area turns dark and mummified in a week or two. The margins separate and the scab falls off, leaving an open ulcer. Secondary infection and regional swollen lymph glands usually become visible at this stage. The outstanding characteristic of the brown recluse bite is an ulcer that does not heal but persists for weeks or months. In addition to the ulcer, there is often a systemic reaction that is serious and may lead to death.

Reactions (fever, chills, joint pain, vomiting, and a generalized rash) occur chiefly in children or debilitated persons.

Tarantulas are large, hairy spiders found mainly in the tropics. Most do not inject venom, out some south American species do. They have large fangs. If bitten, pain and bleeding are certain, and infection is likely. Treat a tarantula bite as for any open wound, and try to prevent infection. If symptoms of poisoning appear, treat as for the bite of the black widow spider.

Scorpions are all poisonous to a greater or lesser degree. There are two different reactions, depending on the species:

- Severe local reaction only, with pain and swelling around the area of the sting. Possible prickly sensation around the mouth and a thick-feeling tongue.
- Severe systemic reaction, with little or no visible local reaction. Local pain may be present. Systemic reaction includes respiratory difficulties, thicK-feeling tongue, body spasms, drooling, gastric distention, double vision, blindness, involuntary rapid movement of the eyeballs, involuntary urination and defecation, and heart failure. Death is rare, occurring mainly in children and adults with high blood pressure or illnesses.

Treat scorpion stings as you would a black widow bite.

Snakebites

The chance of a snakebite in a survival situation is rather small, if you are familiar with the various types of snakes and their habitats. However, it could happen and you should know how to treat a snakebite. Deaths from snakebites are rare. More than one-half of the snakebite victims have little or no poisoning, and only about one-quarter develop serious systemic poisoning. However, the chance of a snakebite in a survival situation can affect morale, and failure to take preventive measures or failure to treat a snakebite properly can result in needless tragedy.

The primary concern in the treatment of snakebite is to limit the amount of eventual tissue destruction around the bite area.

A bite wound, regardless of the type of animal that inflicted it, can become infected from bacteria in the animal's mouth. With nonpoisonous as well as poisonous snakebites, this local infection is responsible for a large part of the residual damage that results.

Snake venoms not only contain poisons that attack the victim's central nervous system (neurotoxins) and blood circulation (hemotoxins), but also digestive enzymes (cytotoxins) to aid in digesting their prey. These poisons can cause a very large area of tissue death, leaving a large open wound. This condition could lead to the need for eventual amputation if not treated.

Shock and panic in a person bitten by a snake can also affect the person's recovery. Excitement, hysteria, and panic can speed up the circulation, causing the body to absorb the toxin quickly. Signs of shock occur within the first 30 mmutes after the bite.

Before you start treating a snakebite, determine whether the snake was poisonous or nonpoisonous. Bites from a nonpoisonous snake will show rows of teeth. Bites from a poisonous snake may have rows of teeth showing, but will have one or more distinctive puncture marks caused by fang penetration. Symptoms of a poisonous bite may be spontaneous bleeding from the nose and anus, blood in the urine, pain at the site of the bite, and swelling at the site of the bite within a few minutes or up to 2 hours later.

Breathing difficulty, paralysis, weakness, twitching, and numbness are also signs of neurotoxic venoms. These signs usually appear 1.5 to 2 hours after the bite.

If you determine that a poisonous snake bit an individual, take the following steps:

- Reassure the victim and keep him still.
- Set up for shock and force fluids or give an intravenous (IV).
- Remove watches, rings, bracelets, or other constricting items.
- Clean the bite area.
- Maintain an airway (especially if bitten near the face or neck) and be prepared to administer mouth-to-mouth resuscitation or CPR.
- Use a constricting band between the wound and the heart.
- Immobilize the site.
- Remove the poison as soon as possible by using a mechanical suction device or by squeezing.

Do not-

- Give the victim alcoholic beverages or tobacco products.

- Give morphine or other central nervous system (CNS) depressors.
- Make any deep cuts at the bite site. Cutting opens capillaries that in turn open a direct route into the blood stream for venom and infection.

Note *If medical treatment is over one hour away, make an incision (no longer than 6 millimeters and no deeper than 3 millimeter) over each puncture, cutting just deep enough to enlarge the fang opening, but only through the first or second layer of skin. Place a suction cup over the bite so that you have a good vacuum seal Suction the bite site 3 to 4 times. Use mouth suction* ***only as a last resort and only if you do not have open sores in your mouth.*** *Spit the envenomed blood out and rinse your mouth with water. This method will draw out 25 to 30 percent of the venom.*

- Put your hands on your face or rub your eyes, as venom may be on your hands. Venom may cause blindness.
- Break open the large blisters that form around the bite site.

After caring for the victim as described above, take the following actions to minimize local effects:

- If infection appears, keep the wound open and clean.
- Use heat after 24 to 48 hours to help prevent the spread of local infection. Heat also helps to draw out an infection.
- Keep the wound covered with a dry, sterile dressing.
- Have the victim drink large amounts of fluids until the infection is gone.

WOUNDS

An interruption of the skin's integrity characterizes wounds. These wounds could be open wounds, skin diseases, frostbite, trench foot, and burns.

Open Wounds

Open wounds are serious in a survival situation, not only because of tissue damage and blood loss, but also because they may become infected. Bacteria on the object that made the wound, on the individual's skin and clothing, or on other foreign material or dirt that touches the wound may cause infection.

By taking proper care of the wound you can reduce further contamination and promote healing. Clean the wound as soon as possible after it occurs by—

- Removing or cutting clothing away from the wound.
- Always looking for an exit wound if a sharp object, gun shot, or projectile caused a wound.
- Thoroughly cleaning the skin around the wound.
- Rinsing (not scrubbing) the wound with large amounts of water under pressure. You can use fresh urine if water is not available.

The "open treatment" method is the safest way to manage wounds in survival situations. Do not try to close any wound by suturing or similar procedures. Leave the wound open to allow the drainage of any pus resulting from infection. As long as the wound can dram, it generally will not Decome life-threatening, regardless of how unpleasant it looks or smells.

Cover the wound with a clean dressing. Place a bandage on the dressing to hold it in place. Change the dressing daily to check for infection.

If a wound is gaping, you can bring the edges together with adhesive tape cut in the form of a "butterfly" or "dumbbell" (Figure 4-7).

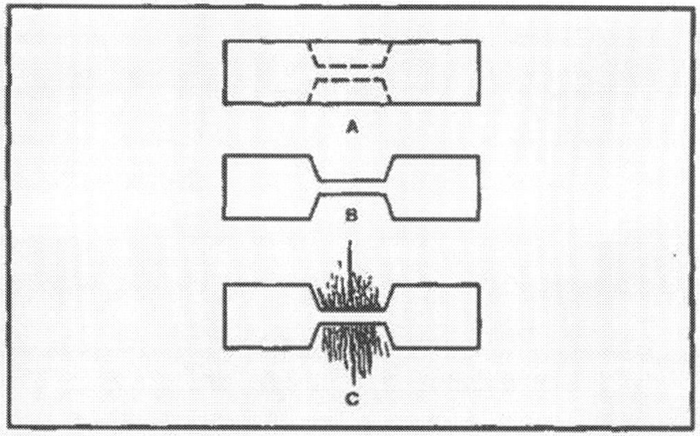

Figure 4-7. Butterfly closure.

In a survival situation, some degree of wound infection is almost inevitable. Pain, swelling, and redness around the wound, increased temperature, and pus in the wound or on the dressing indicate infection is present.

To treat an infected wound—

- Place a warm, moist compress directly on the infected wound. Change the compress when it cools, keeping a warm compress on the wound for a total of 30 minutes. Apply the compresses three or four times daily.
- Drain the wound. Open and gently probe the infected wound with a sterile instrument.
- Dress and bandage the wound.
- Drink a lot of water.

Continue this treatment daily until all signs of infection have disappeared.

If you do not have antibiotics and the wound has become severely infected, does not heal, and ordinary debridement is impossible, consider maggot therapy, despite its hazards:

- Expose the wound to flies for one day and then cover it.
- Check daily for maggots.
- Once maggots develop, keep wound covered but check daily.
- Remove all maggots when they have cleaned out all dead tissue and before they start on healthy tissue. Increased pain and bright red blood in the wound indicate that the maggots have reached healthy tissue.
- Flush the wound repeatedly with sterile water or fresh urine to remove the maggots.
- Check the wound every four hours for several days to ensure all maggots have been removed.
- Bandage the wound and treat it as any other wound. It should heal normally.

Skin Diseases and Ailments

Although boils, fungal infections, and rashes rarely develop into a serious health problem, they cause discomfort and you should treat them.

Boils

Apply warm compresses to bring the boil to a head. Then open the boil using a sterile knife, wire, needle, or similar item. Thoroughly clean out the pus using soap and water. Cover the boil site, checking it periodically to ensure no further infection develops.

Fungal Infections

Keep the skin clean and dry, and expose the infected area to as much sunlight as possible. *Do not scratch* the affected area. During the Southeast Asian conflict, soldiers used antifungal powders, lye soap, chlorine bleach, alcohol, vinegar, concentratedsalt water, and iodine to treat fungal infections with varying degrees of success. As *with any "unorthodox" method of treatment, use it with caution.*

Rashes

To treat a skin rash effectively, first determine what is causing it. This determination may be difficult even in the best of situations. Observe the following rules to treat rashes:

- If it is moist, keep it dry.
- If it is dry, keep it moist.
- Do not scratch it.

Use a compress of vinegar or tannic acid derived from tea or from boiling acorns or the bark or a hardwood tree to dry weeping rashes. Keep dry rashes moist by rubbing a small amount of rendered animal fat or grease on the affected area.

Remember, treat rashes as open wounds and clean and dress them daily. There are many substances available to survivors in the wild or in captivity for use as antiseptics to treat wound:

- *Iodine tablets.* Use 5 to 15 tablets in a liter of water to produce a good rinse for wounds during healing.
- *Garlic.* Rub it on a wound or boil it to extract the oils and use the water to rinse the affected area.
- *Salt water.* Use 2 to 3 tablespoons per liter of water to kill bacteria.
- *Bee honey.* Use it straight or dissolved in water.
- *Sphagnum moss.* Found in boggy areas worldwide, it is a natural source of iodine. Use as a dressing.

Again, use noncommercially prepared materials with caution.

Frostbite

This injury results from frozen tissues. Light frostbite involves only the skin that takes on a dull, whitish pallor. Deep frostbite extends to a depth below the skin. The tissues become solid and immovable. Your feet, hands, and exposed facial areas are particularly vulnerable to frostbite.

When with others, prevent frostbite by using the buddy system. Check your buddy's face often and make sure thathe checks yours. If you are alone, periodically cover your nose and lower part of your face with your mittens.

Do not try to thaw the affected areas by placing them close to an open flame. Gently rub them in lukewarm water. Dry the part and place it next to your skin to warm it at body temperature.

Trench Foot

This condition results from many hours or days of exposure to wet or damp conditions at a temperature just above freezing. The nerves and muscles sustain the main damage, but gangrene can occur. In extreme cases the flesh dies and it may become necessary to have the foot or leg amputated. The best prevention is to keep your feet dry. Carry extra socks with you in a waterproof packet. Dry wet socks against your body. Wash your feet daily and put on dry socks.

Burns

The following field treatment for burns relieves the pain somewhat, seems to help speed healing, and offers some protection against infection:

- First, stop the burning process. Put out the fire by removing clothing, dousing with water or sand, or by rolling on the ground. Cool the burning skin with ice or water. For burns caused by white phosphorous, pick out the white phosphorous with tweezers; do not douse with water.
- Soak dressings or clean rags for 10 minutes in a boiling tannic acid solution (obtained from tea, inner bark of hardwood trees, or acorns boiled in water).
- Cool the dressings or clean rags and apply over bums.
- Rest as an open wound.
- Replace fluid loss.
- Maintain airway.
- Treat for shock.
- Consider using morphine, unless the burns are near the face.

ENVIRONMENTAL INJURIES

Heatstroke, hypothermia, diarrhea, and intestinal parasites are environmental injuries you could face.

Heatstroke

The breakdown of the body's heat regulatory system (body temperature more than 40.5 degrees C]105 degrees F]) causes a heatstroke. Other heat injuries, such as cramps or dehydration, do not always precede a heatstroke. Signs and symptoms of heatstroke are—

- Swollen, beet-red face.
- Reddened whites of eyes.
- Victim not sweating.
- Unconsciousness or delirium, which can cause pallor, a bluish color to lips and nail beds (cyanosis), and cool skin.

Note: By this time the victim is in severe shock. Cool the victim as rapidly as possible. Cool him by dipping him in a cool stream. If one is not available, douse the victim with urine, water, or at the very least, apply cool wet compresses to all the joints, especially the neck, armpits, and crotch. Be sure to wet the victim's head. Heat loss through the scalp is great. Administer IVs and provide drinking fluids. You may fan the individual.

Expect, during cooling—

- Vomiting.
- Diarrhea.
- Struggling.
- Shivering.
- Shouting.

- Prolonged unconsciousness.
- Rebound heatstroke within 48 hours.
- Cardiac arrest; *be ready to perform CPR.*

Note: Treat for dehydration with lightly salted water.

Hypothermia

Defined as the body's failure to maintain a temperature of 36 degrees C (97 degrees F). Exposure to cool or cold temperature over a short or long time can cause hypothermia. Dehydration and lack of food and rest predispose the survivor to hypothermia.

Unlike heatstroke, you must gradually warm the hypothermia victim. Get the victim into dry clothing. Replace lost fluids, and warm him.

Diarrhea

A common, debilitating ailment caused by a change of water and food, drinking contaminated water, eating spoiled food, Decoming fatigued, and using dirty dishes. You can avoid most of these causes by practicing preventive medicine. If you get diarrhea, however, and do not have antt-diarrheal medicine, one of the following treatments may be effective:

- Limit your intake of fluids for 24 hours.
- Drink one cup of a strong tea solution every 2 hours until the diarrhea slows or stops. The tannic acid in the tea helps to control the diarrhea. Boil the inner bark of a hardwood tree for 2 hours or more to release the tannic acid.
- Make a solution of one handful of ground chalk, charcoal, or dried bones and treated water. If you have some apple pomace or the rinds of citrus fruit, add an equal portion to the mixture to make it more effective. Take 2 tablespoons of the solution every 2 hours until the diarrhea slows or stops.

Intestinal Parasites

You can usually avoid worm infestations and other intestinal parasites if you take preventive measures. For example, never go barefoot. The most effective way to prevent intestinal parasites is to avoid uncooked meat and raw vegetables contaminated by raw sewage or human waste used as a fertilizer. However, should you become infested and lack proper medicine, you can use home remedies. Keep in mind that these home remedies work on the principle of changing the environment of the gastrointestinal tract. The following are home remedies you could use:

- *Salt water.* Dissolve 4 tablespoons of salt in 1 liter of water and drink. Do not repeat this treatment.
- *Tobacco.* Eat 1 to 1.5 cigarettes. The nicotine in the cigarette will kill or stun the worms long enough for your system to pass them. If the infestation is severe, repeat the treatment in 24 to 48 hours, *but no sooner.*
- *Kerosene.* Drink 2 tablespoons of kerosene *but no more.* If necessary, you can repeat this treatment in 24 to 48 hours. Be careful not to inhale the fumes. They may cause lung irritation.
- *Hot peppers.* Peppers are effective only if they are a steady part of your diet. You can eat them raw or put them in soups or rice and meat dishes. They create an environment that is prohibitive to parasitic attachment.

HERBAL MEDICINES

Our modern wonder drugs, laboratories, and equipment have obscured more primitive types of medicine involving determination, common sense, and a few simple treatments. In many areas of the world, however, the people still depend on local "witch doctors" or healers to cure their ailments. Many or the herbs (plants) and treatments they use are as effective as the most modern medications available. In fact, many modern medications come from refined herbs.

WARNING

Use herbal medicines with extreme care, however, and only when you lack or have limited medical supplies. Some herbal medicines are dangerous and may cause further damage or even death. See Chapter 9. Survival Use of Plants, for some basic herbal medicine treatments.

CHAPTER 5

SHELTERS

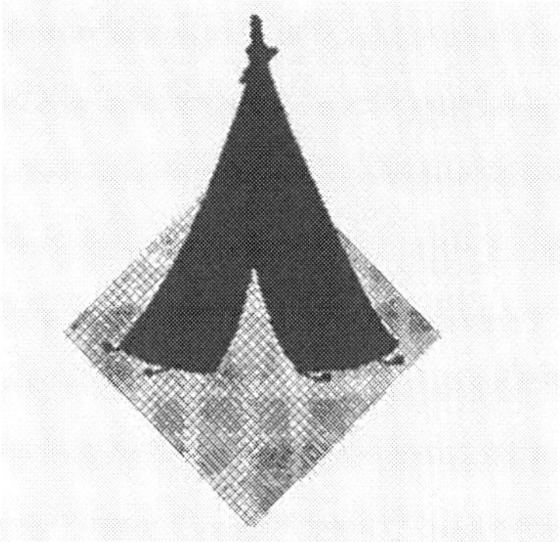

A shelter can protect you from the sun, insects, wind, rain, snow, hot or cold temperatures, and enemy observation. It can give you a feeling of well-being. It can help you maintain your will to survive.

In some areas, your need for shelter may tak precedence over your need for food and possibly even your need for water. For example, prolonged exposure to cold can cause excessive fatigue and weakness (exhaustion). An exhausted person may develop a "passive" outlook, thereby losing the will to survive.

The most common error in making a shelter is to make it too large. A shelter must be large enough to protect you. It must also be small enough to contain your body heat, especially in cold climates.

SHELTER SITE SELECTION

When you are in a survival situation and realize that shelter is a high priority, start looking for shelter as soon as possible. As you do so, remember what you will need at the site. Two requisites are—

- It must contain material to make the type of shelter you need.
- It must be large enough and level enough for you to lie down comfortably.

When you consider these requisites, however, you cannot ignore your tactical situation or your safety. You must also consider whether the site—

- Provides concealment from enemy observation.
- Has camouflaged escape routes.
- Is suitable for signaling, if necessary.
- Provides protection against wild animals and rocks and dead trees that might fall.
- Is free from insects, reptiles, and poisonous plants.

You must also remember the problems that could arise in your environment. For instance—

- Avoid flash flood areas in foothills.
- Avoid avalanche or rockslide areas in mountainous terrain.
- Avoid sites near bodies of water that are below the high water mark.

In some areas, the season of the year has a strong bearing on the site you select. Ideal sites for a shelter differ in winter and summer. During cold winter months you will want a site that will protect you from the cold and wind, but will have a source of fuel and water. During summer months in the same area you will want a source of water, but you will want the site to be almost insect free.

When considering shelter site selection, use the word BLISS as a guide.

B - Blend in with the surroundings.
L - Low silhouette.
I - Irregular shape.
S - Small.
S - Secluded location.

TYPES OF SHELTERS

When looking for a shelter site, keep in mind the type of shelter (protection) you need. However, you must also consider—

- How much time and effort you need to build the shelter.
- If the shelter will adequately protect you from the elements (sun, wind, rain, snow).
- If you have the tools to build it. If not, can you make improvised tools?
- If you have the type and amount of materials needed to build it.

To answer these questions, you need to know how to make various types of shelters and what materials you need to make them.

Poncho Lean-To

It takes only a short time and minimal equipment to build this lean-to (Figure 5-1). You need a poncho, 2 to 3 meters of rope or parachute suspension line, three stakes about 30 centimeters long, and two trees or two poles 2 to 3 meters apart. Before selecting the trees you will use or the location of your poles, check the wind direction. Ensure that the back of your lean-to will be into the wind.

To make the lean-to—

- Tie off the hood of the poncho. Pull the drawstring tight, roll the hood longways, fold it into thirds, and tie it off with the drawstring.
- Cut the rope in half. On one long side of the poncho, tie half of the rope to the corner grommet. Tie the other half to the other corner grommet.
- Attach a drip stick (about a 10-centimeter stick) to each rope about 2.5 centimeters from the grommet. These drip sticks will keep rain-water from running down the ropes into the lean-to. Tying strings (about 10 centimeters long) to each grommet along the poncho's top edge will allow the water to run to and down the lme without drip-ping into the shelter.
- Tie the ropes about waist high on the trees (uprights). Use a round turn and two half hitches with a quick-release knot.
- Spread the poncho and anchor it to the ground, putting sharpened sticks through the grommets and into the ground.

If you plan to use the lean-to for more than one night, or you expect ram, make a center support for the lean-to. Make this support with a line. Attach one end of the line to the poncho hood and the other end to an overhanging branch. Make sure there is no slack in the line.

Another method is to place a stick upright under the center of the lean-to. This method, however, will restrict your space and movements in the shelter.

For additional protection from wind and rain, place some brush, your rucksack, or other equipment at the sides of the lean-to.

To reduce heat loss to the ground, place some type of insulating mate-rial, such as leaves or pine needles, inside your lean-to.

Note: When at rest, you lose as much as 80 percent of your body heat to the ground.

To increase your security from enemy observation, lower the lean-to's silhouette by making two changes. First, secure the support lines to the trees at knee height (not at waist height) using two knee-high sticks in the two center grommets (sides of lean-to). Second, angle the poncho to the ground, securing it with sharpened sticks, as above.

Figure 5-1. Poncho lean-to.

Poncho Tent

This tent (Figure 5-2) provides a low silhouette. It also protects you from the elements on two sides. It has, however, less usable space and observation area than a lean-to, decreasing your reaction time to enemy detection. To make this tent, you need a poncho, two 1.5- to 2.5-meter ropes, six sharpened sticks about 30 centimeters long, and two trees 2 to 3 meters apart.

To make the tent—

- Tie off the poncho hood in the same way as the poncho lean-to.
- Tie a 1.5- to 2.5-meter rope to the center grommet on each side of the poncho.
- Tie the other ends of these ropes at about knee height to two trees 2 to 3 meters apart and stretch the poncho tight.
- Draw one side of the poncho tight and secure it to the ground pushing sharpened sticks through the grommets.
- Follow the same procedure on the other side.

If you need a center support, use the same methods as for the poncho lean-to. Another center support is an A-frame set outside but over the center of the tent (Figure 5-3). Use two 90- to 120-centimeter-long sticks, one with a forked end, to form the A-frame. Tie the hood's drawstring to the A-frame to support the center of the tent.

Figure 5-2. Poncho tent using overhanging branch.

Three-Pole Parachute Tepee

If you have a parachute and three poles and the tactical situation allows, make a parachute tepee. It is easy and takes very little time to make this tepee. It provides protection from the elements and can act as a signaling device by enhancing a small amount of light from a fire or candle. It is large enough to hold several people and their equipment and to allow sleepmg, cooking, and storing firewood.

You can make this tepee using parts of or a whole personnel main or reserve parachute canopy. If using a standard personnel parachute, you need three poles 3.5 to 4.5 meters long and about 5 centimeters in diameter.

To make this tepee (Figure 5-4)—

- Lay the poles on the ground and lash them together at one end.
- Stand the framework up and spread the poles to form a tripod.
- For more support, place additional poles against the tripod. Five or six additional poles work best, but do not lash them to the tripod.

Figure 5-3. Poncho tent with A-frame.

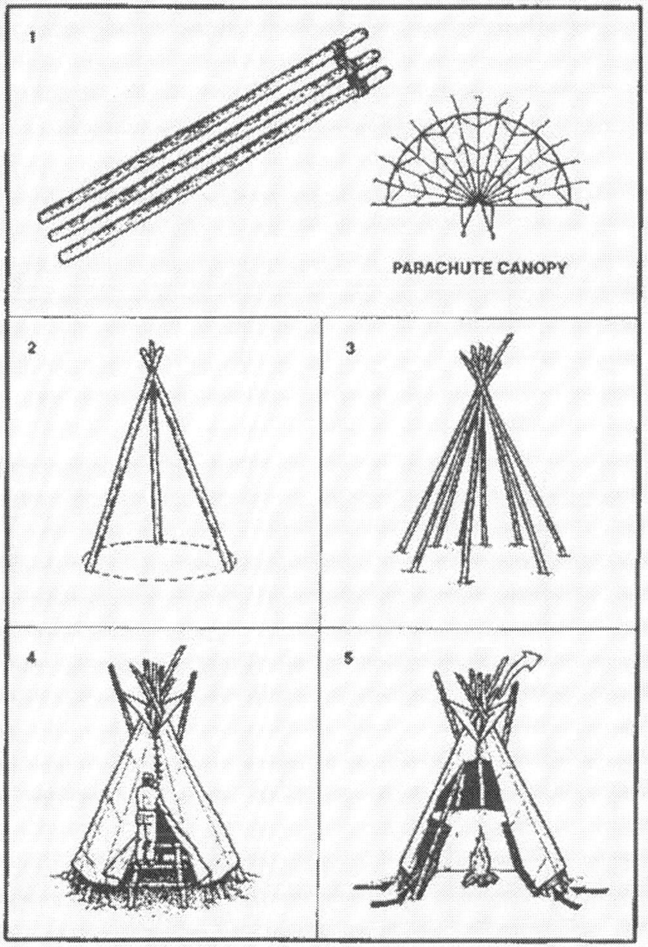

Figure 5-4. Three-pole parachute tepee.

- Determine the wind direction and locate the entrance 90 degrees or more from the mean wind direction.
- Lay out the parachute on the "backside" of the tripod and locate the bridle loop (nylon web loop) at the top (apex) of the canopy.
- Place the bridle loop over the top of a free-standing pole. Then place the pole back up against the tripod so that the canopy's apex is at the same height as the lashing on the three poles.
- Wrap the canopy around one side of the tripod. The canopy should be of double thickness, as you are wrapping an entire parachute. You need only wrap half of the tripod, as the remainder of the canopy will encircle the tripod in the opposite direction.
- Construct the entrance by wrapping the folded edges of the canopy around two free-standing poles. You can then place the poles side by side to close the tepee's entrance.
- Place all extra canopy underneath the tepee poles and inside to create a floor for the shelter.
- Leave a 30- to 50-centimeter opening at the top for ventilation if you intend to have a fire inside the tepee.

One-Pole Parachute Tepee

You need a 14-gore section (normally) of canopy, stakes, a stout center pole, and inner core and needle to construct this tepee. You cut the suspension lines except for 40- to 45-centimeter lengths at the canopy's lower lateral band.

To make this tepee (Figure 5-5)—

- Select a shelter site and scribe a circle about 4 meters in diameter on the ground.
- Stake the parachute material to the ground using the lines remaining at the lower lateral band.
- After deciding where to place the shelter door, emplace a stake and tie the first line (from the lower lateral band) securely to it.
- Stretch the parachute material taut to the next line, emplace a stake on the scribed line, and tie the line to it.
- Continue the staking process until you have tied all the lines.
- Loosely attach the top of the parachute material to the center pole with a suspension line you previously cut and, through trial and error, determine the point at which the parachute material will be pulled tight once the center pole is upright.

- Then securely attach the material to the pole.
- Using a suspension line (or inner core), sew the end gores together leaving 1 or 1.2 meters for a door.

No-Pole Parachute Tepee

You use the same materials, except for the center pole, as for the one-pole parachute tepee.

To make this tepee (Figure 5-6)—

- Tie a line to the top of parachute material with a previously cut suspension line.
- Throw the line over a tree limb, and tie it to the tree trunk.
- Starting at the opposite side from the door, emplace a stake on the scribed 3.5- to 4.3-meter circle.
- Tie the first line on the lower lateral band.
- Continue emplacing the stakes and tying the lines to them.
- After staking down the material, unfasten the line tied to the tree trunk, tighten the tepee material by pulling on this line, and tie it securely to the tree trunk.

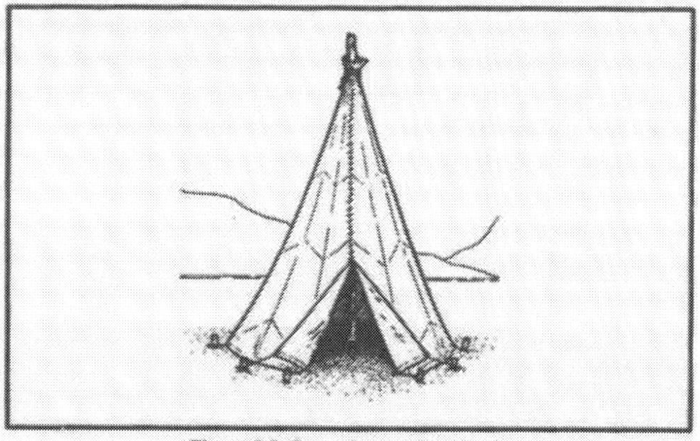

Figure 5-5. One-pole parachute tepee.

One-Man Shelter

A one-man shelter you can easily make using a parachute requires a tree and three poles. One pole should be about 4.5 meters long and the other two about 3 meters long.

To make this shelter (Figure 5-7)—

- Secure the 4.5-meter pole to the tree at about waist height.
- Lay the two 3-meter poles on the ground on either side of and in the same direction as the 4.5-meter pole.
- Lay the folded canopy over the 4.5 meter pole so that about the same amount of material hangs on both sides.
- Tuck the excess material under the 3-meter poles, and spread it on the ground inside to serve as a floor.
- Stake down or put a spreader between the two 3-meter poles at the shelter's entrance so they will not slide inward.
- Use any excess material to cover the entrance.

The parachute cloth makes this shelter wind resistant, and the shelter is small enough that it is easily warmed. A candle, used carefully, can keep the inside temperature comfortable. This shelter is unsatisfactory, however, when snow is falling as even a light snowfall will cave it in.

Figure 5-6. No-pole parachute tepee.

Parachute Hammock

You can make a hammock using 6 to 8 gores of parachute canopy and two trees about 4.5 meters apart (Figure 5-8).

Field-Expedient Lean-To

If you are in a wooded area and have enough natural materials, you can make a field-expedient lean-to (Figure 5-9) without the aid of tools or with only a knife. It takes longer to make this type of shelter than it does to make other types, but it will protect you from the elements.

You will need two trees (or upright poles) about 2 meters apart; one pole about 2 meters long and 2.5 centimeters in diameter; five to eight poles about 3 meters long and 2.5 centimeters in diameter for beams; cord or vines for securing the horizontal support to the trees; and other poles, saplings, or vines to crisscross the beams.

To make this lean-to—

- Tie the 2-meter pole to the two trees at waist to chest height. This is the horizontal support. If a standing tree is not available, construct a biped using Y-shaped sticks or two tripods.
- Place one end of the beams (3-meter poles) on one side of the horizontal support. As with all lean-to type shelters, be sure to place the lean-to's backside into the wind.

Figure 5-7. One-man shelter.

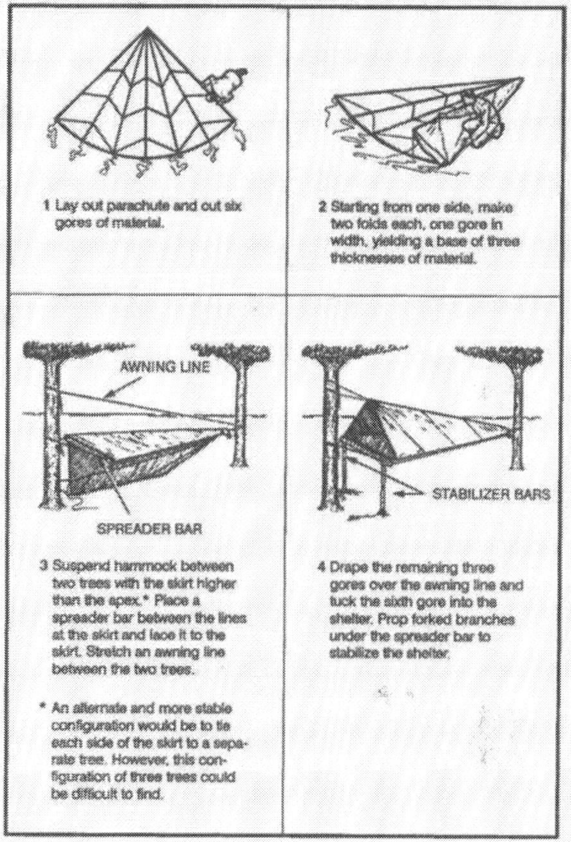

Figure 5-8. Parachute hammock.

- Crisscross saplings or vines on the beams.
- Cover the framework with brush, leaves, pine needles, or grass, starting at the bottom and working your way up like shingling.
- Place straw, leaves, pine needles, or grass inside the shelter for bedding.

In cold weather, add to your lean-to's comfort by building a fire reflector wall (Figure 5-9). Drive four 1.5-meter-long stakes into the ground to support the wall. Stack green logs on top of one another between the support stakes. Form two rows of stacked logs to create an inner space within the wall that you can fill with dirt. This action not only strengthens the wall but makes it more heat reflective. Bind the top of the support stakes so that the green logs and dirt will stay in place.

With just a little more effort you can have a drying rack. Cut a few 2-centimeter-diameter poles (length depends on the distance between the lean-to's horizontal support and the top of the fire reflector wall). Lay one end of the poles on the lean-to support and the other end on top of the reflector wall. Place and tie into place smaller sticks across these poles. You now have a place to dry clothes, meat, or fish.

Figure 5-9. Field-expedient lean-to and fire reflector.

Swamp Bed

In a marsh or swamp, or any area with standing water or continually wet ground, the swamp Ded (Figure 5-10) keeps you out of the water. When selecting such a site, consider the weather, wind, tides, and available materials.

To make a swamp bed—

- Look for four trees clustered in a rectangle, or cut four poles (bamboo is ideal) and drive them firmly into the ground so they form a rectangle. They should be far enough apart and strong enough to support your height and weight, to include equipment.
- Cut two poles that span the width of the rectangle. They, too, must be strong enough to support your weight.
- Secure these two poles to the trees (or poles). Be sure they are high enough above the ground or water to allow for tides and high water.
- Cut additional poles that span the rectangle's length. Lay them across the two side poles, and secure them.
- Cover the top of the bed frame with broad leaves or grass to form a soft sleeping surface.
- Build a fire pad by laying clay, silt, or mud on one comer of the swamp bed and allow it to dry.

Figure 5-10. Swamp bed.

Another shelter designed to get you above and out of the water or wet ground uses the same rectangular configuration as the swamp bed. You very simply lay sticks and branches lengthwise on the inside of the trees (or poles) until there is enough material to raise the sleeping surface above the water level.

Natural Shelters

Do not overlook natural formations that provide shelter. Examples are caves, rocky crevices, clumps of bushes, small depressions, large rocks on leeward sides of hills, large trees with low-hanging limbs, and fallen trees with thick branches. However, when selecting a natural formation—

- Stay away from low ground such as ravines, narrow valleys, or creek beds. Low areas collect the heavy cold air at night and are therefore colder than the surrounding high ground. Thick, orushy, low ground also harbors more insects.
- Check for poisonous snakes, ticks, mites, scorpions, and stinging ants.
- Look for loose rocks, dead limbs, coconuts, or other natural growth than could fall on your shelter.

Debris Hut

For warmth and ease of construction, this shelter is one of the best. When shelter is essential to survival, build this shelter.

To make a debris hut (Figure 5-11)—

- Build it by making a tripod with two short stakes and a long ridgepole or by placing one end of a long ridgepole on top of a sturdy base.
- Secure the ridgepole (pole running the length of the shelter) using the tripod method or by anchoring it to a tree at about waist height.
- Prop large sticks along both sides of the ridgepole to create a wedge-shaped ribbing effect. Ensure the ribbing is wide enough to accommodate your body and steep enough to shed moisture.
- Place finer sticks and brush crosswise on the ribbing. These form a latticework that will keep the insulating material (grass, pine needles, leaves) from falling through the ribbing into the sleeping area.
- Add light, dry, if possible, soft debris over the ribbing; until the insulating material is at least 1 meter thick—the thicker the better.
- Place a 30-centimeter layer of insulating material inside the shelter.
- At the entrance, pile insulating material that you can drag to you once inside the shelter to close the entrance or build a door.
- As a final step in constructing this shelter, add shingling material or branches on top of the debris layer to prevent the insulating material from blowing away in a storm.

Tree-Pit Snow Shelter

If you are in a cold, snow-covered area where evergreen trees grow and you have a digging tool, you can make a tree-pit shelter (Figure 5-12).

To make this shelter—

- Find a tree with bushy branches that provides overhead cover.
- Dig out the snow around the tree trunk until you reach the depth and diameter you desire, or until you reach the ground.

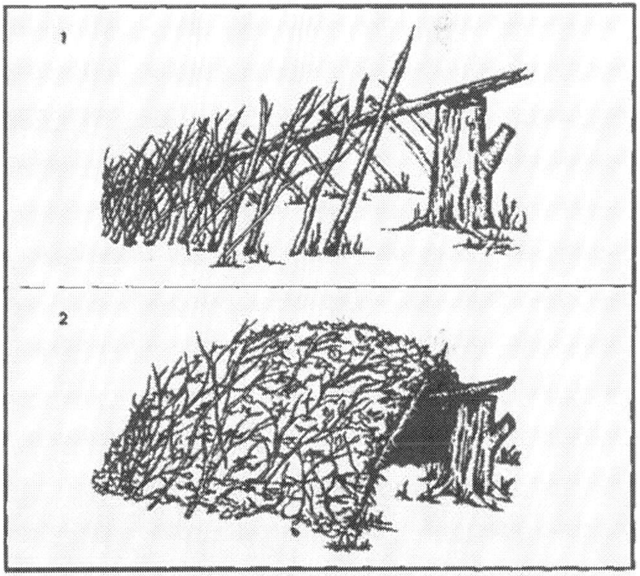

Figure 5-11. Debris hut.

- Pack the snow around the top and the inside of the hole to provide support.
- Find and cut other evergreen boughs. Place them over the top of the pit to give you additional overhead cover. Place evergreen boughs in the bottom of the pit for insulation.

See Chapter 15 for other arctic or cold weather shelters.

Beach Shade Shelter

This shelter protects you from the sun, wind, rain, and heat. It is easy to make using natural materials.

To make this shelter (Figure 5-13)—

- Find and collect driftwood or other natural material to use as support beams and as a digging tool.
- Select a site that is above the high water mark.
- Scrape or dig out a trench running north to south so that it receives the feast amount of sunlight. Make the trench long and wide enough for you to lie down comfortably.

Figure 5-12. Tree-pit snow shelter.

- Mound soil on three sides of the trench. The higher the mound, the more space inside the shelter.
- Lay support beams (driftwood or other natural material) that span the trench on top of the mound to form the framework for a roof.
- Enlarge the shelter's entrance by digging out more sand in front of it.
- Use natural materials such as grass or leaves to form a bed inside the shelter.

Desert Shelters

In an arid environment, consider the time, effort, and material needed to make a shelter. If you have material such as a poncho, canvas, or a parachute, use it along with such terrain features as rock outcropping, mounds of sand, or a depression between dunes or rocks to make your shelter.

Using rock outcroppings—

- Anchor one end of your poncho (canvas, parachute, or other material) on the edge of the outcrop using rocks or other weights.
- Extend and anchor the other end of the poncho so it provides the best possible shade.

Figure 5-13. Beach shade shelter.

In a sandy area—

- Build a mound of sand or use the side of a sand dune for one side of the shelter.
- Anchor one end of the material on top of the mound using sand or other weights.
- Extend and anchor the other end of the material so it provides the best possible shade.

Note: If you have enough material, fold it in half and form a 30-centimeter to 45-centimeter airspace between the two halves. This airspace will reduce the temperature under the shelter.

A belowground shelter (Figure 5-14) can reduce the midday heat as much as 16 to 22 degrees C (30 to 40 degrees F). Building it, however, requires more time and effort than for other shelters. Since your physical effort will make you sweat more and increase dehydration, construct it before the heat or the day.

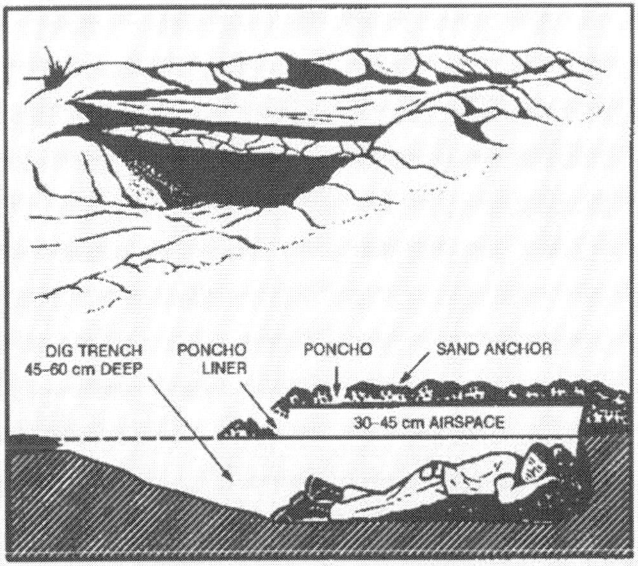

Figure 5-14. Belowground desert shelter.

To make this shelter—

- Find a low spot or depression between dunes or rocks. If necessary, dig a trench 45 to 60 centimeters deep and long and wide enough for you to lie in comfortably.
- Pile the sand you take from the trench to form a mound around three sides.
- On the open end of the trench, dig out more sand so you can get in and out of your shelter easily.
- Cover the trench with your material.
- Secure the material in place using sand, rocks, or other weights.

If you have extra material, you can further decrease the midday temperature in the trench by securing the material 30 to 45 centimeters above the other cover. This layering of the material will reduce the inside temperature 11 to 22 degrees C (20 to 40 degrees F).

Another type of belowground shade shelter is of similar construction, except all sides are open to air currents and circulation. For maximum protection, you need a minimum of two layers of parachute material (Figure 5-15). White is the best color to reflect heat; the innermost layer should be of darker material.

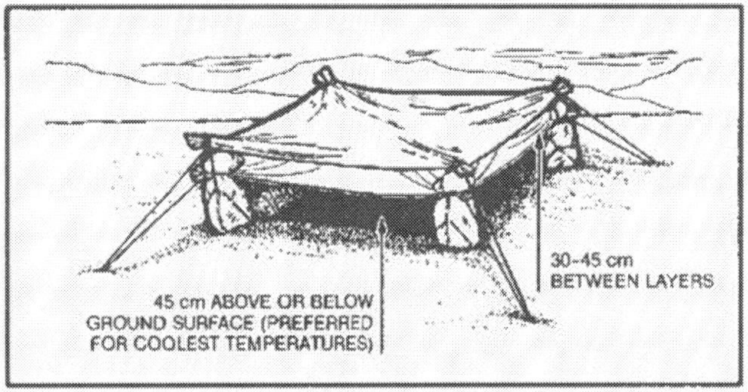

Figure 5-15. Open desert shelter.

CHAPTER 6

WATER PROCUREMENT

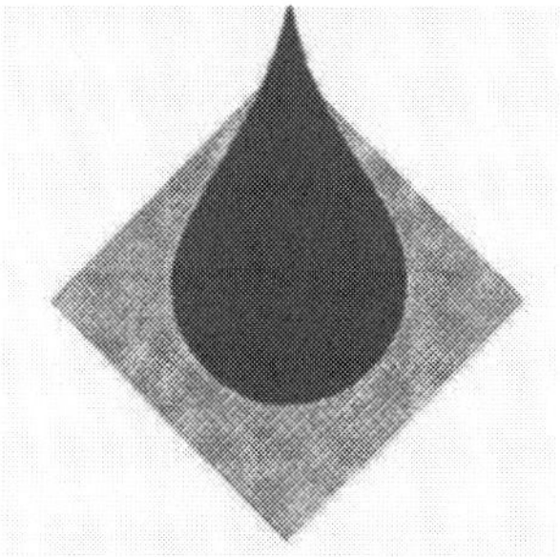

Water is one of your most urgent needs in a survival situation. You can't live long without it, especially in hot areas where you lose water rapidly through perspiration. Even in cold areas, you need a minimum of! liters of water each day to maintain efficiency.

More than three-fourths of your body is composed of fluids. Your body loses fluid as a result of heat, cold, stress, and exertion. To function effectively, you must replace the fluid your body loses. So, one of your first goals is to obtain an adequate supply of water.

WATER SOURCES

Almost any environment has water present to some degree. Figure 6-1 lists possible sources of water in various environments. It also provides information on how to make the water potable.

Note: If you do not have a canteen, a cup, a can, or other type of container, improvise one from plastic or water-resistant cloth. Shape the plastic or cloth into a bowl by pleating it. Use pins or other suitable items—even your hands—to hold the pleats.

Environment	Source of Water	Means of Obtaining and/or Making Potable	Remarks
Frigid areas	Snow and ice	Melt and purify.	**Do not eat** without melting! Eating snow and ice can reduce body temperature and will lead to more dehydration. Snow and ice are no purer than the water from which they come. Sea ice that is gray in color or opaque is salty. Do not use it without desalting it. Sea ice that is crystalline with a bluish cast has little salt in it.
At sea	Sea	Use desalter kit.	**Do not** drink seawater without desalting.
	Rain	Catch rain in tarps or in other water-holding material or containers.	If tarp or water-holding material has become encrusted with salt, wash it in the sea before using (very little salt will remain on it).
	Sea ice		See remarks above for frigid areas.

Figure 6-1. Water sources in different environments.

Environment	Source of Water	Means of Obtaining and/or Making Potable	Remarks
Beach	Ground	Dig hole deep enough to allow water to seep in; obtain rocks, build fire, and heat rocks; drop hot rocks in water; hold cloth over hole to absorb steam; wring water from cloth.	Alternate method if a container or bark pot is available: Fill container or pot with seawater; build fire and boil water to produce steam; hold cloth over container to absorb steam; wring water from cloth.
Desert	Ground • in valleys and low areas • at foot of concave banks of dry river beads • at foot of cliffs or rock outcrops • at first depression behind first sand dune of dry desert lakes • wherever you find damp surface sand • wherever you find green vegetation	Dig holes deep enough to allow water to seep in.	In a sand dune belt, any available water will be found beneath the original valley floor at the edge of dunes.
	Cacti	Cut off the top of a barrel cactus and mash or squeeze the pulp. **CAUTION: Do not eat pulp. Place pulp in mouth, suck out juice, and discard pulp.**	Without a machete, cutting into a cactus is difficult and takes time since you must get past the long, strong spines and cut through the tough rind.

Figure 6-1. Water sources in different environments (continued).

Environment	Source of Water	Means of Obtaining and/or Making Potable	Remarks
Desert (continued)	Depressions or holes in rocks		Periodic rainfall may collect in pools, seep into fissures, or collect in holes in rocks.
	Fissures in rock	Insert flexible tubing and siphon water. If fissure is large enough, you can lower a container into it.	
	Porous rock	Insert flexible tubing and siphon water.	
	Condensation on metal	Use cloth to absorb water, then wring water from cloth.	Extreme temperature variations between night and day may cause condensation on metal surfaces. Following are signs to watch for in the desert to help you find water: • All trails lead to water. You should follow in the direction in which the trails converge. Signs of camps, campfire ashes, animal droppings, and trampled terrain may mark trails. • Flocks of birds will circle over water holes. Some birds fly to water holes at dawn and sunset. Their flight at these times is generally fast and close to the ground. Bird tracks or chirping sounds in the evening or early morning sometimes indicate that water is nearby.

Figure 6-1. Water sources in different environments (continued).

If you do not have a reliable source to replenish your water supply, stay alert for ways in which your environment can help you.

CAUTION
Do not substitute the fluids listed in Figure 6-2 for water.

Heavy dew can provide water. Tie rags or tufts of fine grass around your ankles and walk through dew-covered grass before sunrise. As the rags or grass tufts absorb the dew, wring the water into a container. Repeat the process until you have a supply of water or until the dew is gone. Australian natives sometimes mop up as much as a liter an hour this way.

Bees or ants going into a hole in a tree may point to a water-filled hole. Siphon the water with plastic tubing: or scoop it up with an improvised dipper. You can also stuff cloth in the hole to absorb the water and then wring it from the cloth.

Water sometimes gathers in tree crotches or rock crevices. Use the above procedures to get the water. In arid areas, bird droppings around a crack in the rocks may indicate water in or near the crack.

Green bamboo thickets are an excellent source of fresh water. Water from green bamboo is clear and odorless. To get the water, bend a green

Fluid	Remarks
Alcoholic beverages	Dehydrate the body and cloud judgment.
Urine	Contains harmful body wastes. Is about 2 percent salt.
Blood	Is salty and considered a food; therefore, requires additional body fluids to digest. May transmit disease.
Seawater	Is about 4 percent salt. It takes about 2 liters of body fluids to rid the body of waste from 1 liter of seawater. Therefore, by drinking seawater you deplete your body's water supply, which can cause death.

Figure 6-2. The effects of substitute fluids.

bamboo stalk, tie it down, and cut off the top (Figure 6-3). The water will drip freely during the night. Old, cracked bamboo may contain water.

CAUTION
Purify the water before drinking it.

Wherever you find banana or plantain trees, you can get water. Cut down the tree, leaving about a 30-centimeter stump, and scoop out the center of the stump so that the hollow is bowl-shaped. Water Trom the roots will immediately start to fill the hollow. The first three fillings of water will be bitter, but succeeding fillings will be palatable. The stump (Figure 6-4) will supply water for up to four days. Be sure to cover it to keep out insects.

Some tropical vines can give you water. Cut a notch in the vine as high as you can reach, then cut the vine off close to the ground. Catch the dropping liquid in a container or in your mouth (Figure 6-5).

CAUTION
Do not drink the liquid if it is sticky, milky, or bitter tasting.

Figure 6-3. Water from green bamboo.

The milk from green (unripe) coconuts is a good thirst quencher. However, the mirk from mature coconuts contains an oil that acts as a laxative. Drink in moderation only.

In the American tropics you may find large trees whose branches support air plants. These air plants may holcfa considerable amount of

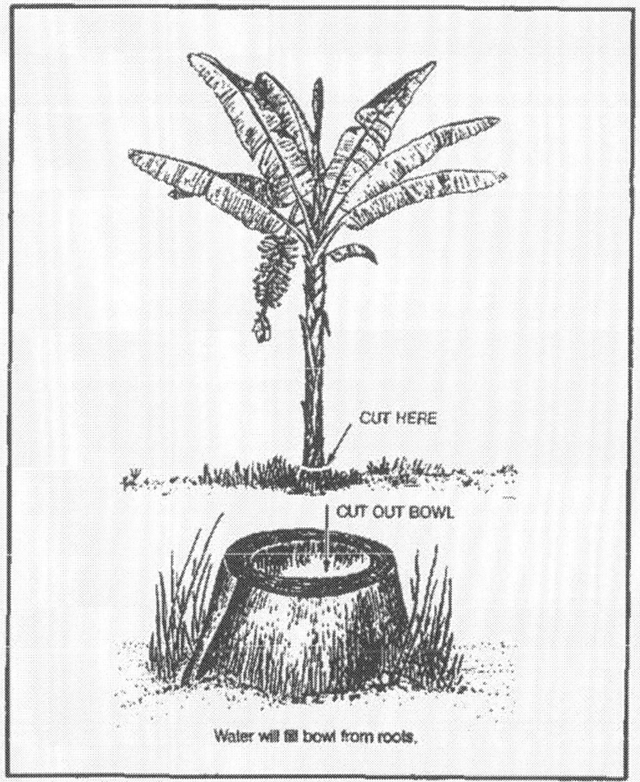

Figure 6-4. Water from plantain or banana tree stump.

rainwater in their overlapping, thickly growing leaves. Strain the water through a cloth to remove insects ana debris.

You can get water from plants with moist pulpy centers. Cut off a section of the plant and squeeze or smash the pulp so that the moisture runs out. Catch the liquid in a container.

Plant roots may provide water. Dig or pry the roots out of the ground, cut them into short pieces, and smash the pulp so that the moisture runs out. Catch the liquid in a container.

Fleshy leaves, stems, or stalks, such as bamboo, contain water. Cut or notch the stalks at the base of a joint to drain out the liquid.

The following trees can also provide water:

- *Palms.* Palms, such as the buri, coconut, sugar, rattan, and nips, contain liquid. Bruise a lower frond and pull it down so the tree will "bleed" at the injury.
- *Traveler's tree.* Found in Madagascar, this tree has a cuplike sheath at the base of its leaves in which water collects.

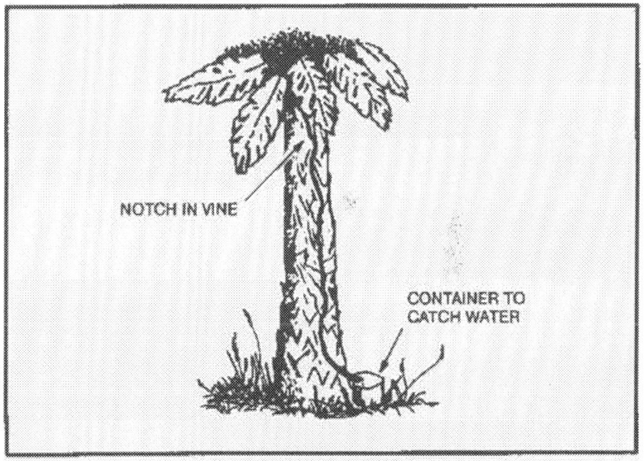

Figure 6-5. Water from a vine.

- *Umbrella tree.* The leaf bases and roots of this tree of western tropical Africa can provide water.
- *Baobab tree.* This tree of the sandy plains of northern Australia and Africa collects water in its bottlelike trunk during the wet season. Frequently, you can find clear, fresh water in these trees after weeks of dry weather.

CAUTION

Do not keep the sap from plants longer than 24 hours. It begins fermenting, becoming dangerous as a water source.

STILL CONSTRUCTION

You can use stills in various areas of the world. They draw moisture from the ground and from plant material. You need certain materials to build a still, and you need time to let it collect the water. It takes about 24 hours to get 05 to 1 liter of water.

Aboveground Still

To make the aboveground still, you need a sunny slope on which to place the still, a clear plastic bag, green leafy vegetation, and a small rock (Figure 6-6).

To make the still—

- Fill the bag with air by turning the opening into the breeze or by "scooping air into the bag.
- Fill the plastic bag half to three-fourths full of green leafy vegetation. Be sure to remove all hard sticks or sharp spines that might puncture the bag.

CAUTION

Do not use poisonous vegetation. It will provide poisonous liquid.

- Place a small rock or similar item in the bag.
- Close the bag and tie the mouth securely as close to the end of the bag as possible to keep the maximum amount of air space. If you have a piece of tubing, a small straw, or a hollow reed, insert one end in the mouth of the bag before you tie it securely. Then tie off or plug the tubing so that air will not escape. This tubing will allow you to drain out condensed water without untying the bag.
- Place the bag, mouth downhill, on a slope in full sunlight. Position the mouth or the bag slightly higher than the low point in the bag.
- Settle the bag in place so that the rock works itself into the low point in the bag.

To get the condensed water from the still, loosen the tie around the bag s mouth and tip the bag so that the water collected around the rock will drain out. Then retie the mouth securely and reposition the still to allow further condensation.

Change the vegetation in the bag after extracting most of the water from it. This will ensure maximum output of water.

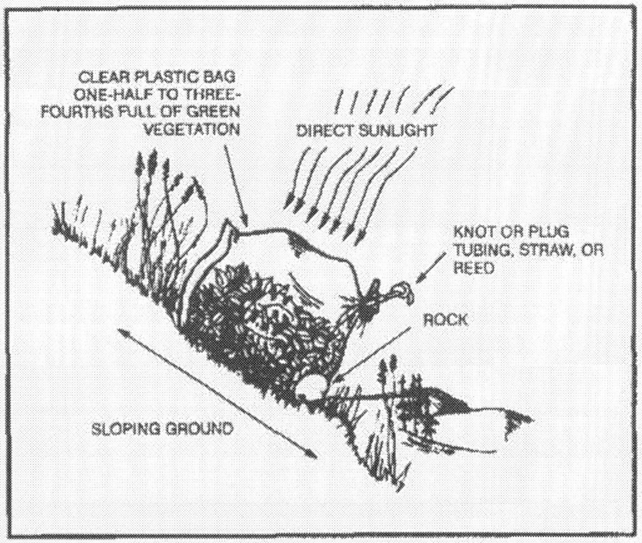

Figure 6-6. Aboveground solar water still.

Belowground Still

To make a belowground still, you need a digging tool, a container, a clear plastic sheet, a drinking tube, and a rock (Figure 6-7).

Select a site where you believe the soil will contain moisture (such as a dry stream bed or a low spot where rainwater has collected). The soil at this site should be easy to dig, and sunlight must hit the site most of the day.

To construct the still—

- Dig a bowl-shaped hole about 1 meter across and 60 centimeters deep.
- Dig a sump in the center of the hole. The sump's depth and perimeter will depend on the size of the container that you have to place in it. The bottom of the sump should allow the container to stand upright.
- Anchor the tubing to the container's bottom by forming a loose overhand knot in the tubing.
- Place the container upright in the sump.

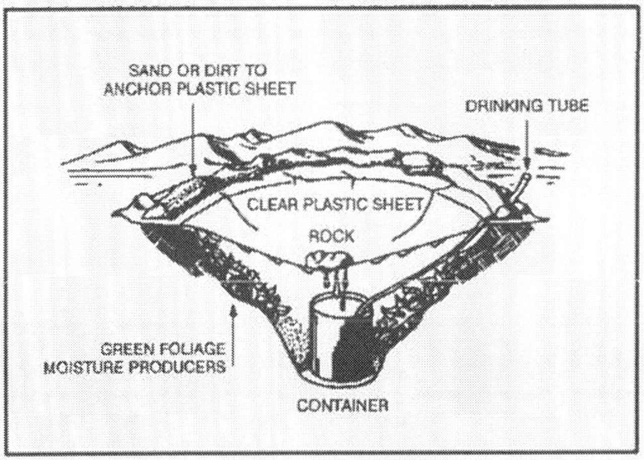

Figure 6-7. Belowground still.

- Extend the unanchored end of the tubing up, over, and beyond the hp of the hole.
- Place the plastic sheet over the hole, covering its edges with soil to hold it in place.
- Place a rock in the center of the plastic sheet.
- Lower the plastic sheet into the hole until it is about 40 centimeters below ground level. It now forms an inverted cone with the rock at its apex. Make sure that the cone's apex is directly over your container. Also make sure the plastic cone does not touch the sides of the hole because the earth will absorb the condensed water.
- Put more soil on the edges of the plastic to hold it securely in place and to prevent the loss of moisture.
- Plug the tube when not in use so that the moisture will not evaporate.

You can drink water without disturbing the still by using the tube as a straw.

You may want to use plants in the hole as a moisture source. If so, dig out additional soil from the sides of the hole to form a slope on which to place the plants. Then proceed as above.

If polluted water is your only moisture source, dig a small trough outside the hole about 25 centimeters from the still's lip (Figure 6-8). Dig the trough about 25 centimeters deep and 8 centimeters wide. Pour the polluted water in the trough. Be sure you do not spill any polluted water around the rim of the hole where the plastic sheet touches the soil. The trough holds the polluted water and the soil filters it as the still draws it. The water then condenses on the plastic and drains into the container. This process works extremely well when your only water source is salt water.

You will need at least three stills to meet your individual daily water intake needs.

WATER PURIFICATION

Rainwater collected in clean containers or in plants is usually safe for drinking. However, purify water from lakes, ponds, swamps, springs, or streams, especially the water near human settlements or m the tropics.

When possible, purify all water you got from vegetation or from the ground by using iodine or chlorine, or by boiling.

Purify water by—

- Using water purification tablets. (Follow the directions provided.)
- Placing 5 drops of 2 percent tincture of iodine in a canteen full of clear water. If the canteen is full of cloudy or cold water, use 10 drops. (Let the canteen of water stand for 30 minutes before drinking.)
- Boiling water for 1 minute at sea level, adding 1 minute for each additional 300 meters above sea level, or boil for 10 minutes no matter where you are.

By drinking nonpotable water you may contract diseases or swallow organisms that can harm you. Examples of such diseases or organisms are—

- *Dysentery.* Severe, prolonged diarrhea with bloody stools, fever, and weakness.
- *Cholera and tuphoid.You* may be susceptible to these diseases regardless of inoculations.

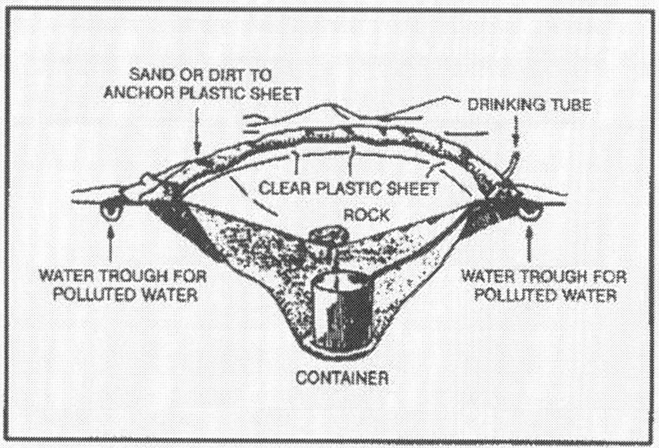

Figure 6-8. Belowground still to get potable water from polluted water.

- *Flukes.* Stagnant, polluted water—especially in tropical areas—often contains blood flukes. If you swallow flukes, they will bore into the bloodstream, live as parasites, and cause disease.
- *Leeches.* If you swallow a leech, it can hook onto the throat passage or inside the nose. It will suck blood, create a wound, and move to another area. Each bleeding wound may become infected.

WATER FILTRATION DEVICES

If the water you find is also muddy, stagnant, and foul smelling, you can clear the water—

- By placing it in a container and letting it stand for 12 hours.
- By pouring it through a filtering system.

Note: These procedures only clear the water and make it more palatable. You will have to purify it.

To make a filtering system, place several centimeters or layers of filtering material such as sand, crushed rock, charcoal, or cloth in bamboo, a hollow log, or an article of clothing (Figure 6-9).

Remove the odor from water by adding charcoal from your fire. Let the water stand for 45 minutes before drinking it.

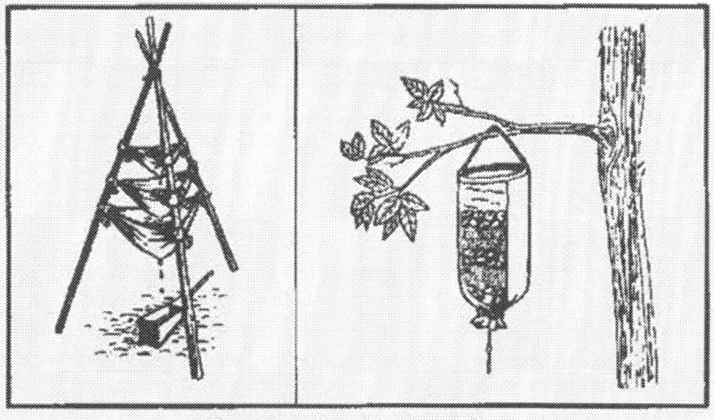

Figure 6-9. Water filtering systems.

CHAPTER 7

FIRECRAFT

In many survival situations, the ability to start a fire can make the difference between living and dying. Fire can fulfill many needs. It can provide warmth and comfort. It not only cooks and preserves food, it also provides warmth in the form of heated food that saves calories our body normally uses to produce body heat. You can use fire to purify water, sterilize bandages, signal for rescue, and provide protection from animals. It can be a psychological boost by providing peace of mind and companionship. You can also use fire to produce tools and weapons.

Fire can cause problems, as well. The enemy can detect the smoke and light it produces. It can cause forest fires or destroy essential equipment. Fire can also cause burns and carbon monoxide poisoning when used in shelters.

Remember weigh your need for fire against your need to avoid enemy detection.

BASIC FIRE PRINCIPLES

To build a fire, it helps to understand the basic principles of a fire. Fuel (in a nongaseous state) does not burn directly. When you apply heat to a fuel, it produces a gas. This gas, combined with oxygen m the air, burns.

Understanding the concept of the fire triangle is very important in correctly constructing ana maintaining a fire. The three sides of the triangle represent *air, heat,* and/we/. If you remove any of these, the fire will go out. The correct ratio of these components is very important for a fire to burn at its greatest capability. The only way to learn this ratio is to practice.

SITE SELECTION AND PREPARATION

You will have to decide what site and arrangement to use. Before building a fire consider—

- The area (terrain and climate) in which you are operating.
- The materials and tools available.
- Time: how much time you have?
- Need: why you need a fire?
- Security: how close is the enemy?

Look for a dry spot that—

- Is protected from the wind.
- Is suitably placed in relation to your shelter (if any).
- Will concentrate the heat in the direction you desire.
- Has a supply of wood or other fuel available. (See page 7-6 for types of material you can use.)

If you are in a wooded or brush-covered area, clear the brush and scrape the surface soil from the spot you have selected. Clear a circle at least 1 meter in diameter so there is little chance of the fire spreading.

If time allows, construct a fire wall using logs or rocks. This wall will help to reflector direct the heat where you want it (Figure 7-1). It will also reduce flying sparks and cut down on the amount of wind blowing into the nre/However, you will need enough wind to keep the fire burning.

CAUTION

Do not use wet or porous rocks as they may explode when heated.

In some situations, you may find that an underground fireplace will best meet your needs. It conceals the fire and serves well for cooking food. To make an underground fireplace or Dakota fire hole (Figure 7-2)-

- Dig a hole in the ground.
- On the upwind side of this hole, poke or dig a large connecting hole for ventilation.
- Build your fire in the hole as illustrated.

Figure 7-1. Types of fire walls.

If you are in a snow-covered area, use green logs to make a dry base for your fire (Figure 7-3). Trees with wrist-sized trunks are easily broken in extreme cold. Cut or break several green logs and lay them side by side on top of the snow. Add one or two more layers. Lay the top layer of logs opposite those below it.

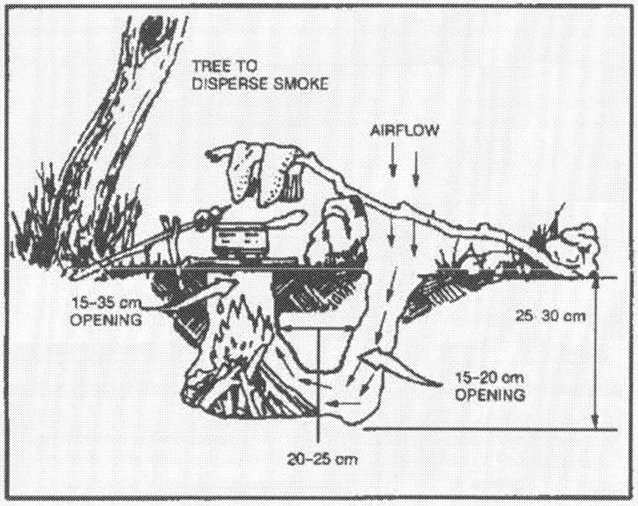

Figure 7-2. Dakota fire hole.

Figure 7-3. Base for fire in snow-covered area.

FIRE MATERIAL SELECTION

You need three types of materials (Figure 7-4) to build a fire—tinder, kindling, and fuel.

Tinder is dry material that ignites with little heat—a spark starts a fire. The tinder must be absolutely dry to be sure just a spark will ignite it. If you only have a device that generates sparks, charred cloth will be almost essential. It holds a spark for long periods, allowing you to put tinder on the hot area to generate a small flame. You can make charred cloth by heating cotton cloth until it turns black, but does not burn. Once it is black, you must keep it in an airtight container to keep it dry. Prepare this cloth well in advance of any survival situation. Ada it to your individual survival kit.

Kindling is readily combustible material that you add to the burning tinder. Again, this material should be absolutely dry to ensure rapid burning. Kindling increases the fire's temperature so that it will ignite less combustible material.

Fuel is less combustible material that burns slowly and steadily once ignited.

HOW TO BUILD A FIRE

There are several methods for laying a fire, each of which has advantages. The situation you find yourself in will determine which fire to use.

Tepee

To make this fire (Figure 7-5), arrange the tinder and a few sticks of kindling in the shape of a tepee or cone. Light the center. As the tepee burns, the outside logs will fall inward, feeding the fire. This type or fire bums well even with wet wood.

Lean-To

To lay this fire (Figure 7-5), push a green stick into the ground at a 30-degree angle. Point the end of the stick in the direction of the wind. Place some tinder deep under this lean-to stick. Lean pieces of kindling against the lean-to stick. Light the tinder. As the kindling catches fire from the tinder, add more kindling.

Tinder	Kindling	Fuel
• Birch bark • Shredded inner bark from cedar, chestnut, red elm trees • Fine wood shavings • Dead grass, ferns, moss, fungi • Straw • Sawdust • Very fine pitchwood scrapings • Dead evergreen needles • Punk (the completely rotted portions of dead logs or trees) • Evergreen tree knots • Bird down (fine feathers) • Down seed heads (milkweed, dry cattails, bulrush, or thistle) • Fine, dried vegetable fibers • Spongy threads of dead puffball • Dead palm leaves • Skinlike membrane lining bamboo • Lint from pocket and seams • Charred cloth • Waxed paper • Outer bamboo shavings • Gunpowder • Cotton • Lint	• Small twigs • Small strips of wood • Split wood • Heavy cardboard • Pieces of wood removed from the inside of larger pieces • Wood that has been doused with highly flammable materials, such as gasoline, oil, or wax	• Dry, standing wood and dry, dead branches • Dry inside (heart) of fallen tree trunks and large branches • Green wood that is finely split • Dry grasses twisted into bunches • Peat dry enough to burn (this may be found at the top of undercut banks) • Dried animal dung • Animal fats • Coal, oil shale, or oil lying on the surface

Figure 7-4. Materials for building fires.

Cross-Ditch

To use this method (Figure 7-5), scratch a cross about 30 centimeters in size in the ground. Dig the cross 7.5 centimeters deep. Put a large wad of tinder in the middle of the cross. Build a kindling pyramid above the tinder. The shallow ditch allows air to sweep under the tinder to provide a draft.

Pyramid

To lay this fire (Figure 7-5), place two small logs or branches parallel on the ground. Place a solid layer of small logs across the parallel logs. Add three or four more layers of logs or branches, each layer smaller than and at a right angle to the layer below it. Make a starter fire on top of the pyramid. As the starter fire burns, it will ignite the logs below it. This gives you a fire that burns downward, requiring no attention during the night.

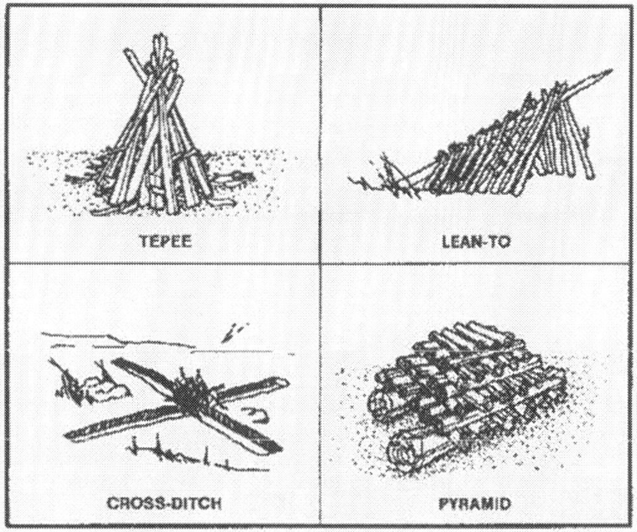

Figure 7-5. Methods for laying fires.

There are several other ways to lay a fire that are quite effective. Your situation and the material available in the area may make another method more suitable.

HOW TO LIGHT A FIRE

Always light your fire from the upwind side. Make sure to lay your tinder, kindling, and fuel so that your fire will burn as long as you need it. Igniters provide the initial heat required to start the tinder burning, they fall into two categories: modem methods and primitive methods.

Modern Methods

Modem igniters use modem devices—items we normally think of to start a fire.

Matches

Make sure these matches are waterproof. Also, store them in a waterproof container along with a dependable striker pad.

Convex lens

Use this method (Figure 7-6) only on bright, sunny days. The lens can come from binoculars, camera, telescopic sights, or magnifying glasses. Angle the lens to concentrate the sun s rays on the tinder. Hold the lens over the same spot until the tinder begins to smolder. Gently blow or fan the tinder into flame, and apply it to the fire lay.

Metal Match

Place a flat, dry leaf under your tinder with a portion exposed. Place the tip of the metal match on the dry leaf, holding the metal match in one hand and a knife in the other. Scrape your knife against the metal match to produce sparks. The sparks will hit the tinder. When the tinder starts to smolder, proceed as above.

Battery

Use a battery to generate a spark. Use of this method depends on the type of battery available. Attach a wire to each terminal. Touch the ends of the bare wires together next to the tinder so the sparks will ignite it.

Gunpowder

Often, you will have ammunition with your equipment. If so, carefully extract the bullet from the shell casing, and use the gunpowder as tinder. A spark will ignite the powder. Be extremely careful when extracting the bullet from the case.

Primitive Methods

Primitive igniters are those attributed to our early ancestors.

Flint and Steel

The direct spark method is the easiest of the primitive methods to use. The flint and steel method is the most reliable of the direct spark methods. Strike a flint or other hard, sharp-edged rock edge with a piece of carbon steel (stainless steel will not produce a good spark). This method requires a loose-jointed wrist and practice. When a spark nas caught in the tinder, blow on it. The spark will spread and burst into flames.

Fire-Plow

The fire-plow (Figure 7-7) is a friction method of ignition. You rub a hardwood shaft against a softer wood base. To use this method, cut a straight groove in the base and plow the blunt tip of the shaft up and down the groove. The plowing action of the shaft pushes out small

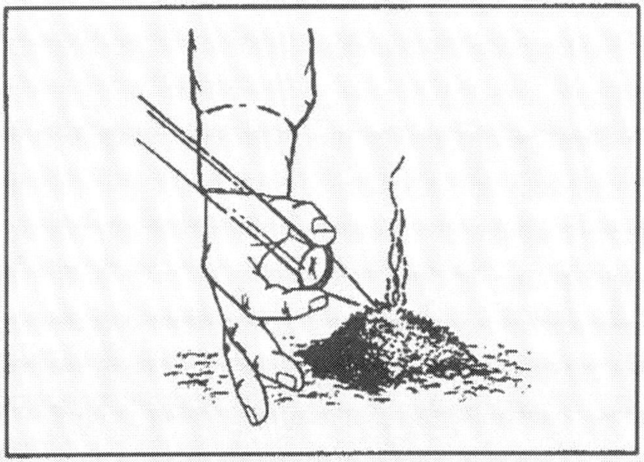

Figure 7-6. Lens method.

particles of wood fibers. Then, as you apply more pressure on each stroke, the friction ignites the wood particles.

Bow and Drill

The technique of starting a fire with a bow and drill (Figure 7-8) is simple, but you must exert much effort and be persistent to produce a fire. You need the following items to use this method:

- *Socket.* The socket is an easily grasped stone or piece of hardwood or bone with a slight depression m one side. Use it to hold the drill in place and to apply downward pressure.
- *Drill.* The drill should be a straight, seasoned hardwood stick about 2 centimeters in diameter and 25 centimeters long. The top end is round and the low end blunt (to produce more friction).
- *Fire board.* Its size is up to you. A seasoned softwood board about 2.5 centimeters thick and 10 centimeters wide is preferable. Cut a depression about 2 centimeters from the edge on one side of the board. On the underside, make a V-shaped cut from the edge of the board to the depression.

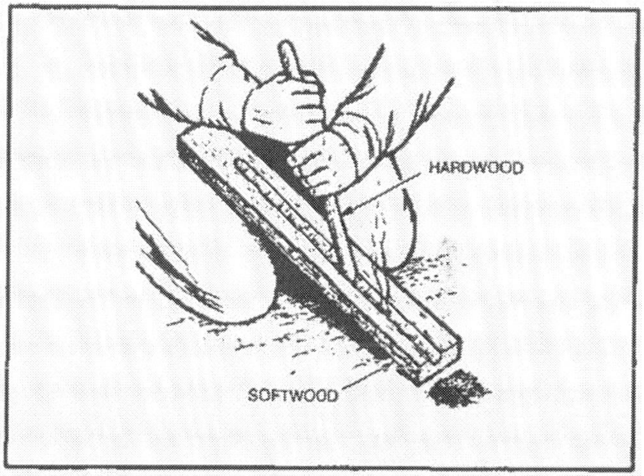

Figure 7-7. Fire-plow.

- *Bow.* The bow is a resilient, green stick about 2.5 centimeters in diameter and a string. The type of wood is not important. The bowstring can be any type of cordage. You tie the bowstring from one end of the bow to the other, without any slack.

To use the bow and drill, first prepare the fire lay. Then place a bundle of tinder under the V-shaped cut in the fire board. Place one foot on the fire board. Loop the bowstring over the drill and place the drill in the precut depression on the fire board. Place the socket, held in one hand, on the top of the drill to hold it in position. Press down on the drill and saw the bow back and forth to twirl the drill (Figure 7-8). Once you have established a smooth motion, apply more downward pressure and work the bow faster. This action will grmd hot black powder into the tinder, causing a spark to catch. Blow on the tinder until it ignites.

Note: Primitive fire-building methods are exhaustive and require practice to ensure success.

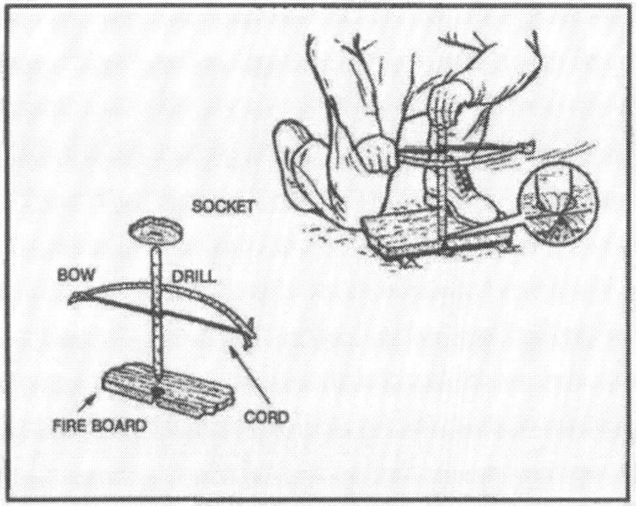

Figure 7-8. Bow and drill.

HELPFUL HINTS
Use nonaromatic seasoned hardwood for fuel, if possible.
Collect kindling and tinder along the trail.
Add insect repellent to the tinder.
Keep the firewood dry.
Dry damp firewood near the fire.
Bank the fire to keep the coals alive overnight,
Carry lighted punk, when possible.
Be sure the fire is out before leaving camp.
Do not select wood lying on the ground. It may appear to be dry but generally doesn't provide enough friction.

CHAPTER 8

FOOD PROCUREMENT

After water, man's most urgent requirement is food. In contemplating virtually any hypothetical survival situation, the mind immediately turns to thoughts of food. Unless the situation occurs in an arid environment, even water, which is more important to maintaining body functions, will almost always follow food in our initial thoughts. The survivor must remember that the three essentials of survival—water, food, and shelter-are prioritized according to the estimate of the actual situation. This estimate must not only be timely but accurate as well. Some situations may well dictate that shelter precede both food and water.

ANIMALS FOR FOOD

Unless you have the chance to take large game, concentrate your efforts on the smaller animals, due to their abundance. The smaller animal

species are also easier to prepare. You must not know all the animal species that are suitable as food. Relatively few are poisonous, and they make a smaller list to remember. What is important is to learn the habits and behavioral patterns of classes of animals. For example, animals that are excellent choices for trapping, those that inhabit a particular range and occupy a den or nest, those that have somewhat fixed feeding areas, and those that have trails leading from one area to another. Larger, herding animals, such as elk or caribou, roam vast areas and are somewhat more difficult to trap. Also, you must understand the food choices of a particular species.

You can, with relatively few exceptions, eat anything that crawls, swims, walks, or flies. The first obstacle is overcoming your natural aversion to a particular food source. Historically, people in starvation situations have resorted to eating everything imaginable for nourishment. A person who ignores an otherwise healthy food source due to a personal bias, or because he feels it is unappetizing, is risking his own survival. Although it may prove difficult at first, a survivor must eat what is available to maintain his health.

Insects

The most abundant life-form on earth, insects are easily caught. Insects provide 65 to 80 percent protein compared to 20 percent forbeef. This tact makes insects an important, if not overly appetizing, food source. Insects to avoid include all adults that sting or bite, hairy or brightly colored insects, and caterpillars and insects that have a pungent odor. Also avoid spiders and common disease carriers such as ticks, flies, and mosquitoes.

Rotting logs lying on the ground are excellent places to look for a variety of msects including ants, termites, beetles, and grubs, which are beetle larvae. Do not overlook insect nests on or in the ground. Grassy areas, such as fields, are good areas to search because the insects are easily seen. Stones, boards, or other materials lying on the ground provide the insects with good nesting sites. Check these sites. Insect larvae are also edible. Insects such as beetles and grasshoppers that have a hard outer shell will have parasites. Cook them before eating. Remove any wings and barbed legs also. You can eat most insects raw. The taste varies from one species to another. Wood grubs are bland, while some species of ants store honey in their bodies, giving them a sweet taste.

You can grind a collection of insects into a paste. You can mix them with edible vegetation. You can cook them to improve their taste.

Worms

Worms *(Annelidea)* are an excellent protein source. Dig for them in damp humus soil or watch for them on the ground after a rain. After capturing them, drop them into clean, potable water for a few minutes. The worms will naturally purge or wash themselves out, after which you can eat them raw.

Crustaceans

Freshwater shrimp range in size from 0.25 centimeter up to 2.5 centimeters. They canlorm rather large colonies in mats of floating algae or in mud bottoms of ponds and lakes.

Crayfish are akin to marine lobsters and crabs. You can distinguish them by their hard exoskeleton and five pairs of legs, the front pair having oversized pincers. Crayfish are active at night, but you can locate them in the daytime by looking under and around stones in streams. You can also find them by looking in the soft mud near the chimneylike breathing holes of their nests, rou can catch crayfish by tying bits of offal or internal organs to a string. When the crayfish grabs the bait, pull it to shore before it has a chance to release the bait.

You find saltwater lobsters, crabs, and shrimp from the surf's edge out to water 10 meters deep. Shrimp may come to a light at night where you can scoop them up with a net. You can catch lobsters and crabs with a baited trap or a baited hook. Crabs will come to bait placed at the edge of the surf, where you can trap or net them. Lobsters and crabs are nocturnal and caught best at night.

Mollusks

This class includes octopuses and freshwater and saltwater shellfish such as snails, clams, mussels, bivalves, barnacles, periwinkles, chitons, and sea urchins (Figure 8-1). You find bivalves similar to our freshwater mussel and terrestrial and aquatic snails worldwide under all water conditions.

River snails or freshwater periwinkles are plentiful in rivers, streams, and lakes of northern coniferous forests. These snails may be pencil point or globular in shape.

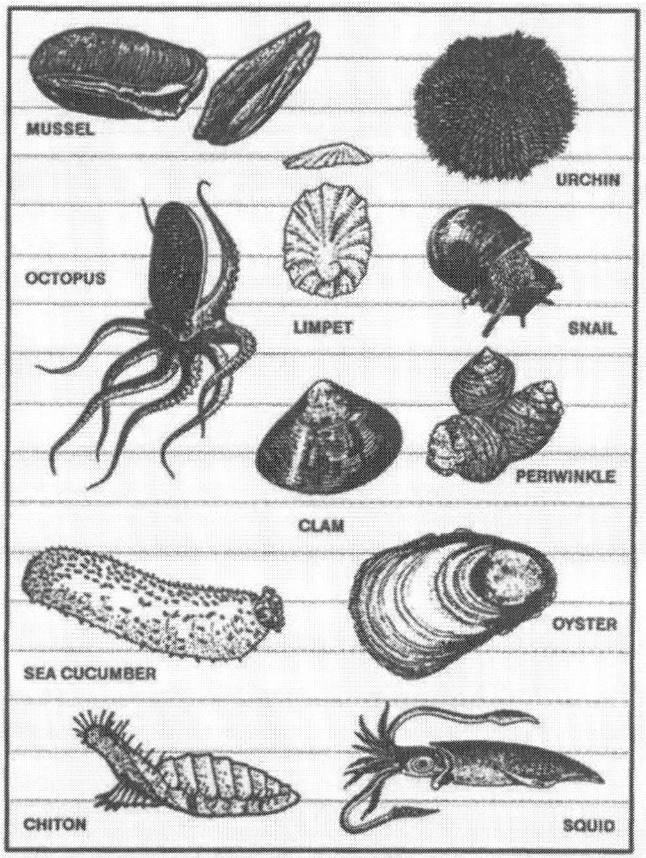

Figure 8-1. Edible mollusks.

In fresh water, look for mollusks in the shallows, especially in water with a sandy or muddy bottom. Look for the narrow trails they leave in the mud or for the dark elliptical slit of their open valves.

Near the sea, look in the tidalpools and the wet sand. Rocks along beaches or extending as reefs into deeper water often bear clinging shellfish. Snails andlimpets cling to rocks and seaweed from the low water mark upward. Large snails, called chitons, adhere tightly to rocks above the surf line.

Mussels usually form dense colonies in rock pools, on logs, or at the base of boulders.

CAUTION
Mussels may be poisonous in tropical zones during the summer!

Steam, boil, or bake mollusks in the shell. They make excellent stews in combination with greens and tubers.

CAUTION
Do not eat shellfish that are not covered by water at high tide!

Fish

Fish represent a good source of protein and fat. They offer some distinct advantages to the survivor or evader. They are usually more abundant than mammal wildlife, and the ways to get them are silent. To be successful at catching fish, you must know their habits. For instance, fish tend to feed heavily before a storm. Fish are not likely to feed after a storm when the water is muddy and swollen. Light often attracts fish at night. When there is a heavy current, fish will rest in places where there is an eddy, such as near rocks. Fish will also gather where there are deep pools, under overhanging brush, and in and around submerged foliage, logs, or other objects that offer them shelter.

There are no poisonous freshwater fish. However, the catfish species has sharp, needlefike protrusions on its dorsal fins and barbels. These can inflict painful puncture wounds that quickly become infected.

Cook all freshwater fish to kill parasites. Also cook saltwater fish caught within a reef or within the influence of a freshwater source as a precaution. Any marine life obtained farther out in the sea will not contain parasites because of the saltwater environment. You can eat these raw.

Certain saltwater species of fish have poisonous flesh. In some species the poison occurs seasonally in others, it is permanent. Examples of poisonous saltwater fish are the porcupine fish, triggerfish, cowfish, thorn fish, oilfish, red snapper, jack, and puffer (Figure 8-2). The barracuda, while not actually poisonous itself, may transmit ciguatera (fish poisoning) if eaten raw.

Amphibians

Frogs and salamanders are easily found around bodies of fresh water. Frogs seldom move from the safety of the water's edge. At the first sign of danger, they plunge into the water and bury themselves in the mud and debris. There are few poisonous species of frogs. Avoid any brightly colored frog or one that has a distinct " X " mark on it's back. Do not confuse toads with frogs. You normally find toads in drier environments. Several species of toads secrete a poisonous substance through their skin as a defense against attack. Therefore, to avoid poisoning, do not handle or eat toads.

Salamanders are nocturnal. The best time to catch them is at night using a light. They can range in size from a few centimeters to well over 60 centimeters in length. Look in water around rocks and mud banks for salamanders.

Reptiles

Reptiles are a good protein source and relatively easy to catch. You should cook them, but in an emergency, you can eat them raw. Their raw flesh may transmit parasites, but because reptiles are cold-blooded, they do not carry the blood diseases of the warm-blooded animals.

The box turtle is a commonly encountered turtle that you should not eat. It feeds on poisonous mushrooms and may build up a highly toxic poison in its flesh. Cooking does not destroy this toxin. Avoid the hawksbill turtle, found in the Atlantic Ocean, because of its poisonous thorax gland. Poisonous snakes, alligators, crocodiles, and large sea turtles present obvious hazards to the survivor.

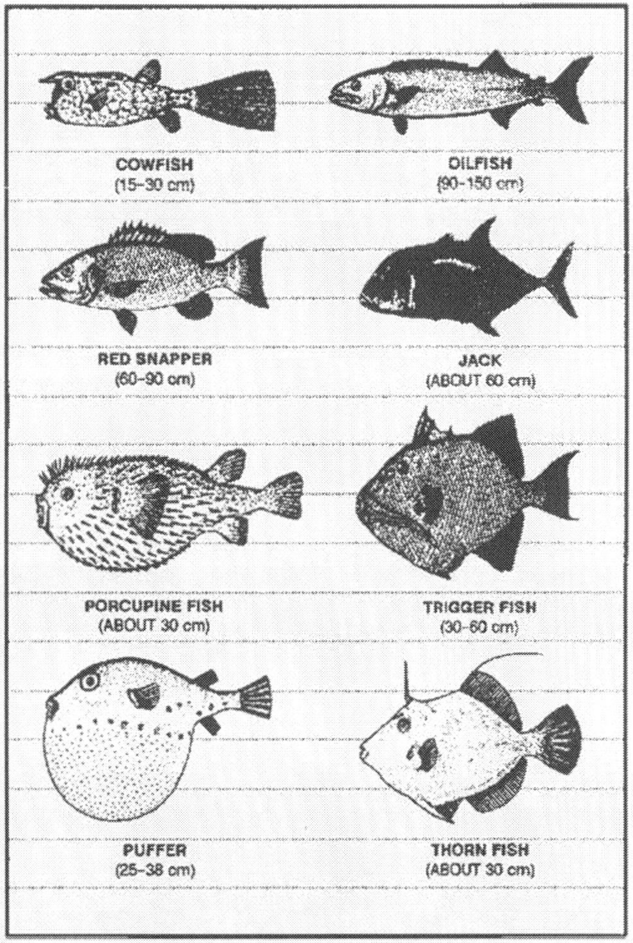

Figure 8-2. Fish with poisonous flesh.

Birds

All species of birds are edible, although the flavor will vary considerably. You may skin fish-eating birds to improve their taste. As with any wild animal, you must understand birds' common habits to have a realistic chance of capturing them. You can take pigeons, as well as some other species, from their roost at night by hand. During the nesting season, some species will not leave the nest even when approached. Knowing where and when the birds nest makes catching them easier (Figure 8-3). Birds tend to have regular flyways going from the roost to a feeding area, to water, and so forth. Careful observation should reveal where these flyways are and indicate good areas for catching birds in nets stretched across the flyways (Figure 8-4). Roosting sites and waterholes are some of the most promising areas for trapping or snaring.

Nesting birds present another food source—eggs. Remove all but two or three eggs from the clutch, marking the ones at you leave. The bird will continue to lay more eggs to fill the clutch. Continue removing the fresh eggs, leaving the ones you marked.

Types of Birds	Frequent Nesting Places	Nesting Periods
Inland birds	Trees, woods, or fields	Spring and early summer in temperate and arctic regions; year round in the tropics
Cranes and herons	Mangrove swamps or high trees near water	Spring and early summer
Some species of owls	High trees	Late December through March
Ducks, geese, and swans	Tundra areas near ponds, rivers, or lakes	Spring and early summer in arctic regions
Some sea birds	Sandbars or low sand islands	Spring and early summer in temperate and arctic regions
Gulls, auks, murres, and cormorants	Steep rocky coasts	Spring and early summer in temperate and arctic regions

Figure 8-3. Bird nesting places.

Mammals

Mammals are excellent protein sources and, for Americans, the most tasty food source. There are some drawbacks to obtaining mammals. In a hostile environment, the enemy may detect any traps or snares placed on land. The amount of injury an animal can inflict is in direct proportion to its size. All mammals nave teeth and nearly all will bite m self-defense. Even a squirrel can inflict a serious wound and any bite presents a serious risk of infection. Also, a mother can be extremely aggressive in defense of her young. Any animal with no route of escape will fight when cornered.

All mammals are edible; however, the polar bear and bearded seal have toxic levels of vitamin A in their livers. The platypus, native to Australia and Tasmania, is an egg-laying, semiaquatic mammal that has poisonous glands. Scavenging mammals, such as the opossum, may carry diseases.

TRAPS AND SNARES

For an unarmed survivor or evader, or when the sound of a rifle shot could be a problem, trapping or snaring wild game is a good alternative.

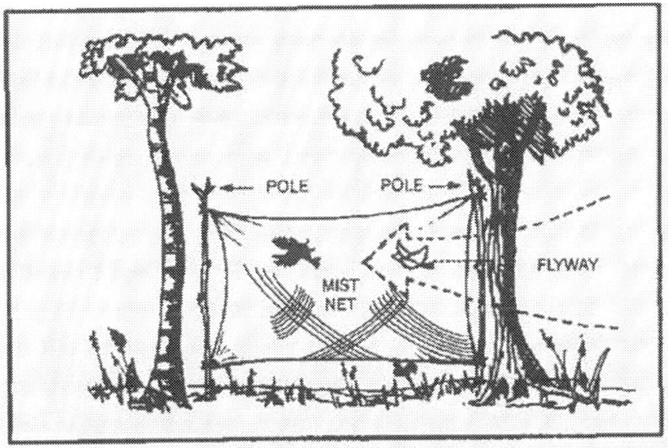

Figure 8-4. Catching birds in a net.

Several well-placed traps have the potential to catch much more game than a man with a rifle is likely to snoot. To be effective with any type of trap or snare, you must—

- Be familiar with the species of animal you intend to catch.
- Be capable of constructing a proper trap.
- Not alarm the prey by leaving signs of your presence.

There are no catchall traps you can set for all animals. You must determine what species are in a given area and set your traps specifically with those animals in mind. Look for the following

- Runs and trails.
- Tracks.
- Droppings.
- Chewed or rubbed vegetation.
- Nesting or roosting sites.
- Feeding and watering areas.

Position your traps and snares where there is proof that animals pass through. You must determine if it is a "run" or a "trail." A trail will show signs of use by several species and will be rather distinct. A run is usually smaller and less distinct and will only contain signs of one species. You may construct a perfect snare, but it will not catch anything if haphazardly placed in the woods. Animals have bedding areas, waterholes, and feedmg areas with trails leading from one to another. You must place snares and traps around these areas to be effective.

For an evader in a hostile environment, trap and snare concealment is important. It is equally important, however, not to create a disturbance that will alarm the animal and cause it to avoid the trap. Therefore, if you must dig, remove all fresh dirt from the area. Most animals will instinctively avoid a pitfall-type trap. Prepare the various parts of a trap or snare away from the site, carry them in, and set them up. Such actions make it easier to avoid disturbing the local vegetation, thereby alerting the prey. Do not use freshly cut, live vegetation to construct a trap or snare. Freshly cut vegetation will "bleed' sap that has an odor the prey will be able to smell. It is an alarm signal to the animal.

You must remove or mask the human scent on and around the trap you set. Although birds do not have a developed sense of smell, nearly all mammals depend on smell even more than on sight. Even the slightest human scent on a trap will alarm the prey and cause it to avoid the area. Actually removing the scent from a trap is difficult but masking it is relatively easy. Use the fluid from the gall and urine bladders of previous kills. Do not use human urine. Mud, particularly from an area with plenty of rotting vegetation, is also good. Use it to coat your hands when handling the trap and to coat the trap when setting it. In nearly all parts of the world, animals know the smell of burned vegetation and smole. It is only when a fire is actually burning that they become alarmed. Therefore, smoking the trap parts is an effective means to mask your scent. If one of the above techniques is not practical, and if time permits, allow a trap to weather for a few days and then set it. Do not handle a trap while it is weathering. When you position the trap, camouflage it as naturally as possible to prevent detection by the enemy and to avoid alarming the prey.

Traps or snares placed on a trail or run should use channelization. To build a channel, construct a funnel-shaped barrier extending from the sides of the trail toward the trap, with the narrowest part nearest the trap. Channelization should be inconspicuous to avoid alerting the prey. As the animalgets to the trap, it cannot turn left or right anctcontrnues into the trap. Few wild animals will back up, preferring to face the direction of travel. Channelization does not have to be an impassable barrier. You only have to make it inconvenient for the animal to go over or through the barrier. For best effect, the channelization should reduce the trail's width to just slightly wider than the targeted animal's body. Maintain this constriction at least as far back from the trap as the animal's body length, then begin the widening toward the mouth of the funnel.

Use of Bait

Baiting a trap or snare increases your chances of catching an animal. When catching fish, you must bait nearly all the devices. Success with an unbaited trap depends on its placement in a good location. A baited trap can actually draw animals to it. The bait should be something the animal knows. This bait, however, should not be so readily available in the immediate area that the animal can get it close by. For example, baiting a trap with corn in the middle of a corn field would not be likely to work. Likewise, if corn is not grown in the region, a corn-baited trap may arouse an animal's curiosity and keep it alerted while it ponders the strange food. Under such circumstances it may not go for the bait. One bait that works well on small mammals is the peanut butter from a meal, ready-to-eat (MRE) ration. Salt is also a good bait. When using such baits, scatter bits of it around the trap to give the prey a chance to sample it and develop a craving for it. The ammal will then overcome some of its caution before it gets to the trap.

If you set and bait a trap for one species but another species takes the bait without being caught, try to determine what the animal was. Then set a proper traplor that animal, using the same bait.

Note: Once you have successfully trapped an animal, you will not only gain confidence in your ability, you also will have resupplied yourself withbait for several more traps.

Trap and Snare Construction

Traps and snares *crush, choke, hang,* or *entangle* the prey. A single trap or snare will commonly incorporate two or more of these principles. The mechanisms that provide power to the trap are almost always very simple. The struggling victim, the force of gravity, or a bent sapling's tension provides the power.

The heart of any trap or snare is the trigger. When planning a trap or snare, ask yourself how it should affect the prey, what is the source of power, and what will be the most efficient trigger. Your answers will help you devise a specific trap for a specific species. Traps are designed to catch and hold or to catch and kill. Snares are traps that incorporate a noose to accomplish either function.

Simple Snare

A simple snare (Figure 8-5) consists of a noose placed over a trail or den hole and attached to a firmly planted stake. If the noose is some type of cordage placed upright on a game trail, use small twigs or blades of grass to hotel it up. Filaments from spider webs are excellent for holding nooses open. Make sure the noose is large enough to pass freely over the animal's head. As the animal continues to move, the noose tightens around its neck. The more the animal struggles, the tighter the noose gets. This type of snare usually does not kill the animal. If you use cordage, it may loosen enough to slip off the animal's neck. Wire is therefore the best choice for a simple snare.

Drag Noose

Use a drag noose on an animal run (Figure 8-6). Place forked sticks on either side of the run and lay a sturdy crossmember across them. Tie the noose to the crossmember and hang it at a height above the animal's head. (Nooses designed to catch by the head should never be low enough for the prey to step into with a foot.) As the noose tightens around the animal's neck, the animal pulls the crossmember from the forked sticks and drags it along. The surrounding vegetation quickly catches the crossmember and the animal becomes entangled.

Twitch-Up

A twitch-up is a supple sapling, which, when bent over and secured with a triggering device, will provide power to a variety of snares. Select a hardwood sapling along the trail. A twitch-up will work much faster and with more force 8 you remove all the branches and foliage.

Twitch- Up Snare

A simple twitch-up snare uses two forked sticks, each with a long and short teg (Figure 8-7). Bend the twitch-up and mark the trail below it. Drive the long leg of one forked stick firmly into the ground at that point. Ensure the cut on the short leg of this stick is parallel to the

Figure 8-5. Simple snare.

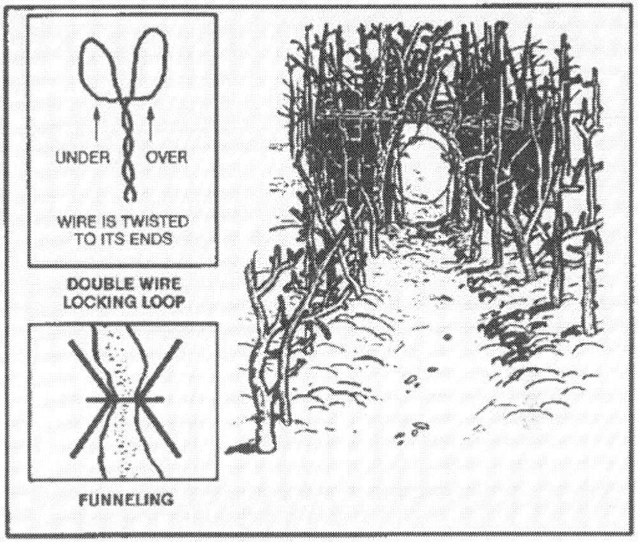

Figure 8-6. Drag noose.

Figure 8-7. Twitch-up snare.

ground. Tie the long leg of the remaining forked stick to a piece of cordage secured to the twitch-up. Cut the short leg so that it catches on the short leg of the other forked stick. Extend a noose over the trail. Set the trap by bending the twitch-up and engaging the short legs of the forked sticks. When an animal catches its head m the noose, it pulls the forked sticks apart, allowing the twitch-up to spring up and hang the prey.

Note: Do not use green stick for the trigger. The sap that oozes out could glue them together.

Squirrel Pole

A squirrel pole is a long pole placed against a tree in an area showing a lot of squirrel activity (Figure 8-8). Place several wire nooses along the top and sides of the pole so that a squirrel frying to go up or down the pole will have to pass through one or more of them. Position the nooses (5 to 6 centimeters in diameter) about 2.5 centimeters off the pole. Place the top and bottom wire nooses 45 centimeters from the top

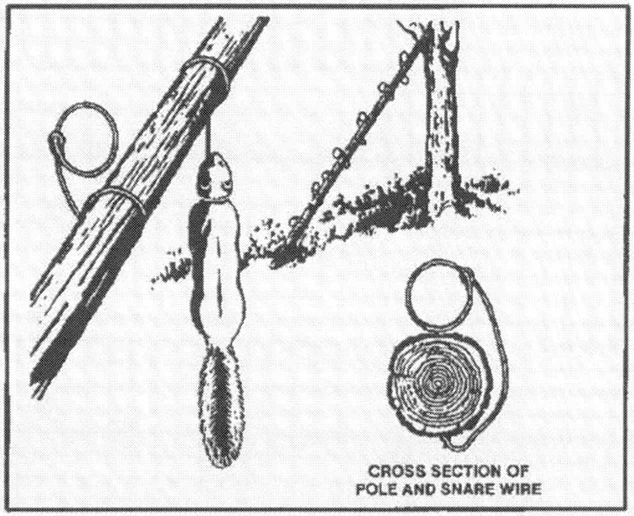

Figure 8-8. Squirrel pole.

and bottom of the pole to prevent the squirrel from getting its feet on a solid surface. If this happens, the squirrel will chew through the wire. Squirrels are naturally curious. After an initial period of caution, they will try to go up or down the pole and will get caught in a noose. The struggling animal will soon fall from the pole and strangle. Other squirrels will soon follow and, in this way, you can catch several squirrels. You can emplace multiple poles to increase the catch.

Ojibwa Bird Pole

An Ojibwa bird pole is a snare used by native Americans for centuries (Figure 8-9). To be effective, place it in a relatively open area away from tall trees. For best results, pick a spot near feeding areas, dusting areas, or watering holes. Cut a pole 1.8 to 2.1 meters long and trim away all limbs and foliage. Do not use resinous wood such as pine. Sharpen the upper end to a point, then drill a small diameter hole 5 to 7.5 centimeters down from the top. Cut a small stick 10 to 15 centimeters long and shape one end so that it will almost fit into the hole. This is the perch. Plant the long pole in the ground with the pointed end up. Tie a small weight, about equal to the weight of the targeted species, to a length of cordage. Pass the free end ofthe cordage through the hole, and tie a slip noose that covers the perch. Tie a single overhand knot in the cordage and place the perch against the hole. Allow the cordage to slip through the hole until the overhand knot rests against the pole and the top ofthe perch. The tension of the overhand knot against the pole

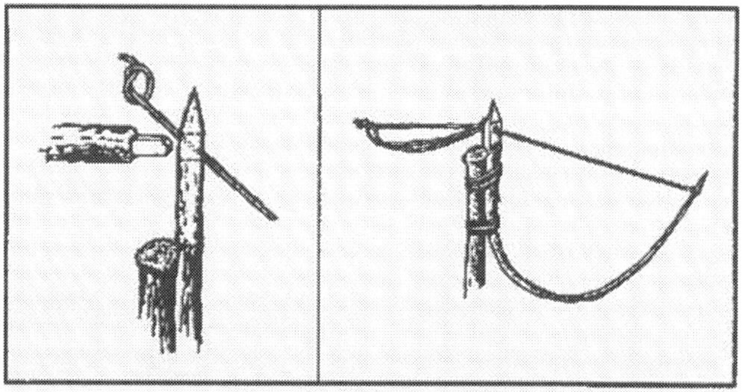

Figure 8-9. Ojibwa bird pole.

and perch will hold the perch in position. Spread the noose over the perch, ensuring it covers the perch and drapes over on both sides. Most birds prefer to rest on something above ground and will land on the perch. As soon as the bird lands, the perch will fall, releasing the overhand knot and allowing the weight to drop. The noose will tighten around the bird's feet, capturing it. If the weight is too heavy, it will cut the bird's feet off, allowing it to escape.

Noosing Wand

A noose stick or "noosing wand" is useful for capturing roosting birds or small mammals (Figure 8-10). It requires a patient operator. This wand is more a weapon than a trap. It consists of a pole (as long as you can effectively handle) with a slip noose of wire or stiff cordage at the small end. To catch an animal, you slip the noose over the neck of a roosting bird and pull it tight. You can also place it over a den hole and hide in a nearby blind. When the animal emerges from the den, you jerk the pole to tighten the noose and thus capture the animal. Carry a stout club to kill the prey.

Treadle Spring Snare

Use a treadle snare against small game on a trail (Figure 8-11). Dig a shallow hole in the trail. Then drive a forked stick (fork down) into the ground on each side of the hole on the same side of the trail. Select two fairly straight sticks that span the two forks. Position these two sticks so that their ends engage the forks. Place several sticks over the hole in the trail by positioning one end over the lower horizontal stick and the other on the ground on the other side of the hole. Cover the hole with enough sticks so that the prey must step on at least one of them to set off the snare. Tie one end of a piece of cordage to a twitch-up or to a weight suspended over a tree limb. Bend the twitch-up or raise the suspended weight to determine where You will tie a 5 centimeter or so long trigger. Form a noose with the other end of the cordage. Route and

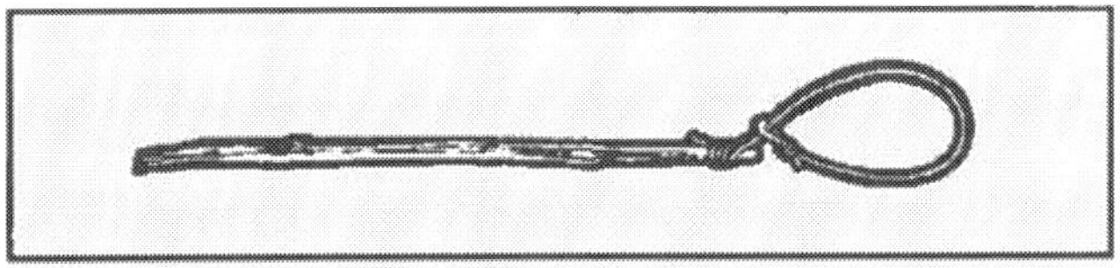

Figure 8-10. Noosing wand.

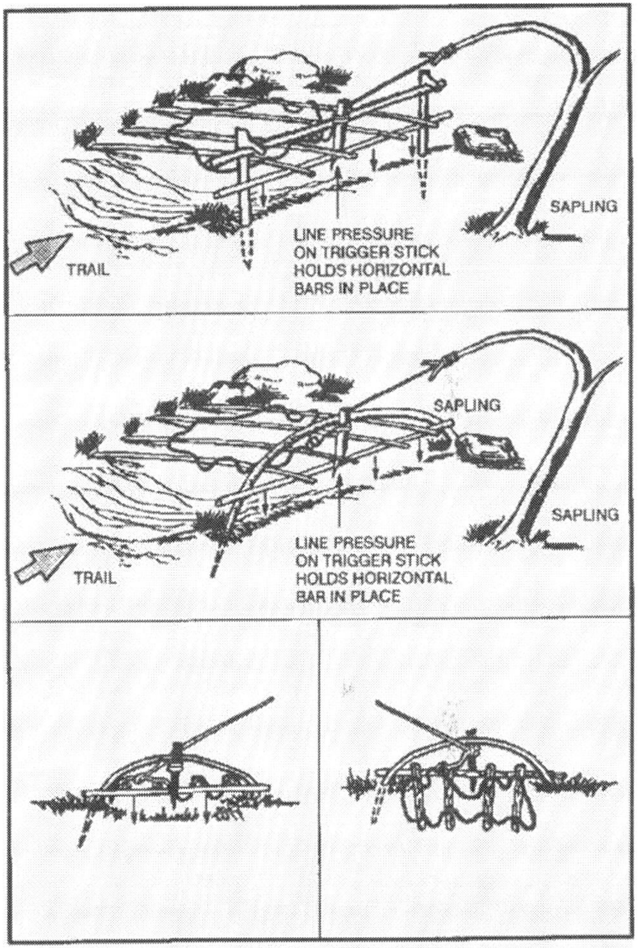

Figure 8-11. Treadle spring snare.

spread the noose over the top of the sticks over the hole. Place the trigger stick against the horizontal sticks and route the cordage behind the sticks so that the tension of the power source will hold it in place. Adjust the bottom horizontal stick so that it will barely hold against the trigger. A the animal places its foot on a stick across the nole, the bottom horizontal stick moves down, releasing the trigger and allowing the noose to catch the animal by the foot. Because ofthe disturbance on the trail, an animal will be wary. You must therefore use channelization.

Figure 4 *Deadfall*

The figure 4 is a trigger used to drop a weight onto a prey and crush it (Figure 8-12). The type of weight used may vary, but it should be heavy enough to kill or incapacitate the prey immediately. Construct the figure 4 using three notched sticks. These notches hold the sticks together m a figure 4 pattern when under tension. Practice making this trigger beforehand; it requires close tolerances and precise angles m its construction.

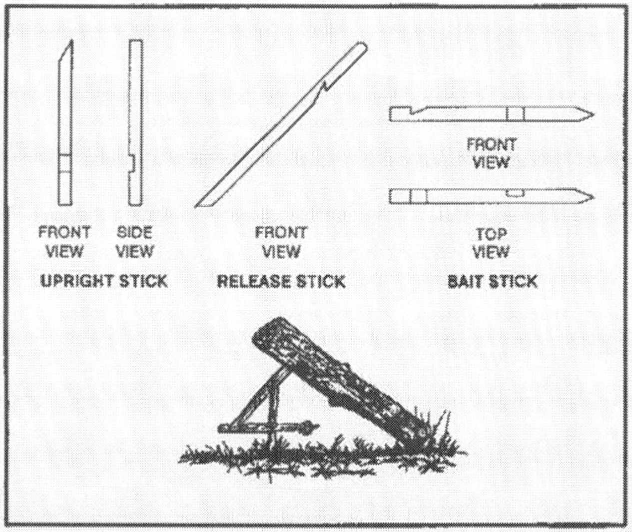

Figure 8-12. Figure 4 deadfall.

Paiute Deadfall

The Paiute deadfall is similar to the figure 4 but uses a piece of cordage and a catch stick (Figure 8-13). It has the advantage of being easier to set than the figure 4. Tie one end of a piece of cordage to the lower end of the diagonal stick. Tie the other end of the cordage to another stick about 5 centimeters long. This 5-centimeter stick is the catch stick. Bring the cord halfway around the vertical stick with the catch stick at a 90-degree angle. Place the bait stick with one end against the drop weight, or a peg driven into the ground, and the other against the catch stick. When a prey disturbs the bait stick, it falls free, releasing the catch stick. As the diagonal stick flies up, the weight falls, crushing the prey.

Bow Trap

A bow trap is one of the deadliest traps. It is dangerous to man as well as animals (Figure 8-14). To construct this trap, build a bow and anchor it to the ground with pegs. Adjust the aiming point as you anchor the bow. Lash a toggle stick to the trigger stick. Two upright sticks driven into the ground hold the trigger stick in place at a point where the toggle stick will engage the pulled bow string. Place a catch stick

Figure 8-13. Paiute deadfall.

between the toggle stick and a stake driven into the ground. Tie a trip wire or cordage to the catch stick and route it around stakes and across the game trail where you tie it off (as in Figure 8-14). When the prey trips the trip wire, the bow looses an arrow into it. A notch in the bow serves to help aim the arrow.

WARNING

This is a lethal trap. Approach it with caution and from the rear only!

Pig *Spear Shaft*

To construct the pig spear shaft, select a stout pole about 2.5 meters long (Figure 8-15). At the smaller end, firmly lash several small stakes. Lasn the large end tightly to a tree along the game trail. Tie a length of cordage to another tree across the trail. Tie a sturdy, smooth stick to the other end of the cord. From the first tree, tie a trip wire or cord low to the ground, stretch it across the trail, and tie it to a catch stick. Make a slip ring from vines or other suitable material. Encircle the trip wire and

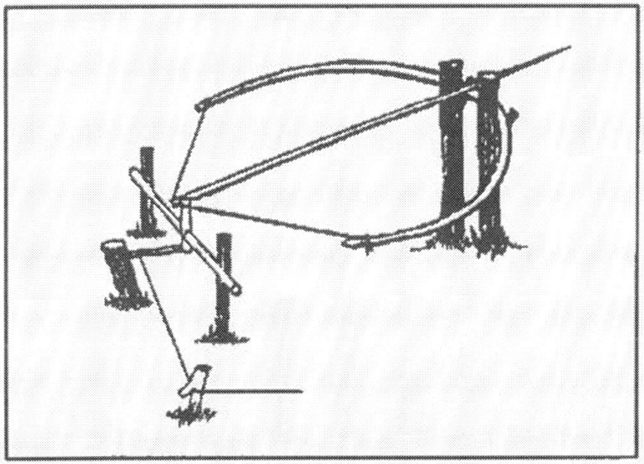

Figure 8-14. Bow trap.

the smooth stick with the slip ring. Emplace one end of another smooth stick within the slip ring and its other end against the second tree. Pull the smaller end of the spear shaft across the trail and position it between the short cord and the smooth stick. As the animal trips the trip wire, the catch stick pulls the slip ring off the smooth sticks, releasing the spear shaft that springs across the trail and impales the prey against the tree.

WARNING

This is a lethal trap. Approach it with caution!

Bottle Trap

A bottle trap is a simple trap for mice and voles (Figure 8-16). Dig a hole 30 to 45 centimeters deep that is wider at the bottom than at the top. Make the top of the hole as small as possible. Place a piece of bark or wood over the hole with small stones under it to hold it up 2.5 to 5 centimeters off the ground. Mice or voles will hide under the cover to escape danger and fall into the hole. They cannot climb out because of the wall's backward slope. Use caution when checking this trap; it is an excellent hiding place for snakes.

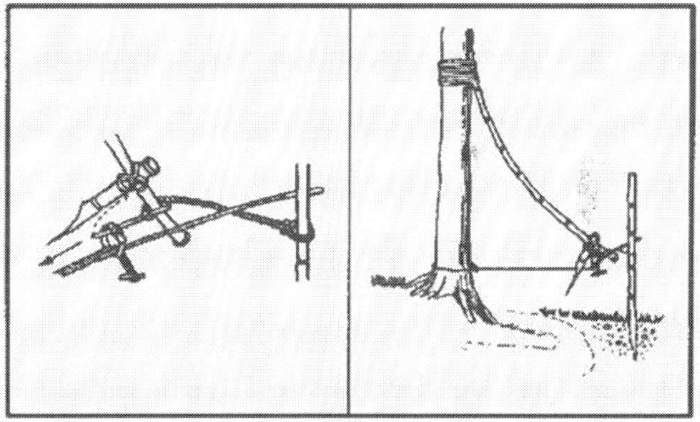

Figure 8-15. Pig spear shaft.

KILLING DEVICES

There are several killing devices that you can construct to help you obtain small game to help you survive. The rabbit stick, the spear, the bow and arrow, and the sling are such devices.

Rabbit Stick

One of the simplest and most effective killing devices is a stout stick as long as your arm, from fingertip to shoulder, called a "rabbit stick." You can throw it either overhand or sidearm and with considerable force. It is very effective against small game that stops and freezes as a defense.

Spear

You can make a spear to kill small game and to fish. Jab with the spear, do not throw it. See spearfishing on page 8-29.

Bow and Arrow

A good bow is the result of many hours of work. You can construct a suitable short-term bow fairly easily. When it loses its spring or breaks, you can replace it. Select a hardwood stick about one meter long that is free of knots or limbs. Carefully scrape the large end down until it has the same pull as the small end. Careful examination will show the natural curve of the stick. Always scrape from the side that faces you, or the bow will break the first time you pull it. Dead, dry wood is preferable to green wood. To increase the pull, lash a second bow to the first, front to front, forming an "X" when viewed from the side. Attach the tips of the bows with cordage and only use a bowstring on one bow.

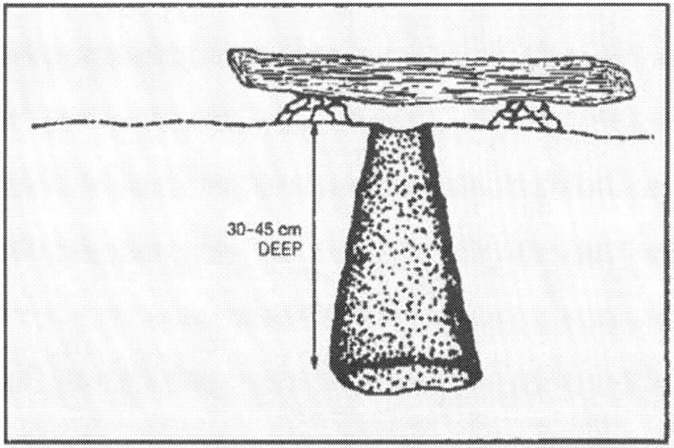

Figure 8-16. Bottle trap.

Select arrows from the straightest dry sticks available. The arrows should be about half as long as the Dow. Scrape each shaft smooth all around. You will probably have to straighten the shaft. You can bend an arrow straight by heating the shaft over hot coals. Do not allow the shaft to scorch orDum. Hold the shaft straight until it cools.

You can make arrowheads from bone, glass, metal, or pieces of rock. You can also sharpen and fire harden the end of the shaft. To fire harden wood, hold it over hot coals, being careful not to bum or scorch the wood.

You must notch the ends of the arrows for the bowstring. Cut or file the notch; do not split it. Fletching (adding feathers to the notched end of an arrow) improves the arrow s flight characteristics, but is not necessary on a field-expedient arrow.

Sling

You can make a sling by tying two pieces of cordage, about sixty centimeters long, at opposite ends of a palm-sized piece of leather or cloth. Place a rock in the cloth and wrap one cord around the middle finger and hold in your palm. Hold the other cord between the forefinger and thumb. To throw the rock, spin the sling several times in a circle and release the cord between the thumb ana forefinger. Practice to gain proficiency. The sling is very effective against small game.

FISHING DEVICES

You can make your own fishhooks, nets and traps and use several methods to obtain fish in a survival situation.

Improvised Fishhooks

You can make field-expedient fishhooks from pins, needles, wire, small nails, or any piece of metal. You can also use wood, bone, coconut shell, thorns, flint, seashell, or tortoise shell. You can also make fishhooks from any combination of these items (Figure 8-17).

To make a wooden hook, cut a piece of hardwood about 2.5 centimeters long and about 6 millimeters in diameter to form the shank. Cut a notch in one end in which to place the point. Place the point (piece of bone, wire, nail) in the notch. Hold the point in the notch and tie securely so that it does not move out of position. This is a fairly large hook. To make smaller hooks, use smaller material.

A gorge is a small shaft of wood, bone, metal, or other material. It is sharp on both ends and notched in the middle where you tie cordage. Bait the gorge by placing a piece of bait on it lengthwise. When the fish swallows the bait, it also swallows the gorge.

Stakeout

A stakeout is a fishing device you can use in a hostile environment (Figure 8-18). To construct a stakeout, drive two supple saplings into the bottom of the lake, pond, or stream with their tops just below the water surface. Tie a cord between them and slightly below the surface. Tie two short cords with hooks or gorges to this cord, ensuring that they cannot wrap around the poles or each other. They should also not slip along the long cord. Bait the hooks or gorges.-

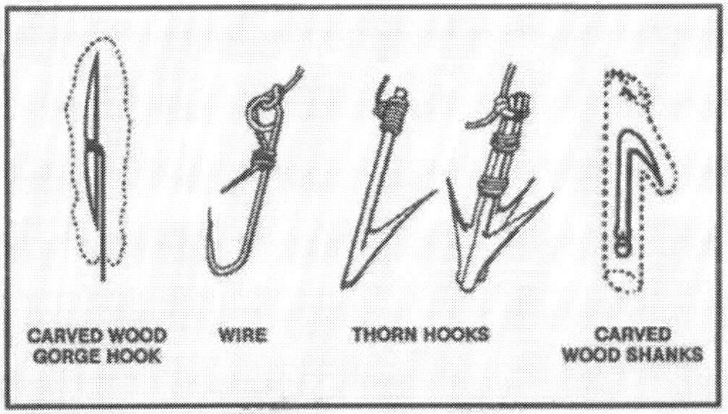

Figure 8-17. Improvised fishhooks.

Gill Net

If a gill net is not available, you can make one using parachute suspensionline or similar material (Figure 8-19). Remove the core lines from the suspension line and tie the easing between two trees. Attach several core lines to the easing by doubling them over and tying them with

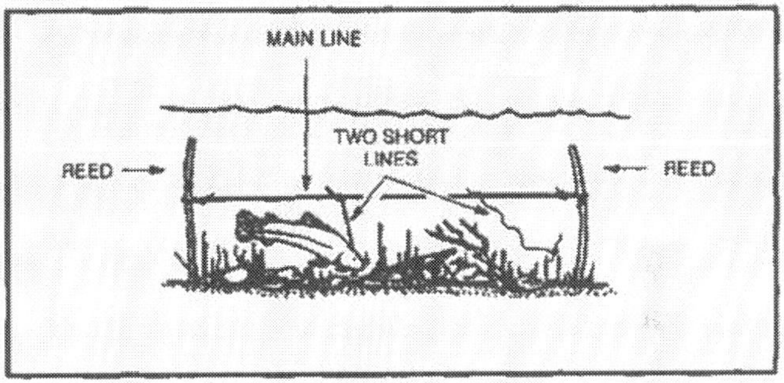

Figure 8-18. Stakeout.

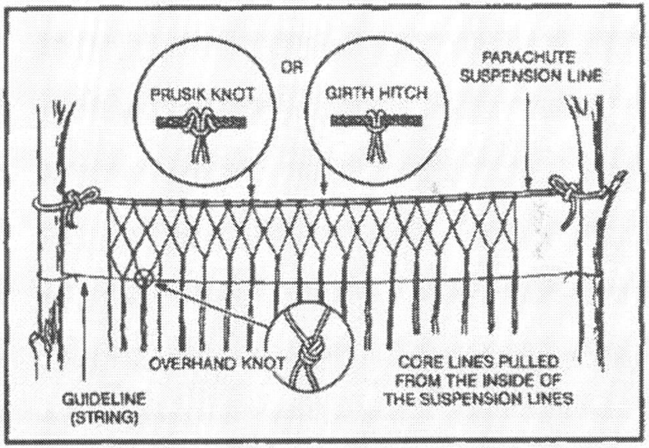

Figure 8-19. Making a gill net.

prusik knots or girth hitches. The length of the desired net and the size of the mesh determine the number of core lines used and the space between them. Starting at one end of the easing, tie the second and the third core lines together using an overhand knot. Then tie the fourth and fifth, sixth and seventh, and so on, until you reach the last core line. You should now have all core lines tied in pairs with a single core line hanging at each end. Start the second row with the first core line, tie it to the second, the third to the fourth, and so on.

To keep the rows even and to regulate the size of the mesh, tie a guideline to the trees. Position the guideline on the opposite side of the net you are working on. Move the guideline down after completing each row. The lines will always hang in pairs and you always tie a cord from one pair to a cord from an adjoining pair. Continue tying rows until the net is the desired width. Thread a suspension line easing along the bottom of the net to strengthen it. Use the gill net as shown in Figure 8-20.

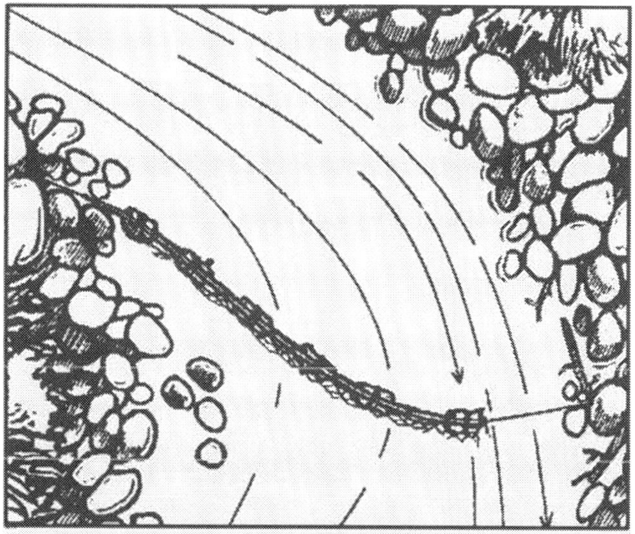

Figure 8-20. Setting a gill net in the stream.

Fish Traps

You may trap fish using several methods (Figure 8-21). Fish baskets are one method. You construct them by lashing several sticks together with vines into a funnel shape. You close the top, leaving a hole large enough for the fish to swim through.

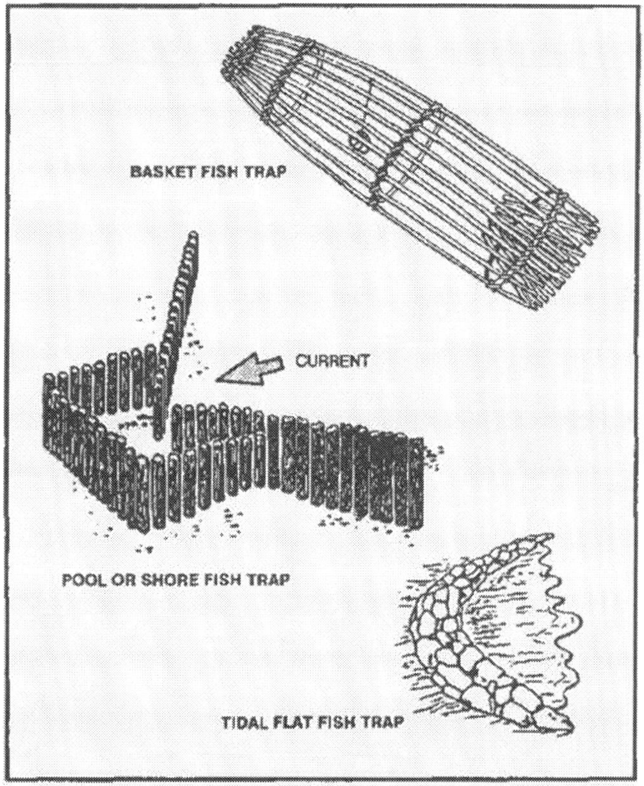

Figure 8-21. Various types of fish traps.

You can also use traps to catch saltwater fish, as schools regularly approach the shore with the incoming tide and often move parallel to the shore. Pick a location at high tide and build the trap at low tide. On rocky shores, use natural rock pools. On coral islands, use natural pools on the surface of reefs by blocking the openings as the tide recedes. On sandy shores, use sandbars and the ditches they enclose. Build the trap as a low stone wall extending outward into the water and forming an angle with the shore.

Spearfishing

If you are near shallow water (about waist deep) where the fish are large and plentiful, you can spear them. To make a spear, cut a long, straight saplmg (Figure 8-22). Sharpen the end to a point or attach a knife, jagged piece of bone, or sharpened metal. You can also make a spear by splitting the shaft a few inches down from the end and inserting a piece or wood to act as a spreader. You then sharpen the two separated halves to points. To spear fish, find an area where fish either gather or where there is a fish run. Place the spear point into the water and slowly move it toward the fish. Then, with a sudden push, impale the fish on the stream bottom. Do not try to lift the fish with the spear, as it with probably slip off and you will lose it; hold the spear with one hand and grab and hold the fish with the other. Do not throw the spear, especially if the point is a knife. You cannot afford to lose a knife in a survival situation. Be alert to the problems caused by light refraction when looking at objects in the water.

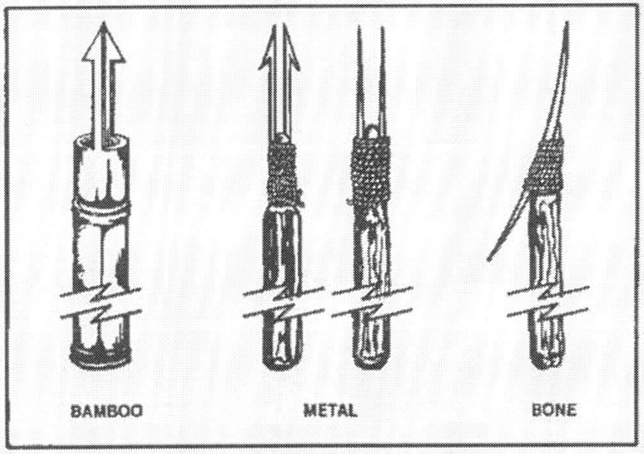

Figure 8-22. Types of spear points.

Chop Fishing

At night, in an area with a good fish density, you can use a light to attract fish. Then, armed with a machete or similar weapon, you can gather fish using the back side of the blade to strike them. Do not use the sharp side as you will cut them in two pieces and end up losing some of the fish.

Fish Poison

Another way to catch fish is by using poison. Poison works quickly. It allows you to remain concealed while it takes effect. It also enables you to catch several fish at one time. When using fish poison, be sure to gather all of the affected fish, because many dead fish floating downstream could arouse suspicion. Some plants that grow in warm regions of the world contain rotenone, a substance that stuns or kills cold-blooded animals but does not harm persons who eat the animals. The best place to use rotenone, or rotenone-producing plants, is in ponds or the headwaiters of small streams containing fish. Rotenone works quickly on fish in water 21 degrees C (70 degrees F) or above. The fish rise helplessly to the surface. It works slowly in water 10 to 21 degrees C (50 to 70 degrees F) and is ineffective m water below 10 degrees C (50 degrees F). The following plants, used as indicated, will stun or kill fish:

- *Anamirta cocculus* (Figure 8-23). This woody vine grows in southern Asia and on islands of the South Pacific. Crush the bean-shaped seeds and throw them in the water.
- *Croton tiglium* (Figure 8-23). This shrub or small tree grows in waste areas on islands of the South Pacific. It bears seeds in three angled capsules. Crush the seeds and throw them into the water.
- *Barringtonia* (Figure 8-23). These large trees grow near the sea in Malaya and parts of Polynesia. They Dear a fleshy one-seeded fruit. Crush the seeds and bark and throw into the water.
- *Derris eliptica* (Figure 8-23). This large genus of tropical shrubs and woody vmes is the main source of commercially produced rotenone.

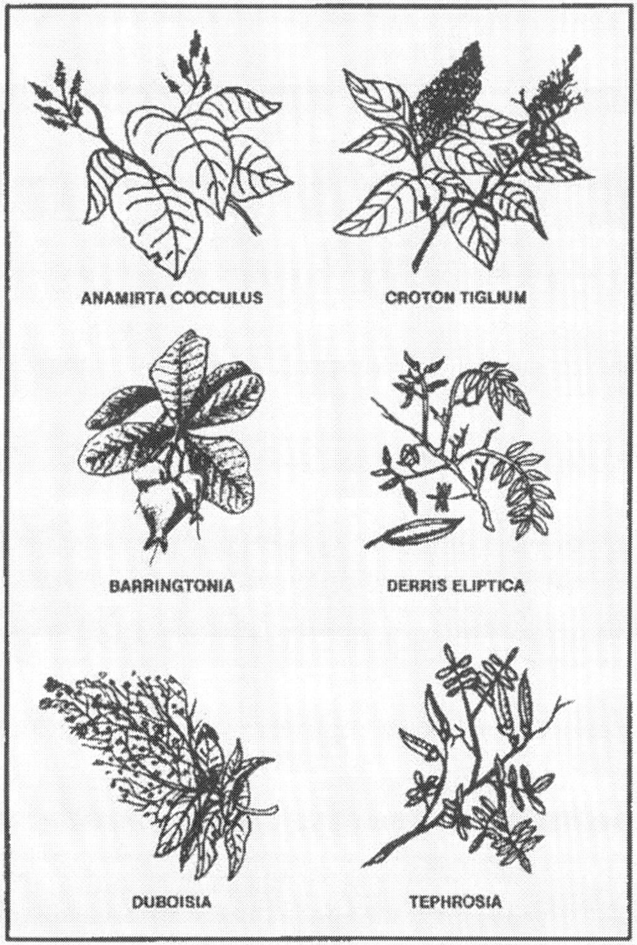

Figure 8-23. Fish-poisoning plants.

Grind the roots into a powder and mix with water. Throw a large quantity of the mixture into the water.

- *Duboisia* (Figure 8-23). This shrub grows in Australia and bears white clusters of flowers and berrylike fruit. Crush the plants and throw them into the water.
- *Tephrosia* (Figure 8-23). This species of small shrubs, which bears beanlike pods, grows throughout the tropics. Crush or bruise bundles of leaves and stems and throw them into the water.
- *Lime.* You can get lime from commercial sources and in agricultural areas that use large quantities of it. You may produce your own by burning coral or seashells. Throw the lime into the water.
- *Nut husks.* Crush green husks from butternuts or black walnuts. Throw the husks into the water.

PREPARATION OF FISH AND GAME FOR COOKING AND STORAGE

You must know how to prepare fish and game for cooking and storage in a survival situation. Improper cleaning or storage can result in inedible fish or game.

Fish

Do not eat fish that appears spoiled. Cooking does not ensure that spoiled fish will be edible. Signs of spoilage are—

- Sunken eyes.
- Peculiar odor.
- Suspicious color. (Gills should be red to pink. Scales should be a pronounced shade of gray, not faded.)
- Dents stay in the fish's flesh after pressing it with your thumb.
- Slimy, rather than moist or wet body.
- Sharp or peppery taste.

Eating spoiled or rotten fish may cause diarrhea, nausea, cramps, vomiting, itching, paralysis, or a metallic taste in the mouth. These symptoms appear suddenly, one to six hours after eating. Induce vomiting if symptoms appear.

Fish spoils quickly after death, especially on a hot day. Prepare fish for eating as soon as possible after catching it. Cut out the gills and large blood vessels that lie near the spine. Gut fish that is more than 10 centimeters long. Scale or skin the fish.

You can impale a whole fish on a stick and cook it over an open fire. However, boiling the fish with the skin on is the best way to get the most food value. The fats and oil are under the skin and, by boiling, you can save the juices for broth. You can use any of the methods used to cook plant food to cook fish. Pack fish into a ball of clay and bury it in the coals of a fire until the clay hardens. Break open the clay ball to get to the cooked fish. Fish is done when the meat flakes off. If you plan to keep the fish for later, smoke or fry it. To prepare fish for smoking, cut off the head and remove the backbone.

Snakes

To skin a snake, first cut off its head and bury it. Then cut the skin down the body 15 to 20 centimeters (Figure 8-24). Peel the skin back, then

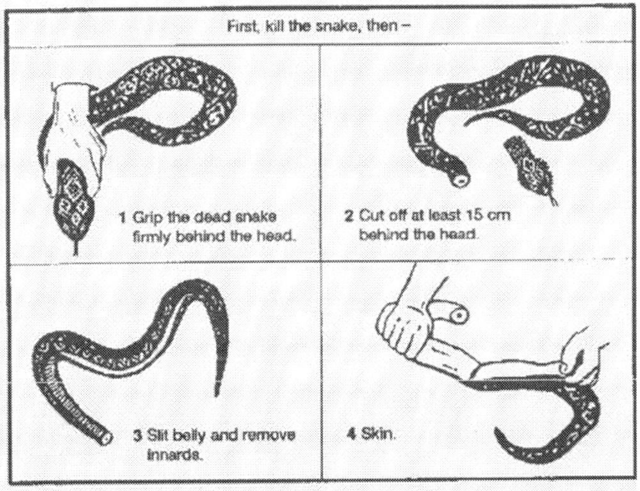

Figure 8-24. Cleaning a snake.

grasp the skin in one hand and the body in the other and pull apart. On Targe, bulky snakes it may be necessary to slit the belly skin. Cook snakes in the same manner as small game. Remove the entrails and discard. Cut the snake into small sections and boil or roast it.

Birds

After killing the bird, remove its feathers by either plucking or skinning. Remember, skinning removes some of the rood value. Open up the body cavity and remove its entrails, saving the craw (in seed-eating birds), heart, and liver. Cut off the feet. Cook by boiling or roasting over a spit. Before cooking scavenger birds, boil them at least 20 minutes to kill parasites.

Skinning and Butchering Game

Bleed the animal by cutting its throat. If possible, clean the carcass near a stream. Place the carcass belly up and split the hide from throat to tail, cutting around all sexual organs (Figure 8-25). Remove the musk glands at points A and B to avoid tainting the meat. For smaller mammals, cut the hide around the body and insert two fingers under the hide on both sides of the cut and pull both pieces off (Figure 8-26).

Note: Wien cutting the hide, insert the knife blade under the skin and turn the blade up so that only the hide gets cut. This mill also prevent cutting hair and getting it on the meat.

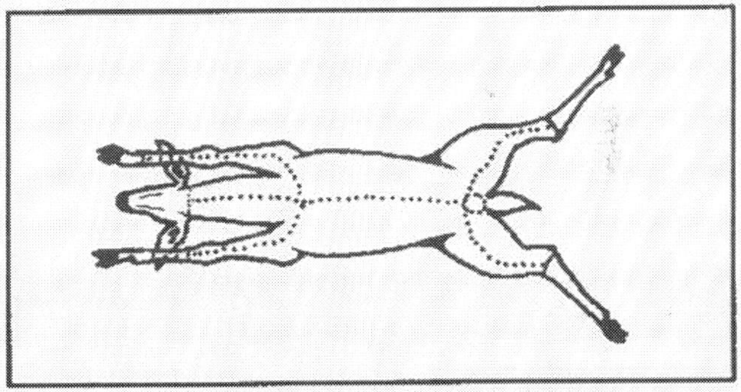

Figure 8-25. Skinning and butchering large game.

Remove the entrails from smaller game by splitting the body open and pulling them out with the fingers. Do not forget the chest cavity. For larger game, cut the gullet away from the diaphragm. Roll the entrails out of the body. Cut around the anus, then reach into the lower abdominal cavity, grasp the lower intestine, and pull to remove. Remove the urine bladder by pinching it off and cutting it below the fingers. If you spill urine on the meat, wash it to avoid tainting the meat. Save the heart and liver. Cut these open and inspect for signs of worms or other parasites. Also inspect the liver's color; it could indicate a diseased animal. The liver's surface should be smooth and wet and its color deep red or purple. If the liver appears diseased, discard it. However, a diseased liver does not indicate you cannot eat the muscle tissue.

Cut along each leg from above the foot to the previously made body cut. Remove the hide by pulling it away from the carcass, cutting the connective tissue where necessary. Cut off the head and feet.

Cut larger game into manageable pieces. First, slice the muscle tissue connecting the front legs to the body. There are no bones or joints connecting the front legs to the body on four-legged animals. Cut the hindquarters off where they join the body. You must cut around a largelDone at the top of the leg and cut to the ball and socket hip joint. Cut the ligaments around the joint and bend it back to separate it. Remove the large muscles (the tenderloin) that he on either side of the spine.

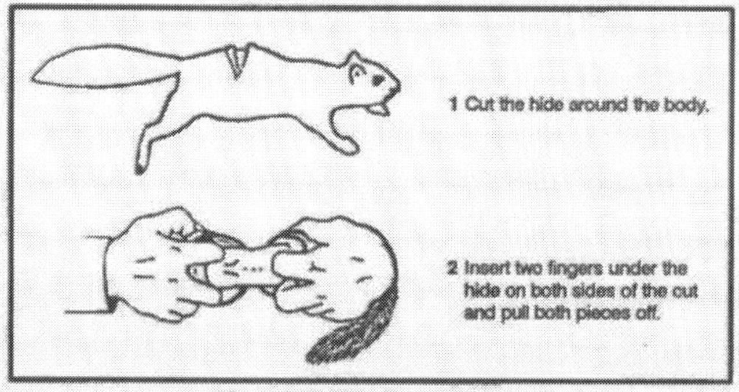

Figure 8-26. Skinning small game.

Separate the ribs from the backbone. There is less work and less wear on your knife if you break the ribs first, then cut through the breaks.

Cook large meat pieces over a spit or boil them. You can stew or boil smaller pieces, particularly those that remain attached to bone after the initial butchering, as soup or broth. You can cook body organs such as the heart, liver, pancreas, spleen, and kidneys using the same methods as for muscle meat. You can also cook and eat the brain. Cut the tongue out, skin it, boil it until tender, and eat it.

Smoking Meat

To smoke meat, prepare an enclosure around a fire (Figure 8-27). Two ponchos snapped together will work. The fire does not need to be big or

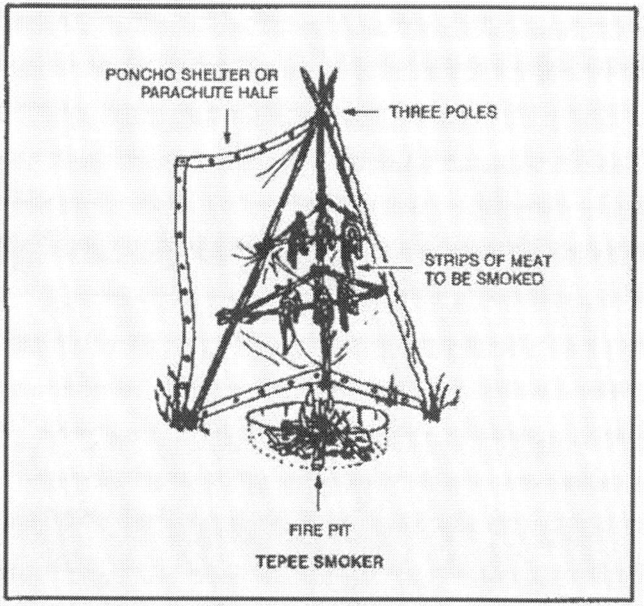

Figure 8-27. Smoking meat.

hot. The intent is to produce smoke, not heat. Do not use resinous wood in the fire because its smoke will ruin the meat. Use hardwoods to produce good smoke. The wood should be somewhat green. If it is too dry, soak it. Cut the meat into thin slices, no more than 6 centimeters thick, and drape them over a framework. Make sure none of the meat touches another piece. Keep the poncho enclosure around the meat to hold the smoke and keep a close watch on the fire. Do not let the fire get too hot. Meat smoked overnight in this manner will last about 1 week. Two days of continuous smoking will preserve the meat for 2 to 4 weeks. Properly smoked meat willlook like a dark, curled, brittle stick and you can eat it without further cooking. You can also use a pit to smoke meat (Figure 8-28).

Drying Meat

To preserve meat by drying, cut it into 6-millimeter strips with the gram. Hang the meat strips on a rack in a sunny location with good air flow. Keep the strips out of the reach of animals and cover them to keep blowflies off. Allow the meat to dry thoroughly before eating. Properly dried meat will have a dry, crisp texture and will not feel cool to the touch.

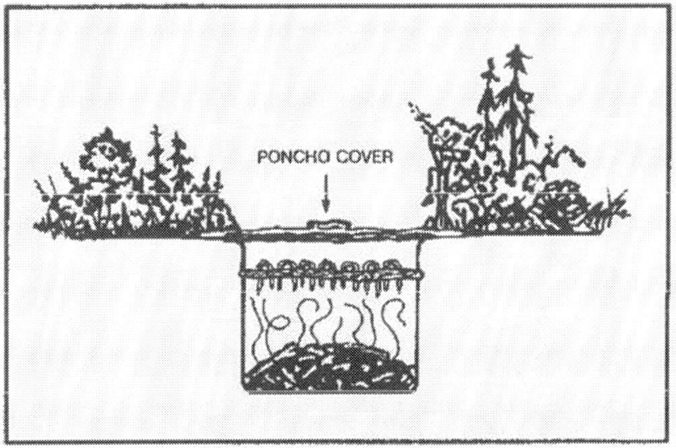

Figure 8-28. Smoking meat over a pit.

Other Preservation Methods

You can also preserve meats using the freezing or brine and salt methods.

Freezing

In cold climates, you can freeze and keep meat indefinitely. Freezing is not a means of preparing meat. You must still cook it before eating.

Brine and Salt

You can preserve meat by soaking it thoroughly in a saltwater solution. The solution must cover the meat. You can also use salt by itself. Wash off the salt before cooking.

C H A P T E R 9

SURVIVAL USE OF PLANTS

After having solved the problems of finding water, shelter, and animal food, you will have to consider the use of plants you can eat. In a survival situation you should always be on the lookout for familiar wild foods and live off the land whenever possible.

You must not count on being able to go for days without food as some sources would suggest. Even in the most static survival situation, maintaining health through a complete and nutritious diet is essential to maintaining strength and peace of mind.

Nature can provide you with food that will let you survive any ordeal, if you don't eat the wrong plant. You must therefore learn as much as possible beforehand about the flora of the region where you will be operating. Plants can provide you with medicines in a survival situation. Plants can supply you with weapons and raw materials to construct shelters and build fires. Plants can even provide you with chemicals for poisoning fish, preserving animal hides, and for camouflaging yourself and your equipment.

Note: You will find illustrations of the plants described in this chapter in Appendixes B and C.

EDIBILITY OF PLANTS

Plants are valuable sources of food because they are widely available, easily procured, and, in the proper combinations, can meet all your nutritional needs.

WARNING

The critical factor in using plants for food is to avoid accidental poisoning. Eat only those plants you can positively identify and you know are safe to eat.

Absolutely identify plants before using them as food. Poison hemlock has killed people who mistook it for its relatives, wild carrots and wild parsnips.

At times you may find yourself in a situation for which you could not plan. In this instance you may not have had the chance to learn the plant life of the region in which you must survive. In this case you can use the Universal Edibility Test to determine which plants you can eat and those to avoid.

It is important to be able to recognize both cultivated and wild edible plants m a survival situation. Most of the information in this chapter is directed towards identifying wild plants because information relating to cultivated plants is more readily available.

Remember the following when collecting wild plants for food:

- Plants growing near homes and occupied buildings or along roadsides may have been sprayed with pesticides. Wash them thoroughly. In more highly developed countries with many automobiles, avoid roadside plants, if possible, due to contamination from exhaust emissions.

- Plants growing in contaminated water or in water containing *Gkrdk lamblia* and other parasites are contaminated themselves. Boil or disinfect them.
- Some plants develop extremely dangerous fungal toxins. To lessen the chance of accidental poisoning, do not eat any fruit that is starting to spoil or showing signs of mildew or fungus.
- Plants of the same species may differ in their toxic or subtoxic compounds content because of genetic or environmental factors. One example of this is the foliage of the common chokecherry. Some chokecherry plants have high concentrations of deadly cyanide compounds while others have low concentrations or none. Horses have died from eating wilted wild cherry leaves. Avoid any weed, leaves, or seeds with an almondlike scent, a characteristic ofthe cyanide compounds.
- Some people are more susceptible to gastric distress (from plants) than others. If you are sensitive in this way, avoid unknown wild plants. If you are extremely sensitive to poison ivy, avoid products from this family, including any parts from sumacs, mangoes, and cashews.
- Some edible wild plants, such as acorns and water lily rhizomes, are bitter. These bitter substances, usually tannin compounds, make them unpalatable. Boiling them in several changes of water will usually remove these bitter properties.
- Many valuable wild plants have high concentrations of oxalate compounds, also known as oxalic acid. Oxalates produce a sharp burning sensation in your mouth and throat and damage the kidneys. Baking, roasting, or drying usually destroys these oxalate crystals. The corm (bulb) of the jack-in-fhe-pulpit is known as the "Indian turnip," but you can eat it only after removing these crystals by slow baking or by drying.

WARNING

Do not eat mushrooms in a survival situation! The only way to tell if a mushroom is edible is by positive identification. There is no room for experimentation. Symptoms of the most dangerous mushrooms affecting the central nervous system may show up after several days have passed when it is too late to reverse their effects.

Plant Identification

You identify plants, other than by memorizing particular varieties through familiarity, by using such factors as leaf shape and margin, leaf arrangements, and root structure.

The basic leaf margins (Figure 9-1) are toothed, lobed, and toothless or smooth.

These leaves maybe lance-shaped, elliptical, egg-shaped, oblong, wedge-shaped, triangular, long-pointed, or top-shaped (Figure 9-2).

The basic types of leaf arrangements (Figure 9-3) are opposite, alternate, compound, simple, ancT basal rosette.

The basic types of root structures (Figure 9-4) are the bulb, clove, taproot, tuber, rhizome, corm, and crown. Bulbs are familiar to us as onions and, when sliced in half, will show concentric rings. Cloves are those bulblike structures that remind us of garlic and will separate into small pieces when broken apart. This characteristic separates wild onions from wild garlic. Taproots resemble carrots and may be single-rooted or branched, but usually only one plant stalk arises from each root. Tubers are like potatoes and daytilies and you will find these structures either on strings or in clusters underneath the parent plants. Rhizomes are

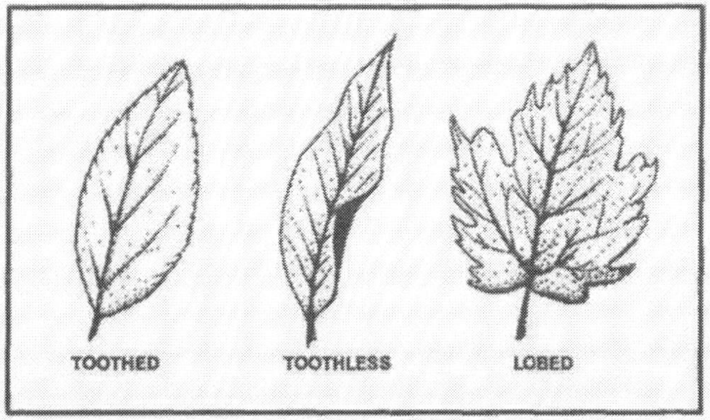

Figure 9-1. Leaf margins.

large creeping rootstock or underground stems and many plants arise from the eyes" of these roots. Corms are similar to bulbs but are solid when cut rather than possessing rings. A crown is the type of root structure found on plants such as asparagus and looks much like a mophead under the soil s surface.

Learn as much as possible about plants you intend to use for food and their unique characteristics. Some plants have both edible and poisonous arts. Many are edible only at certain times of the year. Others may lave poisonous relatives that look very similar to the ones you can eat or use for medicine.

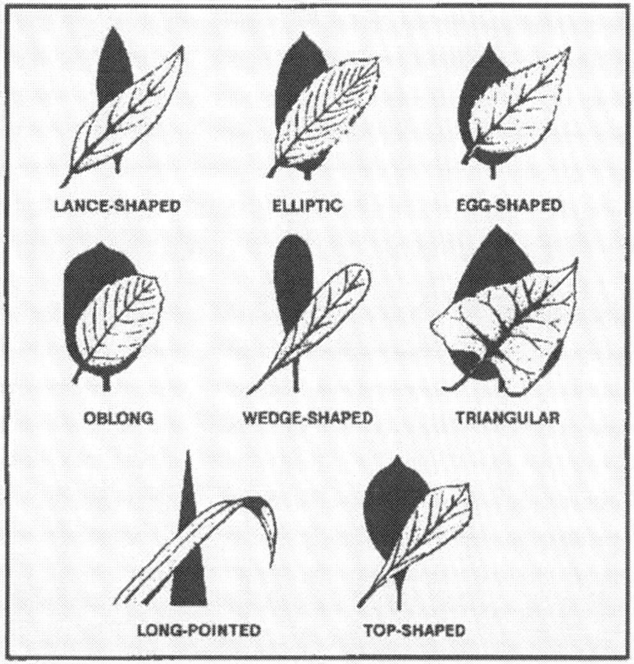

Figure 9-2. Leaf shapes.

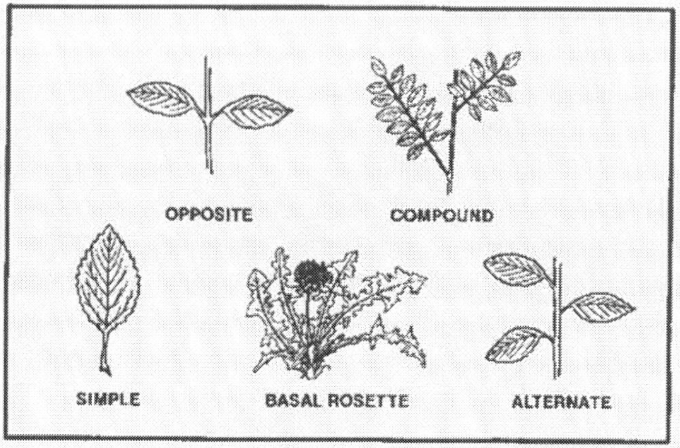

Figure 9-3. Leaf arrangements.

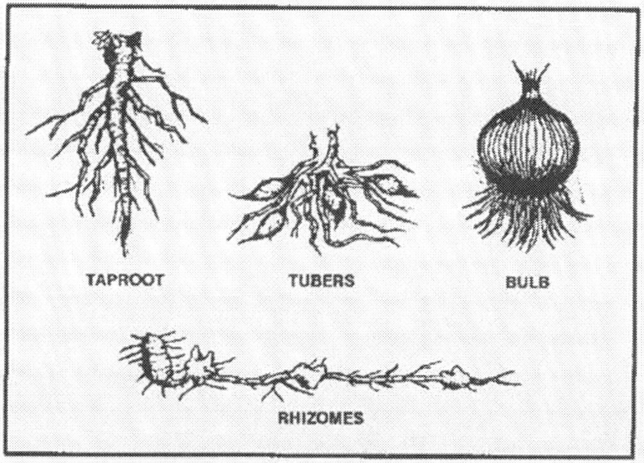

Figure 9-4. Root structures.

Universal Edibility Test

There are many plants throughout the world.Tasting or swallowing even a small portion of some can cause severe discomfort, extreme internal disorders, and even death. Therefore, if you have the slightest doubt about a plant's edibility, apply the Universal Edibility Test (Figure 9-5) before eating any portion of it.

Before testing a plant for edibility, make sure there are enough plants to make the testmg worth your time and effort. Each part of a plant (roots, leaves, flowers, and so on) requires more than 24 hours to test. Do not waste time testing a plant that is not relatively abundant in the area.

Remember, eating large portions of plant food on an empty stomach may cause diarrhea, nausea, or cramps. Two good examples of this are such familiar foods as green apples and wild onions. Even after testing plant food and finding it safe, eat it in moderation.

You can see from the steps and time involved in testing for edibility just how important it is to be able to identify edible plants.

To avoid potentially poisonous plants, stay away from any wild or unknown plants that have—

- Milky or discolored sap.
- Beans, bulbs, or seeds inside pods.
- Bitter or soapy taste.
- Spines, fine hairs, or thorns.
- Dill, carrot, parsnip, or parsleylike foliage.
- "Almond" scent in woody parts and leaves.
- Grain heads with pink, purplish, or black spurs.
- Three-leaved growth pattern.

Using the above criteria as eliminators when choosing plants for the Universal Edibility Test will cause you to avoid some edible plants. More important, these criteria will often help you avoid plants that are potentially toxic to eat or touch.

An entire encyclopedia of edible wild plants could be written, but space limits the number of plants presentedTvere. Learn as much as possible about the plant life of the areas where you train regularly and where

1	Test only one part of a potential food plant at a time.
2	Separate the plant into its basic components—leaves, stems, roots, buds, and flowers.
3	Smell the food for strong or acid odors. Remember, smell alone does not indicate a plant is edible or inedible.
4	Do not eat for 8 hours before starting the test.
5	During the 8 hours you abstain from eating, test for contact poisoning by placing a piece of the plant part you are testing on the inside of your elbow or wrist. Usually 15 minutes is enough time to allow for a reaction.
6	During the test period, take nothing by mouth except purified water and the plant part you are testing.
7	Select a small portion of a single part and prepare it the way you plan to eat it.
8	Before placing the prepared plant part in your mouth, touch a small portion (a pinch) to the outer surface of your lip to test for burning or itching.
9	If after 3 minutes there is no reaction on your lip, place the plant part on your tongue, holding it there for 15 minutes.
10	If there is no reaction, thoroughly chew a pinch and hold it in your mouth for 15 minutes. **Do not swallow.**
11	If no burning, itching, numbing, stinging, or other irritation occurs during the 15 minutes, swallow the food.
12	Wait 8 hours. If any ill effects occur during this period, induce vomiting and drink a lot of water.
13	If no ill effects occur, eat 0.25 cup of the same plant part prepared the same way. Wait another 8 hours. If no ill effects occur, the plant part as prepared is safe for eating.

CAUTION

Test all parts of the plant for edibility, as some plants have both edible and inedible parts. Do not assume that a part that proved edible when cooked is also edible when raw. Test the part raw to ensure edibility before eating raw. The same part or plant may produce varying reactions in different individuals.

Figure 9-5. Universal Edibility Test.

you expect to be traveling or working. Listed below and on the following pages are some of the most common edible and medicinal plants. Detailed descriptions and photographs of these and other common plants are at Appendix B.

TEMPERATE ZONE FOOD PLANTS

- Amaranth *(Amaranthusretrof/exusmA* other species)
- Arrowroot *(SagHlaria species)*
- Asparagus *(Asparagus officinalis)*
- Beechnut *(Fagus* species)
- Blackberries *(Rubus* species)
- Blueberries *(Vaccinium* species)
- Burdock *(Arctium lappa)*
- Cattail *(Typhi* species)
- Chestnut *(Castanea* species)
- Chicory *(Cichorium intybus)*
- Chufa *(Cyperus escukntus)*
- Dandelion *(Taraxacum officinale)*
- Daylily *(Hemerocnllis fulm)*
- Nettle *(Utica* species)
- Oaks *(Quercus* species)
- Persimmon *(Diospyros virginiana)*
- Plantain *(Plantago* species)
- Pokeweed *(Phytolacca americana)*
- Prickly pear cactus *(Opuntia species)*
- Purslane *(Portulaca oleracea)*
- Sassafras *(Sassafras albidum)*
- Sheep sorrel *(Rumex acetosella)*
- Strawberries *(Fragark* species)
- Thistle *(Cirsium* species)
- Water lily and lotus *(Nuphar, Nelumbo,* and other species)
- Wild onion and garlic *(Allium* species)
- Wild rose *(Rosa* species)
- Wood sorrel *(Oxalis* species)

TROPICAL ZONE FOOD PLANTS

- Bamboo *(Bambusa* and other species)
- Bananas (Musa species)
- Breadfruit *(Artocarpus incisa)*
- Cashew nut *(Anacardium occidental)*
- Coconut (Cocos *nucifera)*
- Mango *(Mangifera Mica)*
- Palms (various species)
- Papaya *(Carica* species)
- Sugarcane *(Saccharum officinarum)*
- Taro *(Colocasia* species)

DESERT ZONE FOOD PLANTS

- Acacia *(Acacia farnesiana)*
- Agave *(Agave* species)
- Cactus (various species)
- Date palm *(Phoenix dactylifera)*
- Desert amaranth *(Amaranths palmeri)*

Seaweeds

One plant you should never overlook is seaweed. It is a form of marine algae found on or near ocean shores. There are also some edible freshwater varieties. Seaweed is a valuable source of iodine, other minerals, and vitamin C. Large quantities of seaweed in an unaccustomed stomach can produce a severe laxative effect.

When gathering seaweeds for food, find living plants attached to rocks or floating free. Seaweed washed onshore any length of time may be spoiled or decayed. You can dry freshly harvested seaweeds for later use.

Its preparation for eating depends on the type of seaweed. You can dry thin and tender varieties in the sun or over a fire until crisp. Crush and add these to soups or broths. Boil thick, leathery seaweeds for a short time to soften them. Eat them as a vegetable or with other foods. You can eat some varieties raw after testing for edibility.

SEAWEEDS

- Dulse *(Rhodymenk palmata)*
- Green seaweed *(Ulva lactuca)*
- Irish moss *(Chonirus crispus)*
- Kelp *(Maria esculenta)*
- Laver *(Porphyra* species)
- Mojaban *(Sargassumfulvellum)*
- Sugar wrack *(Laminaria saccharin)*

Preparation of Plant Food

Although some plants or plant parts are edible raw, you must cook others to be edible or palatable. Edible means that a plant or food will provide you with necessary nutrients, while palatable means that it actually is pleasing to eat. Many wild plants are edible but barely palatable. It is a good idea to learn to identify, prepare, and eat wild foods.

Methods used to improve the taste of plant food include soaking, boiling, cooking, or leaching. Leaching is done by crushing the foocf(for example, acorns), placing it in a strainer, and pouring boiling water through it or immersing it in running water.

Boil leaves, stems, and buds until tender, changing the water, if necessary, to remove any bitterness.

Boil, bake, or roast tubers and roots. Drying helps to remove caustic oxalates from some roots like those in the *Arum* family.

Leach acorns in water, if necessary, to remove the bitterness. Some nuts, such as chestnuts, are good raw, but taste better roasted.

You can eat many grains and seeds raw until they mature. When hard or dry, you may have to boil or grind them into meal or flour.

The sap from many trees, such as maples, birches, walnuts, and sycamores, contains sugar. You may boil these saps down to a syrup for sweetening. It takes about 35 liters of maple sap to make one liter of maple syrup!

PLANTS FOR MEDICINE

In a survival situation you will have to use what is available. In using plants and other natural remedies, positive identification of the plants involved is as critical as in using them for food. Proper use of these plants is equally important.

Terms and Definitions

The following terms, and their definitions, are associated with medicinal plant use:

- *Poultice.* The name given to crushed leaves or other plant parts, possibly heated, that you apply to a wound or sore either directly or wrapped in cloth or paper.
- *Infusion or tisane or tea.* The preparation of medicinal herbs for internal or external application. You place a small quantity of a herb in a container, pour hot water over it, and let it steep (covered or uncovered) before use.
- *Decoction.* The extract of a boiled down or simmered herb leaf or root. You add herb leaf or root to water. You bring them to a sustained boil or simmer to draw their chemicals into the water. The average ratio is about 28 to 56 grams (1 to 2 ounces) of herb to 0.5 liter of water.
- *Expressed juice.* Liquids or saps squeezed from plant material and either applied to the wound or made into another medicine.

Many natural remedies work slower than the medicines you know. Therefore, start with smaller doses and allow more time for them to take effect. Naturally, some will act more rapidly than others.

Specific Remedies

The following remedies are for use only in a survival situation, not for routine use:

- *Diarrhea.* Drink tea made from the roots of blackberries and their relatives to stop diarrhea. White oak bark and other barks containing tannin are also effective. However, use them with caution when nothing else is available because of possible negative effects on the kidneys. You can also stop diarrhea by eating white clay or campfire ashes. Tea made from cowberry or cranberry or hazel leaves works too.

- *Antihemorrhagics.* Make medications to stop bleeding from a poultice of the puffball mushroom, from plantain leaves, or most effectively from the leaves of the common yarrow or woundwort *(Achilles millefolium).*
- *Antiseptics.* Use to cleanse wounds, sores, or rashes. You can make them from the expressed juice from wild onion or garlic, or expressed juice from duckweed leaves or the crushed leaves of dock. You can also make antiseptics from a decoction of burdock root, mallow leaves or roots, or white oak bark. All these medications are for external use only.
- *Fevers.* Treat a fever with a tea made from willow bark, an infusion of elder flowers or fruit, linden flower tea, or elm bark decoction.
- *Colds and sore throats.* Treat these illnesses with a decoction made from either plantain leaves or willow bark. You can also use a tea made from burdock roots, mallow or mullein flowers or roots, or mint leaves.
- *Aches, pains, and sprains.* Treat with externally applied poultices of dock, plantain, duckweed, willow bark, garlic, or sorrel. You can also use salves made by mixing the expressed juices of these plants in animal fat or vegetable oils.
- *Itching.* Relieve the itch from insect bites, sunburn, or plant poisoning rashes by applying a poultice of jewelweed *(Impatiens tiflora)* or witch hazel leaves *(Hamamelis virginiana).The* jewelweed juice will help when applied to poison ivy rashes or insect stings. If works on sunburn as well as aloe vera.
- *Sedatives.* Get help in falling asleep by brewing a tea made from mint leaves or passionflower leaves.
- *Hemorrhoids.* Treat them with external washes from elm bark or oak bark tea, from the expressed juice of plantain leaves, or from a Solomon's seal root decoction.
- *Constipation.* Relieve constipation by drinking decoctions from dandelion leaves, rose hips, or walnut bark. Eating raw daylily flowers will also help.
- *Worms or intestinal parasites.* Using moderation, treat with tea made from tansy *(Tanacetum vulgare)* or from wild carrot leaves.
- *Gas and cramps.* Use a tea made from carrot seeds as an antiflatulent; use tea made from mint leaves to settle the stomach.
- *Antifungal washes.* Make a decoction of walnut leaves or oak bark or acorns to treat ringworm and athlete's foot. Apply frequently to the site, alternating with exposure to direct sunlight.

MISCELLANEOUS USES OF PLANTS

Make dyes from various plants to color clothing or to camouflage your skin. Usually, you will have to boil the plants to get the best results. Onion skins produce yellow, walnut hulls produce brown, and pokeberries provide a purple dye.

Make fibers and cordage from plant fibers. Most commonly used are the stems from nettles and milkweeds, yucca plants, and the inner bark of trees like the lindea

Make fish poison by immersing walnut hulls in a small area of quiet water. This poison makes it impossible for the fish to breathe but doesn't adversely affect their edibility.

Make tinder for starting fires from cattail fluff, cedar bark, lighter knot wood from pine trees, or hardened sap from resinous wood trees.

Make insulation by fluffing up female cattail heads or milkweed down.

Make insect repellents by applying the expressed juice of wild garlic or onion to the skin, by placing sassafras leaves in your shelter, or *oy* burning or smudging cattail seed hair fibers.

Plants can be your ally as long as you use them cautiously. *The key to the safe use of plants is positive identification* whether you use them as food or medicine or in constructing shelters or equipment.

CHAPTER 10

POISONOUS PLANTS

Successful use of plants in a survival situation depends on positive identification. Knowing poisonous plants is as important to a survivor as knowing edible plants. Knowing the poisonous plants will help you avoid sustaining injuries from them.

HOW PLANTS POISON

Plants generally poison by—

- *Ingestion.* When a person eats a part of a poisonous plant.
- *Contact.* When a person makes contact with a poisonous plant that causes any type of skin irritation or dermatitis.
- *Absorption or inhalation.* When a person either absorbs the poison through the skin or inhales it into the respiratory system.

Plant poisoning ranges from minor irritation to death. A common question asked is, 'Tlow poisonous is this plant?" It is difficult to say how poisonous plants are because-

- Some plants require contact with a large amount of the plant before noticing any adverse reaction while others will cause death with only a smallamount.
- Every plant will vary in the amount of toxins it contains due to different growing conditions and slight variations in subspecies.
- Every person has a different level of resistance to toxic substances.
- Some persons may be more sensitive to a particular plant.

Some common misconceptions about poisonous plants are—

- *Watch the animals and eat what they* art. Most of the time this statement is true, but some animals can eat plants that are poisonous to humans.
- *Boil the plant in water and any poisons will be removed.* Boiling removes many poisons, but not all.
- *Plants with a red color are poisonous.* Some plants that are red are poisonous, but not all.

The point is there is no one rule to aid in identifying poisonous plants. You must make an effort to learn as much about them as possible.

ALL ABOUT PLANTS

It is to your benefit to learn as much about plants as possible. Many poisonous plants look like their edible relatives or like other edible plants. For example, poison hemlock appears very similar to wild carrot. Certain plants are safe to eat in certain seasons or stages of growth and poisonous in other stages. For example, the leaves of the pokeweed are edible when it first starts to grow, but it soon becomes poisonous. You can eat some plants and their fruits only when they are ripe. For example, the ripe fruit of mayapple is edible, but all other parts and the green fruit are poisonous. Some plants contain both edible and poisonous parts; potatoes and tomatoes are common plant foods, but their green parts are poisonous.

Some plants become toxic after wilting. For example, when the black cherry starts to wilt, hydrocyanic acid develops. Specific preparation methods make some plants edible that are poisonous raw. You can eat the thinly sliced and thoroughly dried corms (drying may take a year) of the jack-in-the-pulpit, but they are poisonous if not thoroughly dried.

Learn to identify and use plants before a survival situation. Some sources of information about plants are pamphlets, books, films, nature trails, botanical gardens, local markets, and focal natives. Gather and cross-reference information from as many sources as possible, because many sources will not contain all the information needed.

RULES FOR AVOIDING POISONOUS PLANTS

Your best policy is to be able to look at a plant and identify it with absolute certainty and to know its uses or dangers. Many times this is not possible. If you have little or no knowledge of the local vegetation, use the rules to select plants for the "Universal Edibility Test. Remember, avoid —

- *All mushrooms.* Mushroom identification is very difficult and must be precise, even more so than with other plants. Some mushrooms cause death very quickly. Some mushrooms have no known antidote. Two general types of mushroom poisoning are gastrointestinal and central nervous system.
- *Contact with or touching plants unnecessarily.*

CONTACT DERMATITIS

Contact dermatitis from plants will usually cause the most trouble in the field. The effects may be persistent, spread by scratching, and are particularly dangerous if there is contact m or around the eyes.

The principal toxin of these plants is usually an oil that gets on the skin upon contact with the plant. The oil can also get on equipment and then infect whoever touches the equipment. Never Dum a contact poisonous plant because the smoke may be as harmful as the plant. There is a greater danger of being affected when overheated and sweating. The infection may be local or it may spread over the body.

Symptoms may take from a few hours to several days to appear. Signs and symptoms can include burning, reddening, itching, swelling, and blisters.

When you first contact the poisonous plants or the first symptoms appear, try to remove the oil by washing with soap and cold water. If water is not available, wipe your skin repeatedly with dirt or sand. Do not use dirt if blisters have developed. The dirt may break open the blisters and leave the body open to infection. After you have removed the oil, dry the area. You can wash with a tannic acid solution and crush and rub jewelweed on the affected area to treat plant-caused rashes. You can make tannic acid from oak bark.

Poisonous plants that cause contact dermatitis are—

- Cowhage.
- Poison ivy.
- Poison oak.
- Poison sumac.
- Rengas tree.
- Trumpet vine.

INGESTION POISONING

hgestion poisoning can be very serious and could lead to death very quickly. Do not eat any plant unless you have positively identified it first. Keep a log of all plants eaten.

Signs and symptoms of ingestion poisoning can include nausea, vomiting, diarrhea, abdominal cramps, depressed heartbeat and respiration, headaches, hallucinations, dry mouth, unconsciousness, coma, and death.

If you suspect plant poisoning, try to remove the poisonous material from the victim's mouth and stomach as soon as possible. Induce vomiting by tickling the back of his throat or by giving nim warm saltwater, if he is conscious. Dilute the poison by administering large quantities of water or milk, if he is conscious.

The following plants can cause ingestion poisoning if eaten:

- Castor bean.
- Chinaberry.
- Death camas.
- Lantana.
- Manchineel.
- Oleander.
- Pangi.
- Physic nut.
- Poison and water hemlocks.
- Rosary pea.
- Strychnine tree.

See Appendix C for photographs and descriptions of these plants.

CHAPTER 11

DANGEROUS ANIMALS

Animals rarely are as threatening to the survivor as the rest ofthe environment. Common sense tells the survivor to avoid encounters with lions, bears, and other large or dangerous animals. You should also avoid large grazing animals with horns, hooves, and great weight. Your actions may prevent unexpected meetings. Move carefully through their environment. Do not attract large predators by leaving food lying around your camp. Carefully survey the scene before entering water or forests.

Smaller animals actually present more of a threat to the survivor than large animals. To compensate for their size, nature has given many small animals weapons such as fangs and stingers to defend themselves. Each year, a few people are bitten by sharks, mauled by alligators, and attacked by bears. Most of these incidents were in some way the victim's fault. However, each year more victims die from bites by relatively small venomous snakes than by large dangerous animals. Even more victims die from allergic reactions to bee stings. For this reason, we will pay more attention to smaller and potentially more dangerous creatures. These are the animals you are more likely to meet as you unwittingly move into their habitat, or they slip into your environment unnoticed.

Keeping a level head and an awareness of your surroundings will keep you alive if you use a few simple safety proce - dures. Do not let curiosity and carelessness kill or injure you.

INSECTS AND ARACHNIDS

You recognize and identify insects, except centipedes and millipedes, by their six legs while arachnids have eight. All these small creatures become pests when they bite, sting, or irritate you.

Although their venom can be quite painful, bee, wasp, and hornet stings rarely kill a survivor unless he is allergic to that particular toxin. Even the most dangerous spiders rarely kill, and the effects of tick-borne diseases are very slow-acting. However, in all cases, avoidance is the best defense. In environments known to have spiders and scorpions, check your footgear and clothing every morning. Also check your bedding and shelter for them. Use care when turning over rocks and logs. See Appendix D for examples of dangerous insects and arachnids.

Scorpions

You find scorpions *(Buthotus* species) in deserts, jungles, and forests of tropical, subtropical, and warm temperate areas of the world. They are mostly nocturnal in habit. You can find desert scorpions from below sea level m Death Valley to elevations as high as 3,60(J meters in the Andes. Typically brown or black in moist areas, they maybe yellow or light green in the desert. Their average size is about 2.5 centimeters. However, there are 20-centimeter giants in the jungles of Central America, New Guinea, and southern Africa. Fatalities from scorpion stings are rare, but they can occur in children, the elderly, and ill persons. Scorpions resemble small lobsters with raised, jointed tails bearing a stinger in the tip. Nature mimics the scorpions with whip scorpions or vinegarroons. These are harmless and have a tail like a wire or whip, rather than the jointed tail and stinger of true scorpions.

Spiders

You recognize the brown recluse or fiddleback spider of North America *(Loxosceles reclusa)*by a prominent violin-shaped light spot on the back of its body. As its name suggests, this spider Tikes to hide in dark places. Though rarely fatal, its bite causes excessive tissue degeneration around the wound and can even lead to amputation of the digits if left untreated.

You find members of the widow family *(Latrodectus specks)* worldwide, though the black widow of North America is perhaps the most well-known. Found in warmer areas of the world, the widows are small, dark spiders with often hourglass-shaped white, red, or orange spots on their abdomens.

Funnelwebs *(Atrax* species) are large, gray or brown Australian spiders. Chunky, with short legs, they are able to move easily up and down the cone-shaped webs from which they get their name. The local populace considers them deadly. Avoid them as they move about, usually at night, in search of prey. Symptoms of their bite are similar to those of the widow's—severe pain accompanied by sweating and shivering, weakness, and disabling episodes that can last a week.

Tarantulas are large, hairy spiders *(Theraphosidae* and *Lycosa* species) best known because they are often sold in pet stores. There is one species in Europe, but most come from tropical America. Some South American species do inject a dangerous toxin, but most simply produce a painful bite. Some tarantulas can be as large as a dinner plate. They all have large fangs for capturing food such as birds, mice, and lizards. If bitten by a tarantula, pain and bleeding are certain, and infection is likely.

Centipedes and Millipedes

Centipedes and millipedes are mostly small and harmless, although some tropical and desert species may reach 25 centimeters. A few varieties of centipedes have a poisonous bite, but infection is the greatest danger, as their sharp claws dig in and puncture the skin. To prevent skin punctures, brush them off in the direction they are traveling, if you find them crawling on your skin.

Bees, Wasps, and Hornets

We are all familiar with bees, wasps, and hornets. They come in many varieties and have a wide diversity of habits and habitats. You recogmze bees by their hairy and usually thick body, while the wasps, hornets, and yellow jackets have more slender, nearly hairless, bodies. Some bees, such as honeybees, live in colonies. They may be either domesticated or living wild in caves or hollow trees. You may find other bees, such as carpenter bees, in individual nest holes in wood, or in the ground, like bumblebees. The main danger from bees is their barbed stinger located on their abdomens. When the bee stings you, it rips its stinger out of its abdomen along with the venom sac, and the bee dies. Except for killer bees, most bees tend to be more docile than wasps, hornets, and yellow jackets that have smooth stingers and are capable of repeated attacks.

Avoidance is the best tactic for self-protection. Watch out for flowers or fruit where bees may be feeding. Be careful of meat-eating yellow jackets when cleaning fish or game. The average person has a relatively minor and temporary reaction to bee stings and recovers in a couple of hours when the pain and headache go away. Those who are allergic to bee venom have severe reactions including anaphylactic shock, coma, and death. If antihistamine medicine is not available and you cannot find a substitute, an allergy sufferer in a survival situation is in grave danger.

Ticks

Ticks are common in the tropics and temperate regions. They are familiar to most of us. Ticks are small round arachnids with eight legs and can have either a soft or hard body. Ticks require a blood host to survive and reproduce. This makes them dangerous because they spread diseases like Lyme disease, Rocky Mountain spotted fever, encephalitis, and others that can ultimately be disabling or fatal. There is little you can do to treat these diseases once contracted, but time is your ally since they are slow-acting ailments. According to most authorities, it takes at least 6 hours of attachment to the host for the tick to transmit the disease organisms. Thus, you have time to thoroughly inspect your body for their presence. Beware of ticks when passing through the thick vegetation they cling to, when cleaning host animals for food, and when gathering natural materials to construct a shelter. Always use insect repellents, if possible.

LEECHES

Leeches are blood-sucking creatures with a wormlike appearance. You find them in the tropics and in temperate zones. You will certainly encounter them when swimming in infested waters or making expedient water crossings. You can find them when passing through swampy, tropical vegetation and bogs. You can also find them while cleaning food animals, such as turtles, found in fresh water. Leeches can crawl into small openings; therefore, avoid camping in their habitats when possible. Keep your trousers tucked in your boots. Check yourself frequently for leecnes. Swallowed or eaten, leeches can be a great hazard. It is therefore essential to treat water from questionable sources by boiling or using chemical water treatments. Survivors have developed severe infections from wounds inside the throat or nose when sores from swallowed leeches became infected.

BATS

Despite the legends, bats (*Desmodus* species) are a relatively small hazard to the survivor. There are many bat varieties worldwide, but you find the true vampire bats only in Central and South America. They are small, agile fliers that land on their sleeping victims, mostly cows and horses, to lap a blood meal after biting their victim. Their saliva contains an anticoagulant that keeps the blood slowly flowing while they feed. Only a small percentage of these bats actually carry rabies; however, avoid any sick or injured bat. They can carry other diseases and infections and will bite readily when handled. Taking shelter in a cave occupied by bats, however, presents the much greater hazard of inhaling powdered bat dung, or guano. Bat dung carries many organisms that can cause diseases. Eating thoroughly cooked flying foxes or other bats presents no danger from rabies and other diseases, Dut again, the emphasis is on thorough cooking.

POISONOUS SNAKES

There are no infallible rules for expedient identification of poisonous snakes in the field, because the guidelines all require close observation or manipulation of the snake's body. The best strategy is to leave all snakes alone. Where snakes are plentiful and poisonous species are present, the risk of their bites negates their food value. Apply the following safety rules when traveling in areas where there are poisonous snakes:

- Walk carefully and watch where you step. Step onto logs rather than over them before looking and moving on.
- Look closely when picking fruit or moving around water.
- Do not tease, molest, or harass snakes. Snakes cannot close their eyes. Therefore, you cannot tell if they are asleep. Some snakes, such as mambas, cobras, and bushmasters, will attack aggressively when cornered or guarding a nest.
- Use sticks to turn logs and rocks.
- Wear proper footgear, particularly at night.
- Carefully check bedding, shelter, and clothing.
- Be calm when you encounter serpents. Snakes cannot hear and you can occasionally surprise them when they are sleeping or sunning. Normally, they will flee if given the opportunity.
- Use extreme care if you must kill snakes for food or safety. Although it is not common, warm, sleeping human bodies occasionally attract snakes.

See Appendix E for detailed descriptions of the snakes listed on the following pages.

Snake-Free Areas

The polar regions are free of snakes due to their inhospitable environments. Other areas considered to be free of poisonous snakes are New Zealand, Cuba, Haiti, Jamaica, Puerto Kico, Ireland, Polynesia, and Hawaii.

POISONOUS SNAKES OF THE AMERICAS

- American Copperhead *(Agkistrodoncontortrix)*
- Bushmaster *(Ladiesis mutus)*
- Coral snake *(Micrurusfulvius)*
- Cottonmouth *(Agkistrodon piscivorus)*
- Fer-de-lance *(Bothrops atrox)*
- Rattlesnake *(Cwtalus* species)

POISONOUS SNAKES OF EUROPE

- Common adder *(Vipers berus)*
- Pallas' viper *(Agkistrodon half)*

POISONOUS SNAKES OF AFRICA AND ASIA

- Boomslang *(Dispholidus typus)*
- Cobra *(Naja* species)
- Gaboon viper *(Bitis gabonica)*
- Green tree pit viper *(Trimeresurus gramineus)*
- Habu pit viper *(Trimeresurus fltvoviridis)*
- Krait *(Bungarus caeruleus)*
- Malayan pit viper *(Callaseksma rhoiostoma)*
- Mamba *(Dendmspis* species)
- Puff adder *(Bitis srietans)*
- Rhinoceros viper *(Bitis nasicornis)*
- Russell's viper *(Vipera russellii)*
- Sand viper (Cerastes *vipera)*
- Saw-scaled viper *(Echis carinatus)*
- Wagler's pit viper *(Trimeresurus wagleri)*

POISONOUS SNAKES OF AUSTRALASIA

- Death adder *(Acanthophis antarcticus)*
- Taipan *(Oxyuranus scutelktus)*
- Tiger snake *(Notechis scutatus)*
- Yellow-bellied sea snake *(Pekmispkturus)*

DANGEROUS LIZARDS

The Gila monster and the Mexican beaded lizard are dangerous and poisonous lizards.

Gila Monster

The Gila monster *(Helodermasuspectrum)oi* the American southwest, including Mexico, is a large lizard with dark, highly textured skin marked by pinkish mottling. It averages 35 to 45 centimeters in length and has a thick, stumpy tail. Unlikely to bite unless molested, it has a poisonous bite.

Mexican Beaded Lizard

The Mexican beaded lizard *(Helodermahorridum)* resembles its relative, the Gila monster. It has more uniform spots rather than bands of color (the Gila monster). It also is poisonous and has a docile nature. You find it from Mexico to Central America.

Komodo Dragon

This giant lizard *(Varanus komodoensis)* grows to more than 3 meters in length and can be dangerous if you try to capture it. This Indonesian lizard can weigh more than 135 kilograms.

DANGERS IN RIVERS

Common sense will tell you to avoid confrontations with hippopotami, alligators, crocodiles, and other large river creatures. There are, however, a few smaller river creatures with which you should be cautious.

Electric Eel

Electric eels *(Ekctrophorus electricus)* may reach 2 meters in length and 20 centimeters in diameter. Avoid them. They are capable of generating up to 500 volts of electricity in certain organs m their body. They use this shock to stun prey and enemies. Normally, you find these eels in the Orinoco and Amazon River systems in South America. They seem to prefer shallow waters that are more highly oxygenated and provide more food. They are bulkier than our native eels. Their upper body is dark gray or black, with a lighter-colored underbelly.

Piranha

Piranhas *(Senasalmo* species) are another hazard of the Orinoco and Amazon River systems, as well as the Paraguay River Basin, where they are native. These fish vary greatly in size and coloration, but usually have a combination of orange undersides and dark tops. They have white, razor-sharp teeth that are clearly visible. They may be as long as 50 centimeters. Use great care when crossing waters where they live. Blood attracts them. They are most dangerous in shallow waters during the dry season.

Turtle

Be careful when handling and capturing large freshwater turtles, such as the snapping turtles and soft-shelled turtles of North America and the matamata and other turtles of South America. All of these turtles will bite in self-defense and can amputate fingers and toes.

Platypus

The platypus or duckbill *(Ornithorhyncus anatinus)* is the only member of its family and is easily recognized. It has a long body covered with grayish, short hair, a tail like a beaver, and a bilflike a duck. Growing up to 60 centimeters in length, it may appear to be a good food source, but this egg-laying mammal, the only one in the world, is very dangerous. The male has a poisonous spur on each hind foot that can inflict intensely painful wounds. You find the platypus only in Australia, mainly along mud banks on waterways.

DANGERS IN BAYS AND ESTUARIES

In areas where seas and rivers come together, there are dangers associated with both fresh and salt water. In shallow salt waters, there are many creatures that can inflict pain and cause infection to develop. Stepping on sea urchins, for example, can produce pain and infection. When moving about in shallow water, wear some form of footgear and shuffle your feet along the bottom, rather than picking up your feet and stepping.

Stingrays *(Dasyatidae* species) are a real hazard in shallow waters, especially tropical waters. The type of bottom appears to be irrelevant. There is a great variance between species, but all have a sharp spike in their tail that may be venomous and can cause extremely painful wounds if stepped on. All rays have a typical shape that resembles a kite. You find them along the coasts ofthe Americas, Africa, and Australasia.

SALTWATER DANGERS

There are several fish that you should not handle, touch, or contact. There are others that you should not eat.

Fish Dangerous to Handle, Touch, or Contact

There are several fish you should not handle, touch, or contact that are identified below.

Shark

Sharks are the most feared animal in the sea. Usually, shark attacks cannot be avoided and are considered accidents. You, as a survivor, should take every precaution to avoid any contact with sharks. There are many shark species, but in general, dangerous sharks have wide mouths and visible teeth, while relatively harmless ones have small mouths on the underside of their heads. However, any shark can inflict painful and often fatal injuries, either through bites or through abrasions from their rough skin.

Rabbitfish

Rabbitfish or spinefoot *(Siganidae* species) occur mainly on coral reefs in the Indian and Pacific oceans. They have very sharp, possibly venomous spines in their fins. Handle them with care, if at all. This fish, like many others of the dangerous fish in this section, is considered edible by native peoples where the fish are found, but deaths occur from careless handling. Seek other nonpoisonous fish to eat if at all possible.

Tang

Tang or surgeonfish *(Acanthuridae* species) average 20 to 25 centimeters in length and often are beautifully colored. They are called surgeonfish because of the scalpellike spines located in the tail. The wounds inflicted by these spines can bring about death through infection, envenomation, and loss or blood, which may incidentally attract sharks.

Toadfish

Toadfish *(Batrachoididae* species) occur in tropical waters off the Gulf Coast of the United States and along both coasts of Central and South America. These dully colored fish average 18 to 25 centimeters in length. They typically bury themselves in the sand to await fish and other prey. They have sharp, very toxic spines along their backs.

Scorpion Fish

Poisonous scorpion fish or zebra fish *(Scorpaenidae*species) are mostly around reefs in the tropical Indian and Pacific oceans and occasionally in the Mediterranean and Aegean seas. They average 30 to 75 centimeters in length. Their coloration is highly variable, from reddish brown to almost purple or brownish yellow. They have long, wavy fins and spines and their sting is intensively painful. Less poisonous relatives live m the Atlantic Ocean.

Stonefish

Stonefish *(Synanceja* species) are in the Pacific and Indian oceans. They can inject a painful venom from their dorsal spines when stepped on or handled carelessly. They are almost impossible to see because of their lumpy shape and drab colors. They range in size up to 40 centimeters.

WeeverFish

Weever fish *(Trachinidae* species) average 30 centimeters long. They are hard to see as they lie buried in the sand off the coasts of Europe, Africa, and the Mediterranean. Their color is usually a dull brown. They have venomous spines on the back and gills.

See Appendix F for more details on these venomous fish.

Animals and Fish Poisonous to Eat

Survival manuals often mention that the livers of polar bears are toxic due to their high concentrations of vitamin A. For this reason, we mention the chance of death after eating this organ. Another toxic meat is the flesh of the hawksbill turtle. You recognize them by their down-turned bill and yellow polka dots on their neck and front flippers. They weigh more than 275 kilograms and are unlikely to be captured.

Many fish living in reefs near shore, or in lagoons and estuaries, are poisonous to eat, though some are only seasonally dangerous. The majority are tropical fish; however, be wary of eating any unidentifiable fish wherever you are. Some predatory fish, such as barracuda and snapper, may become toxic if the fish they feed on in shallow waters are poisonous. The most poisonous types appear to have parrotlike beaks and hard shell-like skins with spines and often can inflate their bodies like balloons. However, at certam times of the year, indigenous populations consider the puffer a delicacy.

Bhwfish

Blowfish or puffer *(Tetraodontidae* species) are more tolerant of cold water. You fmd them along tropical and temperate coasts worldwide, even in some of the rivers of Southeast Asia and Africa. Stout-bodied and round, many of these fish have short spines and can inflate themselves into a ball when alarmed or agitated. Their blood, liver, and gonads are so toxic that as little as 28 milligrams (1 ounce) can be fatal. These fish vary in color and size, growing up to 75 centimeters in length.

Friggerfish

The triggerfish *(Balistidae* species) occur in great variety, mostly in tropical seas.They are deep-bodied and compressed, resembling a seagoing pancake up to 60 centimeters in length, with large and sharp dorsal spines. Avoid them all, as many have poisonous flesh.

Barracuda

Although most people avoid them because of their ferocity, they occasionally eat barracuda *(Sphyraena barracuda).* These predators of mostly tropical seas can reach almost 1.5 meters in length and have attacked humans without provocation. They occasionally carry the poison ciguatera in their flesh, making them deadly if consumed.

See Appendix F for more details on toxic fish and toxic mollusks.

Other Dangerous Sea Creatures

The blue-ringed octopus, jellyfish, and the cone and auger shells are other dangerous sea creatures.

Blue-Ringed Octopus

Most octopi are excellent when properly prepared. However, the blue-ringed octopus *(Hapalochlaena lunulatajcm* mflict a deadly bite from its parrotlike beak. Fortunately, it is restricted to the Great Barrier Reef of Australia and is very small. It is easily recognized by its grayish white overall color and iridescent blue rings. Authorities warn that all tropical octopus species should be treated with caution, since many have poisonous bites, although the flesh is edible.

Jellyfish

Jellyfish-related deaths are rare, but the sting they inflict is extremely painful. The Portuguese man-of-war resembles a large pink or purple balloon floating on the sea. It has poisonous tentacles hanging up to 12 meters below its body. The huge tentacles are actually colonies of stinging cells. Most known deaths from jellyfish are attributed to the man-of-war. Other jellyfish can inflict very painful stings as well. Avoid the long tentacles of any jellyfish, even those washed up on the beach and apparently dead.

Cone Shell

The subtropical and tropical cone shells *(Conidae* species) have a venomous harpoonlike barb. All are cone-shaped and nave a fine netlike pattern on the shell. A membrane may possibly obscure this coloration. There are some very poisonous cone shells, even some lethal ones in the Indian and Pacific oceans. Avoid any shell shaped like an ice cream cone.

Auger Shell

The auger shell or terebra *(Terebridae* species) are much longer and thinner than the cone shells, but can be nearly as deadly as the cone shells. They are found in temperate and tropical seas. Those in the Indian and Pacific oceans have a more toxic venom in their stinging barb. Do not eat these snails, as their flesh may be poisonous.

CHAPTER 12

FIELD-EXPEDIENT WEAPONS, TOOLS, AND EQUIPMENT

As a soldier you know the importance of proper care and use of your weapons, tools, and equipment. This is especially true of your knife. You must always keep it sharp and ready to use. A knife is your most valuable tool in a survival situation. Imagine being in a survival situation without any weapons, tools, or equipment except your knife. It could happen! You might even be without a knife. You would probably feel helpless, but with the proper knowledge and skills, you can easily improvise needed items.

In survival situations, you may have to fashion any number and type of field-expedient tools and equipment to survive. Examples of tools and equipment that could make your life much easier are ropes, rucksacks, clothes, nets, and so on.

Weapons serve a dual purpose. You use them to obtain and prepare food and to provide self-defense. A weapon can also give you a feeling of security and provide you with the ability to hunt on the move.

CLUBS

You hold clubs, you do not throw them. As a field-expedient weapon, the club does not protect you from enemy soldiers. It can, however, extend your area of defense beyond your fingertips. It also serves to increase the force of a blow without injuring yourself. There are three basic types of clubs. They are the simple, weighted, and sling club.

Simple Club

A simple club is a staff or branch. It must be short enough for you to swing easily, but long enough and strong enough for you to damage whatever you hit. Its diameter should fit comfortably in your palm, but it should not be so thin as to allow the club to break easily upon impact. A straight-grained hardwood is best if you can find it.

Weighted Club

A weighted club is any simple club with a weight on one end. The weight may be a natural weight, such as a knot on the wood, or something added, such as a stonelashed to the club.

To make a weighted club, first find a stone that has a shape that will allow you to lash it securely to the club. A stone with a slight hourglass shape works well. If you cannot find a suitably shaped stone, you must fashion a groove or channel into the stone by a technique known as pecking. By repeatedly rapping the club stone with a smaller hard stone, you can get the desired shape.

Next, find a piece of wood that is the right length for you. A straight-grained hardwood is best. The length of the wood should feel comfortable in relation to the weight of the stone. Finally, lash the stone to the handle.

There are three techniques for lashing the stone to the handle: split handle, forked branch, and wrapped handle. The technique you use will depend on the type of handle you choose. See Figure 12-1.

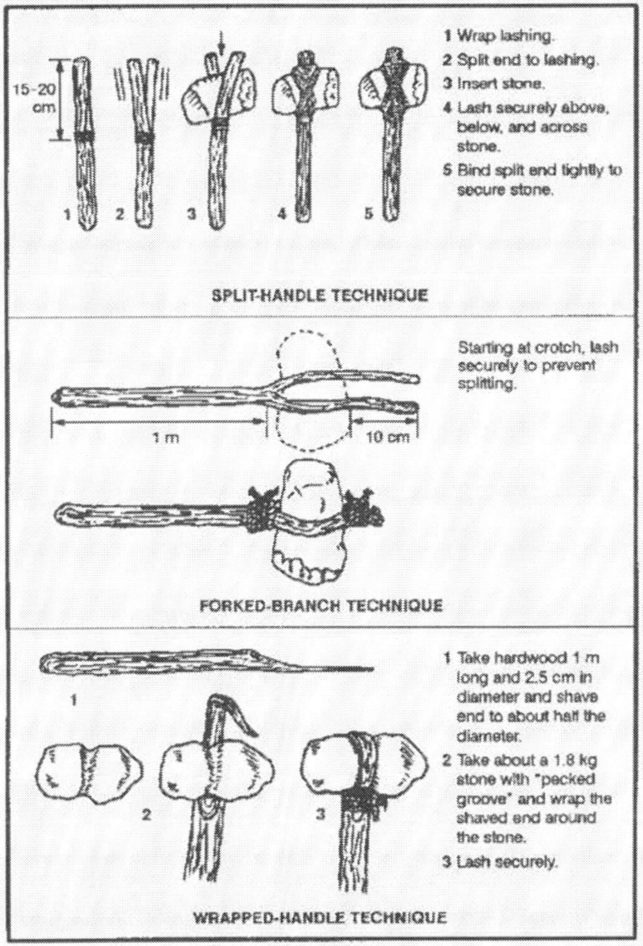

Figure 12-1. Lashing clubs.

Sling Club

A sling club is another type of weighted club. A weight hangs 8 to 10 centimeters from the handle by a strong, flexible lashing (Figure 12-2). This type of club both extends the user s reach and multiplies the force of the blow.

EDGED WEAPONS

Knives, spear blades, and arrow points fall under the category of edged weapons. The following paragraphs will discuss the making of such weapons.

Knives

A knife has three basic functions. It can puncture, slash or chop, and cut. A knife is also an invaluable tool used to construct other survival items. You may find yourself without a knife or you may need another type knife or a spear. To improvise you can use stone, bone, wood, or metal to make a knife or spear blade.

Stone

To make a stone knife, you will need a sharp-edged piece of stone, a chipping tool, and a flaking tool. A chipping toolis a light, blunt-edged

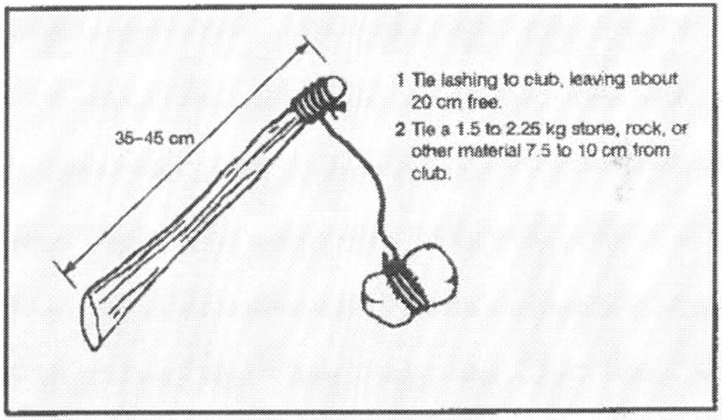

Figure 12-2. Sling club.

tool used to break off small pieces of stone. A flaking tool is a pointed tool used to break off thin, flattened pieces of stone, IOU can make a chipping tool from wood, bone, or metal, and a flaking tool from bone, antler tmes, or soft iron (Figure 12-3).

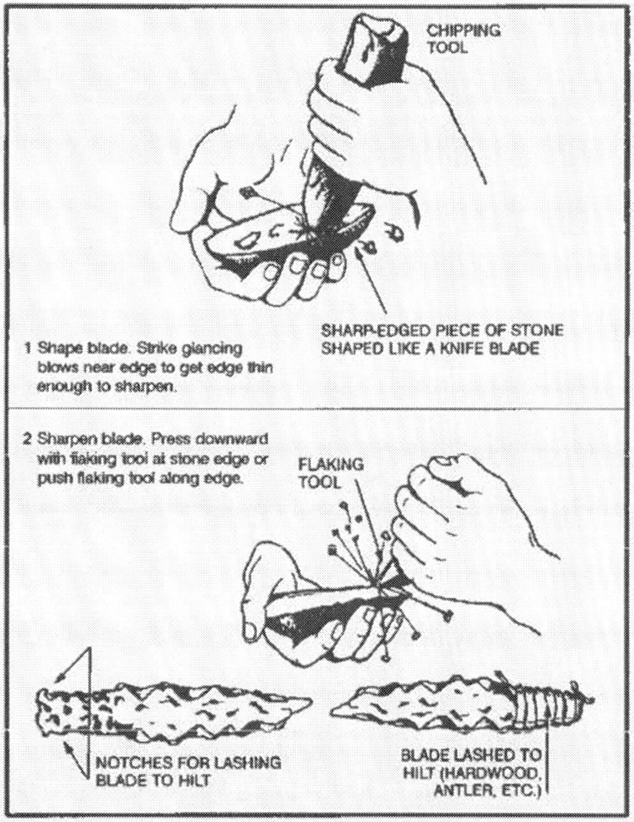

Figure 12-3. Making a stone knife.

Start making the knife by roughing out the desired shape on your sharp piece of stone, using the chipping tool. Try to make the knife fairly thin. Then, using the flaking tool, press it against the edges. This action will cause flakes to come off the opposite side of the edge, leaving a razor sharp edge. Use the flaking tool along the entire length of the edge you need to sharpen. Eventually, you will nave a very sharp cutting edge that you can use as a knife.

Lash the blade to some type of hilt (Figure 12-3).

Note: Stone will make an excellent puncturing tool and a good chopping tool but will not hold a fine edge. Some stones such as chert or flint can nave very fine edges.

Bone

You can also use bone as an effective field-expedient edged weapon. First, you will need to select a suitable bone. The larger Dones, such as the leg bone of a deer or another medium-sized animal, are best. Lay the bone upon another hard object. Shatter the bone by hitting it with a heavy object, such as a rock. From the pieces, select a suitable pointed splinter. You can further shape and sharpen this splinter by rubbing it on a rough-surfaced rock. If the piece is too small to handle, you can still use ithy adding a handle to it. Select a suitable piece of hardwood for a handle and lash the bone splinter securely to it.

Note: Use the bone knife only to puncture. It will not hold an edge and it may flake or break if used differently.

Wood

You can make field-expedient edged weapons from wood. Use these only to puncture. Bamboo is the only wood that will hold a suitable edge. To make a knife using wood, first select a straight-grained piece of hardwood that is about 30 centimeters long and 25 centimeters in diameter. Fashion the blade about 15 centimeters long. Shave it down to a point. Use only the straight-grained portions of the wood. Do not use the core or pith, as it would make a weak point.

Harden the point by a process known as fire hardening. If a fire is possible, dry the blade portion over the fire slowly until lightly charred. The drier the wood, the harder the point. After lightly charring the blade portion, sharpen it on a coarse stone. If using bamboo and after fashioning the blade, remove any other wood to make the blade thinner from the inside portion of the bamboo. Removal is done this way because bamboo's hardest part is its outer layer. Keep as much of this layer as possible to ensure the hardest blade possible. When charring bamboo over a fire, char only the inside wood; do not char the outside.

Metal

Metal is the best material to make field-expedient edged weapons. Metal, when properly designed, can fulfill a knife's three uses—puncture, slice or chop, and cut. First, select a suitable piece of metal, one that most resembles the desired end product. Depending on the size and original shape, you can obtain a point and cutting edge by rubbing the metal on a rough-surfaced stone. If the metal is soft enough, you can hammer out one edge while the metal is cold. Use a suitable flat, hard surface as an anvil and a smaller, harder object of stone or metal as a hammer to hammer out the edge. Make a knife handle from wood, bone, or other material that will protect your hand.

Other Materials

You can use other materials to produce edged weapons. Glass is a good alternative to an edged weapon or tool, if no other material is available. Obtain a suitable piece in the same manner as described for bone. Glass has a natural edge but is less durable for heavy work. You can also sharpen plastic—if it is thick enough or hard enough—into a durable point for puncturing.

Spear Blades

To make spears, use the same procedures to make the blade that you used to make a knife blade. Then select a shaft (a straight sapling) 1.2 to 1.5 meters long. The length should allow you to handle the spear easily and effectively. Attach the spear blade to the shaft using lashing. The preferred method is to split the handle, insert the blade, then wrap or lash it tightly. You can use other materials without adding a blade. Select a 1.2- to 1.5-meter long straight hardwood shaft and shave one end to a point. Ifpossible, fire harden the point. Bamboo also makes an excellent spear. Select a piece 1.2 to 1.5 meters long. Starting 8 to 10 centimeters back from the end used as the point, shave down the end at a 45-degree angle (Figure 12-4). Remember, to sharpen the edges, shave only the inner portion.

Arrow Points

To make an arrow point, use the same procedures for making a stone knife blade. Chert, flint, and shell-type stones are best for arrow points. You can fashion bone like stone—by flaking. You can make an efficient arrow point using broken glass.

OTHER EXPEDIENT WEAPONS

You can make other field-expedient weapons such as the throwing stick, archery equipment, and theoola.

Throwing Stick

The throwing stick, commonly known as the rabbit stick, is very effective against small game (squirrels, chipmunks, and rabbits). The rabbit stick itself is a blunt stick, naturally curved at about a 45-degree angle. Select a stick with the desired angle from heavy hardwood such as oak. Shave off two opposite sides so that the stick is flat like a boomerang (Figure 12-5). You must practice the throwing technique for accuracy and speed. First, align the target by extending the nonthrowing arm in line with the mid to lower section of the target. Slowly and repeatedly raise the throwing arm up and back until the throwing stick crosses the back at about a 45-degree angle or is in line with the nonthrowing hip. Bring the throwing arm forward until it is just slightly above and parallel to the nonthrowing arm. This will be the throwing stick's release point. Practice slowly and repeatedly to attain accuracy.

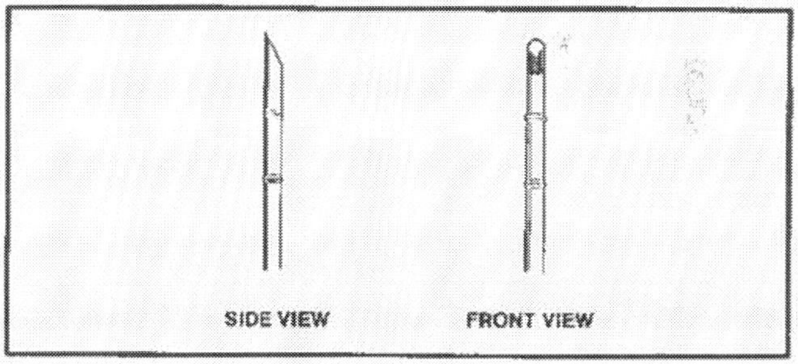

Figure 12-4. Bamboo spear.

Archery Equipment

You can make a bow and arrow (Figure 12-6) from materials available in your survival area. To make a bow, use the procedure described under Killing Devices in Chapter 8.

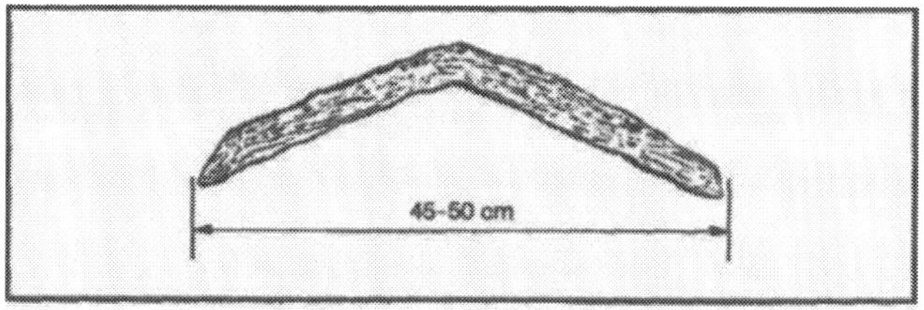

Figure 12-5. Rabbit stick.

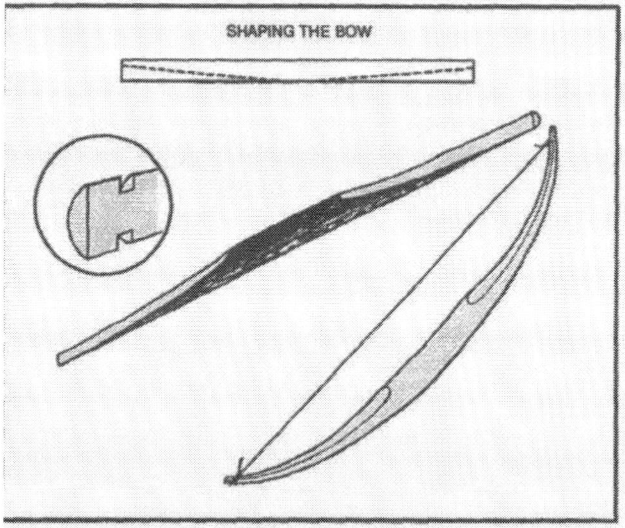

Figure 12-6. Archery equipment.

While it may be relatively simple to make a bow and arrow, it is not easy to use one. You must practice using it a long time to be reasonably sure that you will hit your target. Also, a field-expedient bow will not last very long before you have to make a new one. For the time and effort involved, you may well decide to use another type of field-expedient weapon.

Bola

The bola is another field-expedient weapon that is easy to make (Figure 12-7). It is especially effective for capturing running game or low-flying fowl in a flock. To use the bola, hold it by the centerknot and twirl it above your head. Release the knot so that the bola flies toward your target. When you release the bola, the weighted cords will separate. These cords will wrap around and immobilize the fowl or animal that you hit.

LASHING AND CORDAGE

Many materials are strong enough for use as lashing and cordage. A number of natural and man-made materials are available in a survival

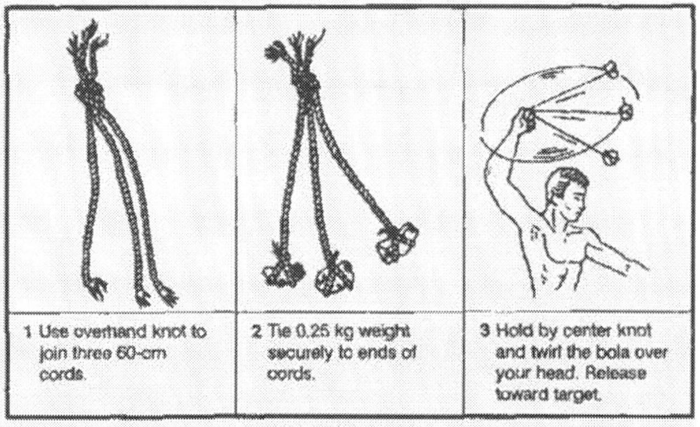

Figure 12-7. Bola.

situation. For example, you can make a cotton web belt much more useful by unraveling it. You can then use the string for other purposes (fishing line, thread for sewing, and lashing).

Natural Cordage Selection

Before making cordage, there are a few simple tests you can do to determine you material s suitability. First, pull on a length of the material to test for strength. Next, twist it between your fingers and roll the fibers together. If it withstands this handling and does not snap apart, tie an overhand knot with the fibers and gently tighten. If the Knot does not break, the material is usable. Figure 12-8 shows various methods of making cordage.

Lashing Material

The best natural material for lashing small objects is sinew. You can make sinew from the tendons of large game, such as deer. Remove the tendons from the game and dry them completely. Smash the dried tendons so that they separate into fibers. Moisten the fibers and twist them into a continuous strand. If you need stronger lashing material, you can braid the strands. When you use sinew for small lashings, you do not need knots as the moistened sinew is sticky and it hardens when dry.

You can shred and braid plant fibers from the inner bark of some trees to make cord. You can use the linden, elm, hickory, white oak, mulberry,

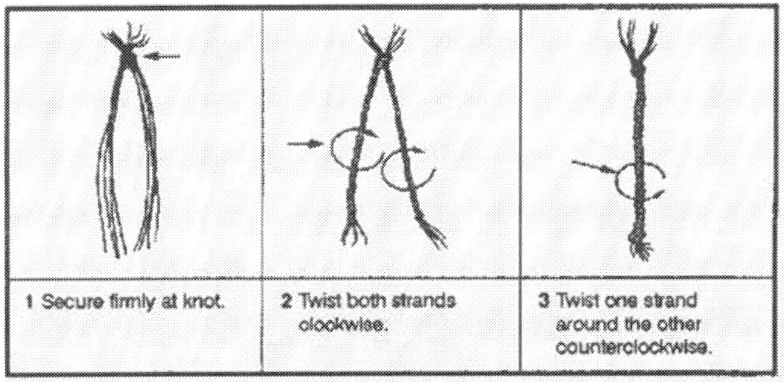

Figure 12-8. Making lines from plant fibers.

chestnut, and red and white cedar trees. After you make the cord, test it to be sure it is strong enough for your purpose. You can make these materials stronger by braiding several strands together.

You can use rawhide for larger lashing jobs. Make rawhide from the skins of medium or large game. After skinning the animal, remove any excess fat and any pieces of meat from the skin. Dry the skin completely. You do not need to stretch it as long as there are no folds to trap moisture. You do not have to remove the hair from the skin. Cut the skin while it is dry. Make cuts about 6 millimeters wide. Start from the center of the hide and make one continuous circular cut, working clockwise to the hide's outer edge. Soak the rawhide for 2 to 4 hours or until it is soft. Use it wet, stretching it as much as possible while applying it. It will be strong and durable when it dries.

RUCKSACK CONSTRUCTION

The materials for constructing a rucksack or pack are almost limitless. You can use wood, bamboo, rope, plant fiber, clothing, animal skins, canvas, and many other materials to make a pack.

There are several construction techniques for rucksacks. Many are very elaborate, but those that are simple and easy are often the most readily made in a survival situation.

Horseshoe Pack

This pack is simple to make and use and relatively comfortable to carry over one shoulder. Lay available square-shaped material, such as poncho, blanket, or canvas, flat on the ground. Lay items on one edge of the material. Pad the hard items. Roll the material (with the items) toward the opposite edge and tie both ends securely. Add extra ties along the length of the bundle. You can drape the pack over one shoulder with a line connecting the two ends (Figure 12-9).

Square Pack

This pack is easy to construct if rope or cordage is available. Otherwise, you must first make cordage. To make this pack, construct a square frame from bamboo, limbs, or sticks. Size will vary for each person and the amount of equipment carried (Figure 12-10).

Figure 12-9. Horseshoe pack.

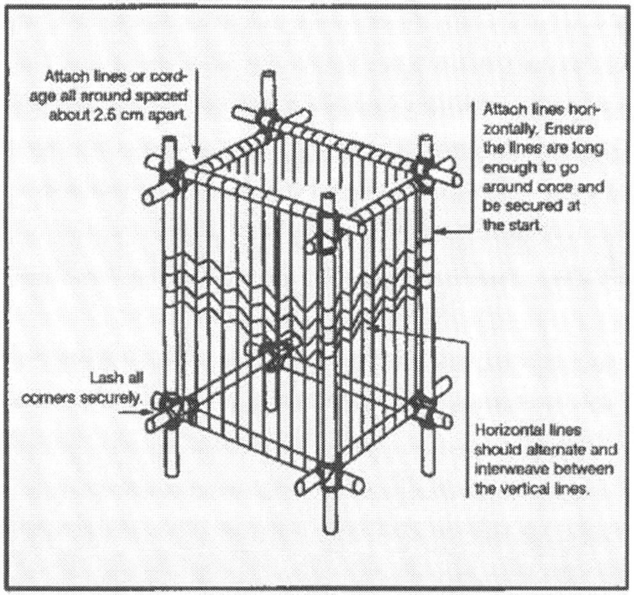

Figure 12-10. Square pack.

CLOTHING AND INSULATION

You can use many materials for clothing and insulation. Both man-made materials, such as parachutes, and natural materials, such as skins and plant materials, are available and offer significant protection.

Parachute Assembly

Consider the entire parachute assembly as a resource. Use every piece of material and hardware, to include the canopy, suspension lines, connector snaps, and parachute harness. Before disassembling the parachute, consider all of your survival requirements and plan to use different portions of the parachute accordingly. For example, consider shelter requirements, need for a rucksack, and so on, in addition to clothing or insulation needs.

Animal Skins

The selection of animal skins in a survival situation will most often be limited to what you manage to trap or hunt. However, if there is an abundance of wildlife, select the hides of larger animals with heavier coats and large fat content. Do not use the skins of infected or diseased animals if at all possible. Since they live in the wild, animals are carriers of pests such as ticks, lice, and fleas. Because of these pests, use water to thoroughly clean any skin obtained from any animal. If water is not available, at least shake out the skin thoroughly. As with rawhide, lay out the skin, and remove all fat and meat. Dry the skin completely. Use the hind quarter joint areas to make shoes and mittens or socks. Wear the hide with the fur to the inside for its insulating factor.

Plant Fibers

Several plants are sources of insulation from cold. Cattail is a marshland plant found along lakes, ponds, and the backwaters of rivers. The fuzz on the tops of the stalks forms dead air spaces and makes a good down-like insulation when placed between two pieces of material. Milkweed has pollenlike seeds that act as good insulation. The husk fibers from coconuts are very good for weaving ropes and, when dried, make excellent tinder and insulation.

COOKING AND EATING UTENSILS

Many materials may be used to make equipment for the cooking, eating, and storing of food.

Bowls

Use wood, bone, horn, bark, or other similar material to make bowls. To make wooden bowls, use a hollowed out piece of wood that will hold your food and enough water to cook it in. Hang the wooden container over the fire and add hot rocks to the water and food. Remove the rocks as they cool and add more hot rocks until your food is cooked.

CAUTION

Do not use rocks with air pockets, such as limestone and sandstone. They may explode while heating in the fire.

You can also use this method with containers made of bark or leaves. However, these containers will burn above the waterline unless you keep them moist or keep the fire low.

A section of bamboo works very well, if you cut out a section between two sealed joints (Figure 12-11).

CAUTION

A sealed section of bamboo will explode if heated because of trapped air and water in the section.

Forks, Knives, and Spoons

Carve forks, knives, and spoons from nonresinous woods so that you do not get a wood resin aftertaste or do not taint the food. Nonresinous woods include oak, birch, and other hardwood trees.

Note: Do not use those trees that secrete a syrup or resinlike liquid on the bark or when cut.

Pots

You can make pots from turtle shells or wood. As described with bowls, using hot rocks in a hollowed out piece of wood is very effective. Bamboo is the best wood for making cooking containers.

To use turtle shells, first thoroughly boil the upper portion of the shell. Then use it to heat food and water over a flame (Figure 12-11).

Water Bottles

Make water bottles from the stomachs of larger animals. Thoroughly flush the stomach out with water, then tie off the bottom. Leave the top open, with some means of fastening it closed.

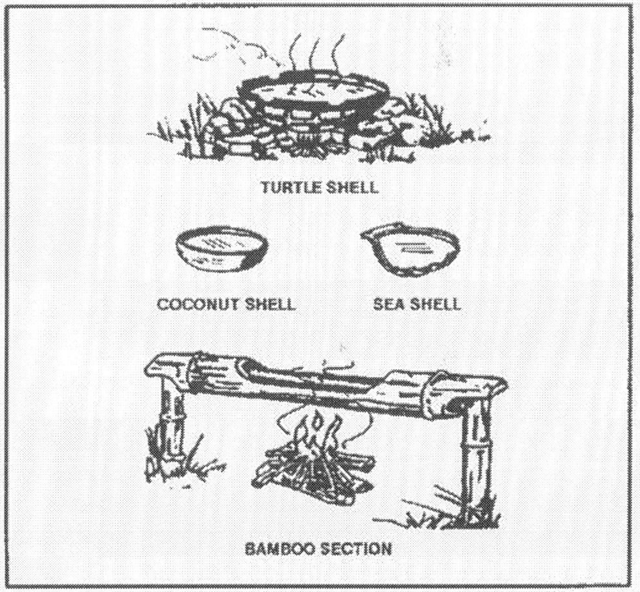

Figure 12-11. Containers for boiling food.

CHAPTER 13

DESERT SURVIVAL

To survive and evade in arid or desert areas, you must understand and prepare for the environment you will face. You must determine your equipment needs, the tactics you will use, and how the environment will affect you and your tactics. Your survival will depend upon your knowledge of the terrain, basic climatic elements, your ability to cope with these elements, and your will to survive.

TERRAIN

Most arid areas have several types of terrain. The five basic desert terrain types are—

- Mountainous (High Altitude).
- Rocky plateau.
- Sand dunes.
- Salt marshes.
- Broken, dissected terrain ("gebel" or "wadi").

Desert terrain makes movement difficult and demanding. Land navigation will be extremely difficult as there may be very fewlandmarks. Cover and concealment may be very limited; therefore, the threat of exposure to the enemy remains constant.

Mountain Deserts

Scattered ranges or areas of barren hills or mountains separated by dry, flat basins characterize mountain deserts. High ground may rise gradually or abruptly from flat areas to several thousand meters above sea level. Most of the infrequent rainfall occurs on high ground and runs off rapidly in the form of flash floods. These floodwaters erode deep gullies and ravines and deposit sand and gravel around the edges of the basins. Water rapidly evaporates, leaving the land as barren as Defore, although there may be short-lived vegetation. If enough water enters the basin to compensate for the rate of evaporation, shallow lakes may develop, such as the Great Salt Lake in Utah, or the Dead Sea. Most of these lakes have a high salt content.

Rocky Plateau Deserts

Rocky plateau deserts have relatively slight relief interspersed with extensive flat areas with quantities of solid or broken rock at or near the surface. There may be steep-walled, eroded valleys, known as wadis in the Middle East and arroyos or canyons in the United States and Mexico. Although their flat bottoms may be superficially attractive as assembly areas, the narrower valleys can be extremely dangerous to men and material due to flash flooding after rains. The Golan Heights is an example of a rocky plateau desert.

Sandy or Dune Deserts

Sandy or dune deserts are extensive flat areas covered with sand or gravel. "Flat" is a relative term, as some areas may contain sand dunes that are over 300 meters high and 16 to 24 kilometers long. Trafficability in such terrain will depend on the windward or leeward slope of the dunes and the texture of the sand. Other areas, however, may be flat for 3,000 meters and more. Plant life may vary from none to scrub over 2 meters high. Examples of this type of desert include the edges of the

Sahara, the empty quarter of the Arabian Desert, areas of California and New Mexico, and the Kalahari in South Africa.

Salt Marshes

Salt marshes are flat, desolate areas, sometimes studded with clumps of grass but devoid of other vegetation. They occur in arid areas where rainwater has collected, evaporated, and left large deposits of alkali salts and water with a high salt concentration. The water is so salty it is undrfnkable. A crust that may be 2.5 to 30 centimeters thick forms over the saltwater.

In arid areas there are salt marshes hundreds of kilometers square. These areas usually support many insects, most of which bite. Avoid salt marshes. This type of terrain is highly corrosive to boots, clothing, and skin. A good example is the Shat-el-Arab waterway along the Iran-Iraq border.

Broken Terrain

All arid areas contain broken or highly dissected terrain. Rainstorms that erode soft sand and carve out canyons form this terrain. A wadi may range from 3 meters wide and 2 meters deep to several hundred meters wide and deep. The direction it takes varies as much as its width and depth. It twists and turns and forms a mazelike pattern. A wadi will give you good cover and concealment, but do not try to move through it because it is very difficult terrain to negotiate.

ENVIRONMENTAL FACTORS

Surviving and evading the enemy in an arid area depends on what you know and how prepared you are for the environmental conditions you will face. Determine what equipment you will need, the tactics you will use, and the environment's impact on them and you.

In a desert area there are seven environmental factors that you must consider—

- Low rainfall.
- Intense sunlight and heat.
- Wide temperature range.
- i Sparse vegetation.
- : High mineral content near ground surface.
- : Sandstorms.
- : Mirages.

Low Rainfall

Low rainfall is the most obvious environmental factor in an arid area. Some desert areas receive less than 10 centimeters of rain annually, and this rain comes in brief torrents that quickly run off the ground surface. You cannot survive long without water in high desert temperatures. In a desert survival situation, you must first consider "How much water do I have?" and "Where are other water sources?"

Intense Sunlight and Heat

Intense sunlight and heat are present in all arid areas. Air temperature can rise as high as 60 degrees t (140 degrees F) during the day. Heat gain results from direct sunlight, hot blowing winds, reflective heat (the sun's rays bouncing off the sand), and conductive heat from direct contact with the desert sand and rock (Figure 13-1).

The temperature of desert sand and rock averages 16 to 22 degrees C (30 to 40 degrees F) more than that of the air. For instance, when the air temperature is 43 degrees C (110 degrees F), the sand temperature may be 60 degrees C (140 degrees F).

Intense sunlight and heat increase the body's need for water. To conserve your body fluids and energy, you will need a shelter to reduce your exposure to the heat of the day. Travel at night to lessen your use of water.

Radios and sensitive items of equipment exposed to direct intense sunlight will malfunction.

Wide Temperature Range

Temperatures in arid areas may get as high as 55 degrees C during the day and as low as 10 degrees C during the night. The drop in temperature at night occurs rapidly and will chill a person who lacks warm clothing and is unable to move about. The cool evenings and nights are the Dest times to work or travel. If your plan is to rest at night, you will find a wool sweater, long underwear, and a wool stocking cap extremely helpful.

Sparse Vegetation

Vegetation is sparse in arid areas. You will therefore have trouble finding shelter and camouflaging your movements. During daylight hours large areas of terrain are visible and easily controlledby a small opposing force.

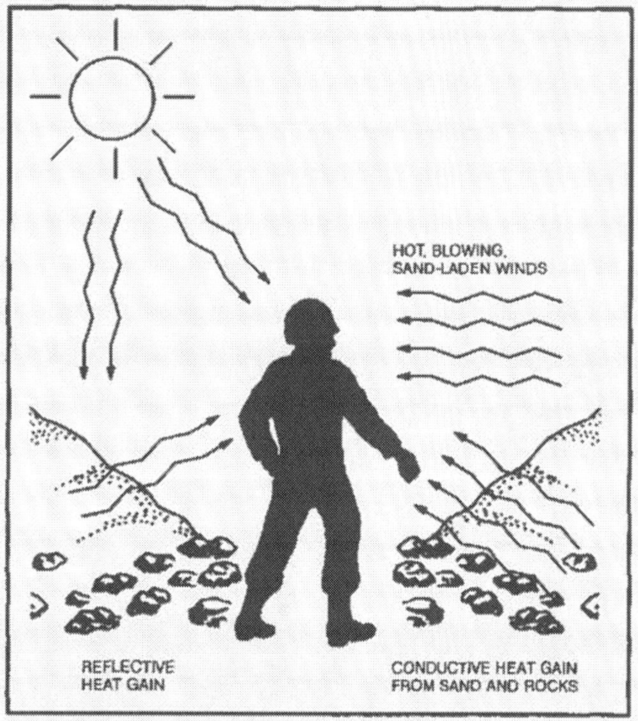

Figure 13-1. Types of heat gain.

If traveling in hostile territory, follow the principles of desert camouflage—

- Hide or seek shelter in dry washes (wadis) with thicker growths of vegetation and cover from oblique observation.
- Use the shadows cast from brush, rocks, or outcropping. The temperature in shaded areas will be 11 to 17 degrees C cooler than the air temperature.
- Cover objects that will reflect the light from the sun.

Before moving, survey the area for sites that provide cover and concealment. You **win** have trouble estimating distance. The emptiness of desert terrain causes most people to underestimate distance Dy a factor of three: What appears to be 1 kilometer away is really 3 kilometers away.

High Mineral Content

All arid regions have areas where the surface soil has a high mineral content (borax, salt, alkali, and lime). Material in contact with this soil wears out quickly, and water in these areas is extremely hard and undrinkable. Wetting your uniform in such water to cool off may cause a skin rash. The Great Salt Lake area in Utah is an example of this type of mineral-laden water and soil. There is little or no plant life; therefore, shelter is hard to find. Avoid these areas if possible.

Sandstorms

Sandstorms (sand-laden winds) occur frequently in most deserts. The "Seistan" desert wind in Iran and Afghanistan blows constantly for up to 120 days. Within Saudi Arabia, winds average 3.2 to 4.8 kilometers per hour (kph) and can reach 112 to 128 kph in early afternoon. Expect major sandstorms and dust storms at least once a week.

The greatest danger is getting lost in a swirling wall of sand. Wear goggles and cover your mouth and nose with cloth. If natural shelter is unavailable, mark your direction of travel, lie down, and sit out the storm.

Dust and wind-blown sand interfere with radio transmissions. Therefore, be ready to use other means for signaling, such as pyrotechnics, signal mirrors, or marker panels, if available.

Mirages

Mirages are optical phenomena caused by the refraction of light through heated air rising from a sandy or stony surface. They occur in the interior of the desert about 10 kilometers from the coast. They make objects that are 1.5 kilometers or more away appear to move.

This mirage effect makes it difficult for you to identify an object from a distance. It also blurs distant range contours so much that you feel surrounded by a sheet of water from which elevations stand out as "islands."

The mirage effect makes it hard for a person to identify targets, estimate range, and see objects clearly. However, if you can get to high ground (3 meters or more above the desert floor), you can get above the superheated air close to the ground and overcome the mirage effect. Mirages make land navigation difficult because they obscure natural features. You can survey the area at dawn, dusk, or by moonlight when there is little likelihood of mirage.

Light levels in desert areas are more intense than in other geographic areas. Moonlit nights are usually crystal clear, winds die down, naze and glare disappear, and visibility is excellent. You can see lights, red flashlights, and blackout lights at great distances. Sound carries very far.

Conversely, during nights with little moonlight, visibility is extremely poor. Traveling is extremely hazardous. You must avoid getting lost, falling into ravmes, or stumbling into enemy positions. Movement during such a night is practical only if you have a compass and have spent the day in a shelter, resting, observing and memorizing the terrain, and selecting your route.

NEED FOR WATER

The subject of man and water in the desert has generated considerable interest and confusion since the early days of World War II when the U.S. Army was preparing to fight in North Africa. At one time the U.S. Army thought it could condition men to do with less water by progressively reducing their water supplies during training. They called it water discipline. It caused hundreds of heat casualties.

A key factor in desert survival is understanding the relationship between physical activity, air temperature, and water consumption. The hody requires a certain amount of water for a certain level of activity at a certain temperature. For example, a person performing hard work in the sun at 43 degrees C requires 19 liters of water daily. Lack of the required amount of water causes a rapid decline in an individual's ability to make decisions and to perform tasks efficiently.

Your body's normal temperature is 36.9 degrees C (98.6 degrees F). Your body gets rid of excess heat (cools off) by sweating. The warmer your body becomes—whether caused by work, exercise, or air temperature—the more you sweat. The more you sweat, the more moisture you lose. Sweating is the principal cause of water loss. If a person stops sweating during periods of high air temperature and heavy work or exercise, he will quickly develop heat stroke. This is an emergency that requires immediate medical attention.

Figure 13-2 shows daily water requirements for various levels of work. Understanding how the air temperature and your physical activity affect your water requirements allows you to take measures to get the most from your water supply. These measures are—

- Find shade! Get out of the sun!
- Place something between you and the hot ground.
- Limit your movements!
- Conserve your sweat. Wear your complete uniform to include T-shirt. Roll the sleeves down, cover your head, and protect your neck with a scarf or similar item. These steps will protect your body from hot-blowing winds and the direct rays of the sun. Your clothing will absorb your sweat, keeping it against your skin so that you gain its full cooling effect. By staying in the shade quietly, fully clothed, not talking, keeping your mouth closed, and breathing through your nose, your water requirement for survival drops dramatically.
- If water is scarce, do not eat. Food requires water for digestion; therefore, eating food will use water that you need for cooling.

Thirst is not a reliable guide for your need for water. A person who uses thirst as a guide will drink only two-thirds of his daily water requirement. To prevent this "voluntary" dehydration, use the following guide:

- At temperatures below 38 degrees C, drink 0.5 liter of water every hour.
- At temperatures above 38 degrees C, drink 1 liter of water every hour.

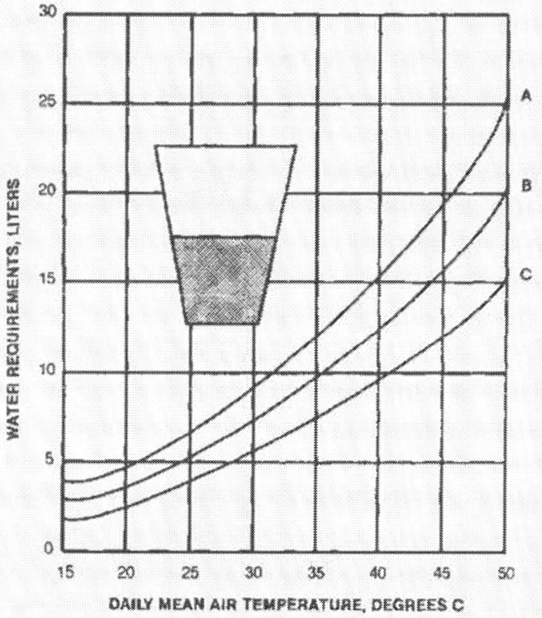

A: Hard work in sun (creeping and crawling with equipment on).
B: Moderate work in sun (cleaning weapons and equipment).
C: Rest in shade.

This graph shows water needs, in liters per day, for men at three activity levels in relation to the daily mean air temperature. For example, if one is doing 8 hours of hard work in the sun (curve A) when the average temperature for the day is 50 degrees C (horizontal scale), one's water requirement for the day will be approximately 25 liters (vertical scale).

From Technical Report E-P118. Southwest Asia: Environment and Its Relationship to Military Activities. July 1959.

Figure 13-2. Daily water requirements for three levels of activity.

Drinking water at regular intervals helps your body remain cool and decreases sweating. Even when your water supply is low, sipping water constantly will keep your body cooler and reduce water loss through sweating. Conserve your fluids by reducing activity during the heat of day. **Do not** ration your water! If you try to ration water, you stand a good chance of becoming a heat casualty.

HEAT CASUALTIES

Your chances of becoming a heat casualty as a survivor are great, due to injury, stress, and lack of :ritical items of equipment. Following are the major types of heat casualties and their treatment *when little water and no medical help are available.*

Heat Cramps

The loss of salt due to excessive sweating causes heat cramps. Symptoms are moderate to severe muscle cramps in legs, arms, or abdomen. These symptoms may start as a mild muscular discomfort. You should now stop all activity, get in the shade, and drink water. If you fail to recognize the early symptoms and continue your physical activity, you will have severe muscle cramps and pain. Treat as for heat exhaustion, below.

Heat Exhaustion

A large loss of body water and salt causes heat exhaustion. Symptoms are headache, mental confusion, irritability, excessive sweating, weakness, dizziness, cramps, and pale, moist, cold (clammy) skin. Immediately get the patient under shade. Make him lie on a stretcher or similar item about 45 centimeters off the ground. Loosen his clothing. Sprinkle him with water and fan him. Have him drink small amounts of water every 3 minutes. Ensure he stays quiet and rests.

Heat Stroke

A severe heat injury caused by extreme loss of water and salt and the body's inability to cool itself. The patient may die if not cooled immediately. Symptoms are the lack of sweat, hot and dry skin, headache, dizziness, fast pulse, nausea and vomiting, and mental confusion leading to unconsciousness. Immediately get the person to shade. Lay him on a stretcher or similar item about 45 centimeters off the ground. Loosen his clothing. Pour water on him (it does not matter if the water is polluted or brackish) and fan him. Massage his arms, legs, and body. If he regains consciousness, let him drink small amounts of water every 3 minutes.

PRECAUTIONS

In a desert survival and evasion situation, it is unlikely that you will have a medic or medical supplies with you to treat heat injuries. Therefore, take extra care to avoid heat injuries. Rest during the day. Work during the cool evenings and nights. Use a buddy system to watch for heat injury, and observe the following guidelines:

- Make sure you tell someone where you are going and when you will return.
- Watch for signs of heat injury. If someone complains of tiredness or wanders away from the group, he may be a heat casualty.
- Drink water at least once an hour.
- Get in the shade when resting; do not lie directly on the ground.
- Do not take off your shirt and work during the day.
- Check the color of your urine. A light color means you are drinking enough water, a dark color means you need to drink more.

DESERT HAZARDS

There are several hazards unique to desert survival. These include insects, snakes, thorned plants and cacti, contaminated water, sunburn, eye irritation, and climatic stress.

Insects of almost every type abound in the desert. Man, as a source of water and food, attracts lice, mites, wasps, and flies. They are extremely unpleasant and may carry diseases. Old ouildings, ruins, and caves are favorite habitats of spiders, scorpions, centipedes, lice, and mites. These areas provide protection from the elements and also attract other wildlife. Therefore, take extra care when staying in these areas. Wear gloves at all times in the desert. Do not place yournands anywhere without first looking to see what is there. Visually inspect an area before sitting or lying down. When you get up, shake out and inspect your boots and clothing.

All desert areas have snakes. They inhabit ruins, native villages, garbage dumps, caves, and natural rock outcropping that offer shade. Never go barefoot or walk through these areas without carefully inspecting them for snakes. Pay attention to where you place your feet ana hands. Most snakebites result from stepping on or handling snakes. Avoid them. Once you see a snake, give it a wide berth.

CHAPTER 14

TROPICAL SURVIVAL

Most people think of the tropics as a huge and forbidding tropical rainforest through which every step taken must be hacked out, and where every inch of the way is crawling with danger. Actually, over half of the land in the tropics is cultivated in some way.

A knowledge of field skills, the ability to improvise, and the application of the principles of survival will increase the prospects of survival. Do not be afraid of being alone in the jungle; fear will lead to panic. Panic will lead to exhaustion and decrease your chance of survival.

Everything in the jungle thrives, including disease germs and parasites that breed at an alarming rate. Nature will provide water, food, and plenty of materials to build shelters.

Indigenous peoples have lived for millennia by hunting and gathering. However, it will take an outsider some time to get used to the conditions and the nonstop activity of tropical survival.

TROPICAL WEATHER

High temperatures, heavy rainfall, and oppressive humidity characterize equatorial and subtropical regions, except at high altitudes. At low altitudes, temperature variation is seldom less than 10 degrees C and is often more than 35 degrees C. At altitudes over 1,500 meters, ice often forms at night. The rain has a cooling effect, but when it stops, the temperature soars.

Rainfall is heavy, often with thunder and lightning. Sudden rain beats on the tree canopy, turning trickles into raging torrents and causing rivers to rise. Just as suddenly, the rain stops, violent storms may occur, usually toward the end of the summer months.

Hurricanes, cyclones, and typhoons develop over the sea and rush inland, causing tidal waves and devastation ashore. In choosing campsites, make sure you are above any potential flooding. Prevailing winds vary between winter and summer. The dry season has rain once a day and the monsoon has continuous rain. In Southeast Asia, winds from the Indian Ocean bring the monsoon, but it is dry when the wind blows from the landmass of China.

Tropical day and night are of equal length. Darkness falls quickly and daybreak is just as sudden.

JUNGLE TYPES

There is no standard jungle. The tropical area may be any of the following:

- Rain forests.
- Secondary jungles.
- Semievergreen seasonal and monsoon forests.

- Scrub and thorn forests.
- Savannas.
- Saltwater swamps.
- Freshwater swamps.

Tropical Rain Forests

The climate varies little in rain forests. You find these forests across the equator in the Amazon and Congo basins, parts of Indonesia, and several Pacific islands. Up to 3.5 meters of rain fall evenly throughout the year. Temperatures range from about 32 degrees C in the day to 21 degrees C at night.

There are five layers of vegetation in this jungle (Figure 14-1). Where untouched by man, jungle trees rise from buttress roots to heights of 60 meters. Below them, smaller trees produce a canopy so thick that little light reaches the jungle floor. Seedlings struggle oeneath them to reach light, and masses of vines and lianas twine up to the sun. Ferns, mosses, and herbaceous plants push through a thick carpet of leaves, and a great variety of fungi grow on leaves and fallen tree trunks.

Because of the lack of light on the jungle floor, there is little undergrowth to hamper movement, but dense growth limits visibility to about 50 meters. You can easily lose your sense of direction in this jungle, and it is extremely hard for aircraft to see you.

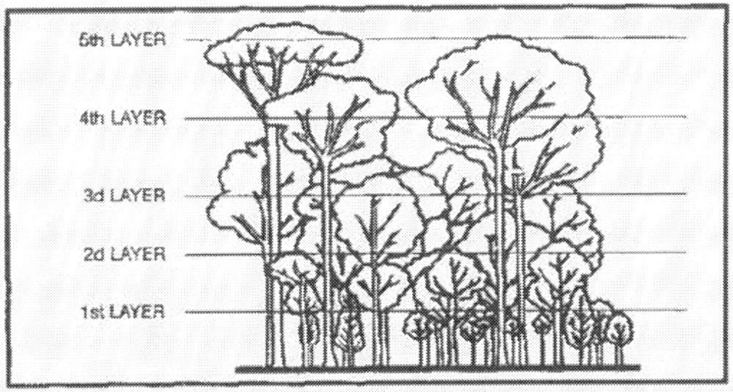

Figure 14-1. Five layers of tropical rain forest vegetation.

Secondary Jungles

Secondary jungle is very similar to rain forest. Prolific growth, where sunlight penetrates to the jungle floor, typifies this type of forest. Such growth happens mainly along river banks, onjungle fringes, and where man has cleared rain forest. When abandoned, tangled masses of vegetation quickly reclaim these cultivated areas. You can often find cultivated food plants among this vegetation.

Semievergreen Seasonal and Monsoon Forests

The characteristics of the American and African semievergreen seasonal forests correspond with those of the Asian monsoon forests. These characteristics are—

- Their trees fall into two stories of tree strata. Those in the upper story average 18 to 24 meters; those in the lower story average 7 to 13 meters.
- The diameter of the trees averages 0.5 meter.
- Their leaves fall during a seasonal drought.

Except for the sage, nipa, and coconut palms, the same edible plants grow in these areas as in the tropical rain forests.

You find these forests in portions of Columbia and Venezuela and the Amazon basin in South America; in portions of southeast coastal Kenya, Tanzania, and Mozambique in Africa; in Northeastern India, much of Burma, Thailand, Indochina, Java, and parts of other Indonesian islands in Asia.

Tropical Scrub and Thorn Forests

The chief characteristics of tropical scrub and thorn forests are—

- There is a definite dry season.
- Trees are leafless during the dry season.
- The ground is bare except for a few tufted plants in bunches; grasses are uncommon.
- Plants with thorns predominate.
- Fires occur frequently.

You find tropical scrub and thorn forests on the west coast of Mexico, Yucatan peninsula, Venezuela, Brazil; on the northwest coast and central parts of Africa; and in Asia, in Turkestan and India.

Within the tropical scrub and thorn forest areas, you will find it hard to obtain food plants during the dry season. During the rainy season, plants are considerably more abundant.

Tropical Savannas

General characteristics of the savanna are—

- It is found within the tropical zones in South America and Africa.
- It looks like a broad, grassy meadow, with trees spaced at wide intervals.
- It frequently has red soil.
- It grows scattered trees that usually appear stunted and gnarled like apple trees. Palms also occur on savannas.

You find savannas in parts of Venezuela, Brazil, and the Guianas in South America. In Africa, you find them in the southern Sahara (north-central Cameroon and Gabon and southern Sudan), Benin, Togo, most of Nigeria, northeastern Zaire, northern Uganda, western Kenya, part of Malawi, part of Tanzania, southern Zimbabwe, Mozambique, and western Madagascar.

Saltwater Swamps

Saltwater swamps are common in coastal areas subject to tidal flooding. Mangrove trees thrive in these swamps. Mangrove trees can reach heights of 12 meters, and their tangled roots are an obstacle to movement. Visibility in this type of swamp is poor, and movement is extremely difficult. Sometimes, streams that you can raft form channels, but you usually must travel on foot through this swamp.

You find saltwater swamps in West Africa, Madagascar, Malaysia, the Pacific islands, Central and South America, and at the mouth of the Ganges River in India. The swamps at the mouths of the Orinoco and Amazon rivers and rivers of Guyana consist of mud and trees that offer little shade. Tides in saltwater swamps can vary as much as 12 meters.

Everything in a saltwater swamp may appear hostile to you, from leeches and insects to crocodiles and caimans. Avoid the dangerous animals in this swamp.

Avoid this swamp altogether if you can. If there are water channels through it, you may be able to use a raft to escape.

Freshwater Swamps

You find freshwater swamps in low-lying inland areas. Their characteristics are masses of thorny undergrowth, reeds, grasses, and occasional short palms that reduce visibility and make travel difficult. There are often islands that dot these swamps, allowing you to get out of the water. Wildlife is abundant in these swamps.

TRAVEL THROUGH JUNGLE AREAS

With practice, movement through thick undergrowth and jungle can be done efficiently. Always wear long sleeves to avoid cuts and scratches.

To move easily, you must develop "jungle eye," that is, you should not concentrate on the pattern of bushes and trees to your immediate front. You must focus on the jungle further out and find natural breaks in the foliage. Look *through* the jungle, not at it. Stop and stoop down occasionally to look along the jungle floor. This action may reveal game trails that you can follow.

Stay alert and move slowly and steadily through dense forest or jungle. Stop periodically to listen and take your bearings. Use a machete to cut through dense vegetation, but do not cut unnecessarily or you will quickly wear yourself out. If using a machete, stroke upward when cutting vines to reduce noise because sound carries long distances in the jungle. Use a stick to part the vegetation. Using a stick will also help dislodge biting ants, spiders, or snakes. Do not grasp at brush or vines when climbing slopes; they may have irritating spines or sharp thorns.

Many jungle and forest animals follow game trails. These trails wind and cross, but frequently lead to water or clearings. Use these trails if they lead in your desired direction of travel.

In many countries, electric and telephone lines run for miles through sparsely inhabited areas. Usually, the right-of-way is clear enough to allow easy travel. When traveling along these lines, be careful as you approach transformer and relay stations. In enemy territory, they may be guarded.

TRAVEL TIPS
Pinpoint your initial location as accurately as possible to determine a general line of travel to safety. If you do not have a compass, use a field-expedient direction finding method.
Take stock of water supplies and equipment.
Move in one direction, but not necessarily in a straight line. Avoid obstacles. In enemy territory, take advantage of natural cover and concealment.
Move smoothly through the jungle. Do not blunder through it since you will get many cuts and scratches. Turn your shoulders, shift your hips, bend your body, and shorten or lengthen your stride as necessary to slide between the undergrowth.

IMMEDIATE CONSIDERATIONS

There is less likelihood of your rescue from beneath a dense jungle canopy than in other survival situations. You will probably have to travel to reach safety.

If you are the victim of an aircraft crash, the most important items to take with you from the crash site are a machete, a compass, a first aid kit, and a parachute or other material for use as mosquito netting and shelter.

Take shelter from tropical rain, sun, and insects. Malaria-carrying mosquitoes and other insects are immediate dangers, so protect yourself against bites.

Do not leave the crash area without carefully blazing or marking your route. Use your compass. Know what direction you are taking.

In the tropics, even the smallest scratch can quickly become dangerously infected. Promptly treat any wound, no matter how minor.

WATER PROCUREMENT

Even though water is abundant in most tropical environments, you may, as a survivor, have trouble finding it. If you do find water, it may not be safe to drink. Some of the many sources are vines, roots, palm trees, and condensation. You can sometimes follow animals to water. Often you can get nearly clear water from muddy streams or lakes by digging a hole in sandy soil about 1 meter from the bank. Water will seep into the hole. You must purify any water obtained in this manner.

Animals as Signs of Water

Animals can often lead you to water. Most animals require water regularly. Grazing animals such as deer, are usually never far from water and usually drink at dawn and dusk. Converging game trails often lead to water. Carnivores (meat eaters) are not reliable indicators of water. They get moisture from the animals they eat and can go without water for long periods.

Birds can sometimes also lead you to water. Grain eaters, such as finches and pigeons, are never tar from water. They drink at dawn and dusk. When they fly straight and low, they are heading for water. When returning from wafer, they are full and will fly from tree to tree, resting frequently. Do not rely on water birds to lead you to water. They fly long distances without stopping. Hawks, eagles, and other birds of prey get liquids from their victims; you cannot use them as a water indicator.

Insects can be good indicators of water, especially bees. Bees seldom range more than 6 kilometers from their nests or "hives. They usually will have a water source in this range. Ants need water. A column of ants marching up a tree is going to a small reservoir of trapped water. You find such reservoirs even in arid areas. Most flies stay within 100 meters of water, especially the European mason fly, easily recognized by its iridescent green body.

Human tracks will usually lead to a well, bore hole, or soak. Scrub or rocks may cover it to reduce evaporation. Replace the cover after use.

Water From Plants

Plants such as vines, roots, and palm trees are good sources of water.

Vines

Vines with rough bark and shoots about 5 centimeters thick can be a useful source of water. You must learn by experience which are the water-bearing vines, because not all have drinkable water. Some may even have a poisonous sap. The poisonous ones yield a sticky, milky sap when cut. Nonpoisonous vines will give a clear fluid. Some vines cause a skin irritation on contact; therefore let the liquid drip into your mouth, rather than put your mouth to the vine. Preferably, use some type of container. Use the procedure described in Chapter 6 to obtain water from a vine.

Roofs

In Australia, the water tree, desert oak, and bloodwood have roots near the surface. Pry these roots out of the ground and cut them into 30-centimeter lengths. Remove the bark and suck out the moisture, or shave the root to a pulp and squeeze it over your mouth.

Palm Trees

The buri, coconut, and nipa palms all contain a sugary fluid that is very good to drink. To obtain the liquid, bend a flowering stalk of one of these palms downward, and cut off its tip. If you cut a thin slice off the stalk every 12 hours, the flow will renew, making it possible to collect up to a liter per day. Nipapalm shoots grow from the base, so that you can work at ground level. On grown trees of other species, you may nave to climb them to reach a flowering stalk. Milk from coconuts has a large water content, but may contain a strong laxative in ripe nuts. Drinking too much of this milk may cause you to lose more fluid than you drink.

Water From Condensation

Often it requires too much effort to dig for roots containing water. It may be easier to let a plant produce water for you in the form of condensation. Tying a clear plastic bag around a green leafy branch will cause water m the leaves to evaporate and condense in the bag. Placing cut vegetation in a plastic bag will also produce condensation. This is a solar still (see Chapter 6).

FOOD

Food is usually abundant in a tropical survival situation. To obtain animal food, use the procedures outlined in Chapter 8.

In addition to animal food, you will have to supplement your diet with edible plants. The best places to forage are the banks of streams and rivers. Wherever the sun penetrates the jungle, there will be a mass of vegetation, but river banks may be the most accessible areas.

If you are weak, do not expend energy climbing or felling a tree for food. There are more easily obtained sources of food nearer the ground. Do not pick more food than you need. Food spoils rapidly in tropical conditions. Leave food on the growing plant until you need it, and eat it fresh.

There are an almost unlimited number of edible plants from which to choose. Unless you can positively identify these plants, it may be safer at first to begin with palms, bamboos, and common fruits. The list below identifies some of the most common foods. Detailed descriptions and photographs are at Appendix B.

TROPICAL ZONE FOOD PLANTS

- Bael fruit (*Aegle marmelos*)
- Bamboo (various species)
- Banana or plantain (*Musa species*)
- Bignay (*Antidesma bunius*)
- Breadfruit (*Artocarpus incisa*)
- Coconut palm (*Cocos nucifera*)
- Fishtail palm (*Caryota urens*)
- Horseradish tree (*Moringa pterygosperma*)
- Lotus (*Nelumbo* species)
- Mango (*Mangifera indica*)
- Manioc (*Manihot utillissima*)
- Nipa palm (*Nipa fruticans*)
- Papaya (*Carica papaya*)
- Persimmon (*Diospyros virginiana*)
- Rattan palm (*Calamus* species)
- Sago palm (*Metroxylon sagu*)
- Sterculia (*Sterculia foetida*)
- Sugarcane (*Saccharum officinarum*)
- Sugar palm (*Arenga pinnata*)
- Sweetsop (*Annona squamosa*)
- Taro (*Colocasia* and *Alocasia* species)
- Water lily (*Nymphaea odorata*)
- Wild fig (*Ficus* species)
- Wild rice (*Zizania aquatica*)
- Yam (*Dioscorea* species)

POISONOUS PLANTS

The proportion of poisonous plants in tropical regions is no greater than in any other area of the world. However, it may appear that most plants in the tropics are poisonous because of the great density of plant growth in some tropical areas. See Appendix C.

CHAPTER 15

COLD WEATHER SURVIVAL

One of the most difficult survival situations is a cold weather scenario. Remember, cold weather is an adversary that can be as dangerous as an enemy soldier. Every lime you venture into the cold, you are pitting yourself against the elements. With a little knowledge of the environment, proper plans, and appropriate equipment, you can overcome the elements. As you remove one or more of these factors, survival becomes increasingly difficult. Remember, winter weather is highly variable. Prepare yourself to adapt to blizzard conditions even during sunny and clear weather.

Cold is a far greater threat to survival than it appears. It decreases your ability to think and weakens your will to do anything except to get warm. Cold is an insidious enemy; as it numbs the mind and body, it subdues the will to survive.

Cold makes it very easy to forget your ultimate goal—to survive.

COLD REGIONS AND LOCATIONS

Cold regions include arctic and subarctic areas and areas immediately adjoining them. You can classify about 48 percent of the northern hemisphere's total landmass as a cold region due to the influence and extent of air temperatures. Ocean currents affect cold weather and cause large areas normally included in the temperate zone to fall within the cold regions during winter periods. Elevation also has a marked effect on defining cold regions.

Within the cold weather regions, you may face two types of cold weather environments—wet or dry.TCnowmg in which environment your area of operations falls will affect planning and execution of a cold weather operation.

Wet Cold Weather Environments

Wet cold weather conditions exist when the average temperature in a 24-hour period is -10 degrees C or above. Characteristics of this condition are freezing during the colder night hours and thawing during the day. Even though the temperatures are warmer during this condition, the terrain is usually very sloppy due to slush and mud. You must concentrate on protecting yourself from the wet ground and from freezing rain or wet snow.

Dry Cold Weather Environments

Dry cold weather conditions exist when the average temperature in a 24-hour period remains below -10 degrees C. Even though the temeratures in this condition are much lower than normal, you do not ave to contend with the freezing and thawing. In these conditions, you need more layers of inner clothing to protect you from temperatures as low as -60 degrees C. Extremely hazardous conditions exist when wind and low temperature combine.

WINDCHILL

Windchill increases the hazards in cold regions. Windchill is the effect of moving air on exposed flesh. For instance, with a 27.8-kph

(15-knot) wind and a temperature of -10 degrees C the equivalent windchill temperature is -23 degrees C. Figure 15-1 gives the windchill factors for various temperatures and wind speeds.

Remember, even when there is no wind, you will create the equivalent wind by skiing, running, being towed on stis behind a vehicle, working around aircraft that produce wind blasts.

BASIC PRINCIPLES OF COLD WEATHER SURVIVAL

It is more difficult for you to satisfy your basic water, food, and shelter needs in a cold environment than m a warm environment. Even if you have the basic requirements, you must also have adequate protective clothing and the will to survive. The will to survive is as important as the basic needs. There have been incidents when trained and well-equipped individuals have not survived cold weather situations because they lacked the will to live. Conversely, this will has sustained individuals less well-trained and equipped.

There are many different items of cold weather equipment and clothing issued by the U.S. Army today. Specialized units may have access to newer, lightweight gear such as polypropylene underwear, GORE-TEX outerwear and boots, and other special equipment. Remember, however, the older gear will keep you warm as long as you apply a few cold weather principles. If the newer types or doming are available, use them. If not, then your clothing should be entirely wool, with the possible exception of a windbreaker.

You must not only have enough clothing to protect you from the cold, ou must also know how to maximize the warmth you get from it. jor example, always keep your head covered. You canlose 40 to 45 percent of body heat from an unprotected head and even more from the unprotected neck, wrist, and ankles. These areas of the body are good radiators of heat and have very little insulating fat. The brain is very susceptible to cold and can stand the least amount of cooling. Because there is much blood circulation in the head, most of which is on the surface, you can lose heat quickly if you do not cover your head.

COOLING POWER OF WIND EXPRESSED AS "EQUIVALENT CHILL TEMPERATURE"

WIND SPEED		TEMPERATURE (DEGREES C)																				
CALM	CALM	4	2	-1	-4	-7	-9	-12	-15	-18	-21	-23	-26	-29	-32	-34	-37	-40	-43	-46	-48	-51
KNOTS	KPH	EQUIVALENT CHILL TEMPERATURE																				
4	8	2	-1	-4	-7	-9	-12	-15	-18	-21	-23	-26	-29	-32	-34	-37	-40	-43	-46	-48	-54	-57
9	16	-1	-7	-9	-12	-15	-18	-23	-26	-29	-32	-37	-40	-43	-46	-51	-54	-57	-59	-62	-68	-71
13	24	-4	-9	-12	-18	-21	-23	-29	-32	-34	-40	-43	-46	-51	-54	-57	-62	-65	-68	-73	-76	-79
17	32	-7	-12	-15	-18	-23	-28	-32	-34	-37	-43	-46	-51	-54	-59	-62	-65	-71	-73	-79	-82	-84
22	40	-9	-12	-18	-21	-26	-29	-34	-37	-43	-46	-51	-54	-59	-62	-68	-71	-76	-79	-84	-87	-93
26	48	-12	-15	-18	-23	-29	-32	-34	-40	-46	-48	-54	-57	-62	-65	-71	-73	-79	-82	-87	-90	-96
30	56	-12	-15	-21	-23	-29	-34	-37	-40	-46	-51	-54	-59	-62	-68	-73	-76	-82	-84	-90	-93	-98
35	64	-12	-18	-21	-26	-29	-34	-37	-43	-48	-51	-57	-59	-65	-71	-73	-79	-82	-87	-90	-96	-101
(Higher winds have little additional effects)		LITTLE DANGER						INCREASING DANGER (Flesh may freeze within 1 minute)						GREAT DANGER (Flesh may freeze within 30 seconds)								
DANGER OF FREEZING EXPOSED FLESH FOR PROPERLY CLOTHED PERSONS																						

Figure 15-1. Windchill table.

There are four basic principles to follow to keep warm. An easy way to remember these basic principles is to use the word COLD—

C - Keep clothing *clean.*

O- Avoid *overheating.*

L - Wear clothes loose and in *layers.*

D- Keep clothing *dry.*

C - *Keep clothing clean.* This principle is always important for sanitation and comfort. In winter, it is also important from the standpoint of warmth. Clothes matted with dirt and grease lose much of their insulation value. Heat can escape more easily from the body through the clothing's crushed or filled up air pockets.

O - *Avoid overheating.* When you get too hot, you sweat and your clothing absorbs the moisture. This affects your warmth m two ways: dampness decreases the insulation quality of clothing, and as sweat evaporates, your body cools. Adjust your clothing so that you do not sweat. Do this by partially opening your parka or jacket, by removing an inner layer of clothing, by removing heavy outer mittens, or by throwing back your parka hood or changing to lighter headgear. The head and hands act as efficient heat dissipaters when overheated.

L - *Wear your clothing loose and in layers.* Wearing tight clothing and footgear restricts blood circulation and invites cold injury. It also decreases the volume of air trapped between the layers, reducing its insulating value. Several layers of lightweight clothing are better than one equally thick layer of clothing, because the layers have dead-air space between them. The deacFair space provides extra insulation. Also, layers of clothing allow you to tale off or add clothing layers to prevent excessive sweating or to increase warmth.

D - *Keep clothing dry.* In cold temperatures, your inner layers of clothing can become wet from sweat and your outer layer, if not water repellent, can become wet from snow and frost melted by body heat. Wear water repellent outer clothing, if available. It will shed most of the water collected from melting snow and frost. Before entering a heated shelter, brush off the snow and frost. Despite the precautions you take, there will be times when you cannot keep from getting wet. At such times, drying your clothing may become a major problem. On the march, hang your damp mittens and socks on your rucksack. Sometimes in freezing temperatures, the wind and sun will dry this clothing. You can also place damp socks or mittens, unfolded, near your body so that your body heat can dry them. In a campsite, hang damp clothing inside the shelter near the top, using drying lines or improvised racks. You may even be able to dry each item by holding it before an open fire. Dry leather items slowly. If no other means are available for drying your boots, put them between your sleeping bag shell and liner. Your body heat will help to dry the leather.

A heavy, down-lined sleeping bag is a valuable piece of survival gear in cold weather. Ensure the down remains dry. If wet, it loses a lot of its insulation value. If you do not have a sleeping bag, you can make one out of parachute cloth or similar material and natural dry material, such as leaves, pine needles, or moss. Place the dry material between two layers of the material.

Other important survival items are a knife; waterproof matches in a waterproof container, preferably one with a flint attached; a durable compass; map; watch; waterproof ground cloth and cover; flashlight; binoculars; dark glasses; fatty emergency foods; food gathering gear; and signaling items.

Remember, a cold weather environment can be very harsh. Give a good deal of thought to selecting the right equipment for survival in the cold. If unsure of an item you have never used, test it in an "overnight backyard" environment before venturing further. Once you have selected items that are essential for your survival, do not lose them after you enter a cold weather environment.

HYGIENE

Although washing yourself may be impractical and uncomfortable in a cold environment, you must do so. Washing helps prevent skin rashes that can develop into more serious problems.

In some situations, you may be able to take a snow bath. Take a handful of snow and wash your body where sweat and moisture accumulate, such as under the arms and between the legs, and then wipe yourself dry. If possible, wash your feet daily and put on clean, dry socks. Change your underwear at least twice a week. If you are unable to wash your underwear, take it off, shake it, and let it air out for an hour or two.

If you are using a previously used shelter, check your body and clothing for lice each night. If your clothing has become infested, use insecticide powder if you have any. Otherwise, hang your clothes in the cold, then beat and brush them. This will help get rid of the lice, but not the eggs.

If you shave, try to do so before going to bed. This will give your skin a chance to recover before exposing it to the elements.

MEDICAL ASPECTS

When you are healthy, your inner core temperature (torso temperature) remains almost constant at 37 degrees C (98.6 degrees F). Since your limbs and head have less protective body tissue than your torso, their temperatures vary and may not reach core temperature.

Your body has a control system that lets it react to temperature extremes to maintain a temperature balance. There are three main factors that affect this temperature balance—heat production, heat loss, and evaporation. The difference between the body's core temperature and the environment's temperature governs the heat production rate. Your body can get rid of heat better than it can produce it. Sweating helps to control the heat balance. Maximum sweating will get rid of heat about as fast as maximum exertion produces it.

Shivering causes the body to produce heat. It also causes fatigue that, in turn, leads to a drop in body temperature. Air movement around your body affects heat loss. It has been calculated that a naked man exposed to still air at or about 0 degrees C can maintain a heat balance if he shivers as hard as he can. However, he can't shiver forever.

It has also been calculated that a man at rest wearing the maximum arctic clothing in a cold environment can keep his internal heat balance during temperatures well below freezing. To withstand really cold conditions for any length of time, however, he will have to become active or shiver.

COLD INJURIES

The best way to deal with injuries and sicknesses is to take measures to prevent them from happening in the first place. Treat any injury or sickness that occurs as soon as possible to prevent it from worsening.

The knowledge of signs and symptoms and the use of the buddy system are critical in maintaining health. Following are cold injuries that can occur.

Hypothermia

Hypothermia is the lowering of the body temperature at a rate faster than the body can produce neat. Causes of hypothermia may be general exposure or the sudden wetting of the body by falling into a lake or spraying with fuel or other liquids.

The initial symptom is shivering. This shivering may progress to the point that it is uncontrollable and interferes with an individual's ability to care for himself. This begins when the body's core (rectal) temperature falls to about 35.5 degrees C (96 degrees F). When the core temperature reaches 35 to 32 degrees C (95 to 90 degrees F), sluggish thinking, irrational reasoning, and a false feeling of warmth may occur. Core temperatures of 32 to 30 degrees C (90 to 86 degrees F) and below result in muscle rigidity, unconsciousness, and barely detectable signs of life. If the victim's core temperature falls below 25 degrees C (77 degrees F), death is almost certain.

To treat hypothermia, rewarm the entire body. If there are means available, rewarm the person by first immersing the trunk area only in warm water of 37.7 to 43.3 degrees C (100 to 110 degrees F).

CAUTION

Rewarming the total body in a warm water bath should be done only in a hospital environment because of the increased risk of cardiac arrest and rewarming shock.

One of the quickest ways to get heat to the inner core is to give warm water enemas. Such an action, however, may not be possible in a survival situation. Another method is to wrap the victim in a warmed sleeping bag with another person who is already warm; both should be naked.

CAUTION

The individual placed in the sleeping bag with victim could also become a hypothermia victim if left in the bag too long.

If the person is conscious, give him hot, sweetened fluids. One of the best sources of calories is honey or dextrose; if unavailable, use sugar, cocoa, or a similar soluble sweetener.

CAUTION

Do not force an unconscious person to drink.

There are two dangers in treating hypothermia-rewarming too rapidly and "after drop." Rewarming too rapidly can cause the victim to have circulatory problems, resulting in heart failure. After drop is the sharp body core temperature drop that occurs when taking the victim from the warm water. Its probable muse is the return of previously stagnant limb blood to the core (inner torso) area as recirculation occurs. Concentrating on warming the core area and stimulating peripheral circulation will lessen the effects of after drop. Immersing the torso in a warm bath, if possible, is the best treatment.

Frostbite

This injury is the result of frozen tissues. Light frostbite involves only the skin that takes on a dull whitish pallor. TJeep frostbite extends to a depth below the skin. The tissues become solid and immovable. Your feet, hands, and exposed facial areas are particularly vulnerable to frostbite.

The best frostbite prevention, when you are with others, is to use the buddy system. Check your buddy's face often and make sure that he checks yours. If you are alone periodically cover your nose and lower part of your face with your mittened hand.

The following pointers will aid you in keeping warm and preventing frostbite when it is extremely cold or when you have less than adequate clothing:

- *Face.* Maintain circulation by twitching and wrinkling the skin on your face making faces. Warm with your hands.
- *Ears.* Wiggle and move your ears. Warm with your hands.
- *Hands.* Move your hands inside your gloves. Warm by placing your hands close to your body.
- *Feet.* Move your feet and wiggle your toes inside your boots.

A loss of feeling in your hands and feet is a sign of frostbite. If you have lost feeling for only a short time, the frostbite is probably light. Otherwise, assume the frostbite is deep. To rewarm a light frostbite, use your hands or mittens to warm your face and ears. Place your hands under your armpits. Place your feet next to your buddy's stomach. A deep frostbite injury, if thawed and refrozen, will cause more damage than a nonmedically trained person can handle. Figure 15-2 lists some do's and don'ts regarding frostbite.

Trench Foot and Immersion Foot

These conditions result from many hours or days of exposure to wet or damp conditions at a temperature just above freezing. The symptoms are a sensation of pins and needles, tingling, numbness, and then pain. The skin will initially appear wet, soggy, white, and shriveled. As it progresses and damage appears, the skin will take on a red and then a bluish or black discoloration. The feet become cold, swollen, and have a waxy appearance. Walking becomes difficult and the feet feel heavy and numb. The nerves and muscles sustain the main damage, but gangrene can occur. In extreme cases, the flesh dies and it may become necessary to have the foot or leg amputated. The best prevention is to keep your feet dry. Carry extra socks with you in a waterproof packet. You can dry wet socks against your torso (back or chest). Wash your feet and put on dry socks daily.

Do	Don't
• Periodically check for frostbite.	• Rub injury with snow.
• Rewarm light frostbite.	• Drink alcoholic beverages.
• Keep injured areas from refreezing.	• Smoke.
	• Try to thaw out a deep frostbite injury if you are away from definitive medical care.

Figure 15-2. Frostbite do's and don'ts.

Dehydration

When bundled up in many layers of clothing during cold weather, you may be unaware that you are losing body moisture. Your heavy clothing absorbs the moisture that evaporates in the air. You must drink water to replace this loss of fluid. Your need for water is as great in a cold environment as it is in a warm environment (Chapter 13). One way to tell if you are becoming dehydrated is to check the color of your urine on snow. If your urine makes the snow dark yellow, you are becoming dehydrated and need to replace body fluids. If it makes the snow light yellow to no color, your body fluids have a more normal balance.

Cold Diuresis

Exposure to cold increases urine output. It also decreases body fluids that you must replace.

Sunburn

Exposed skin can become sunburned even when the air temperature is below freezing. The sun's rays reflect at all angles from snow, ice, and water, hitting sensitive areas of skin—lips, nostrils, and eyelids. Exposure to the sun results in sunburn more quickly at high altitudes than at low altitudes. Apply sunburn cream or hp salve to your face when in the sun.

Snow Blindness

The reflection of the sun's ultraviolet rays off a snow-covered area causes this condition. The symptoms of snow blindness are a sensation of grit in the eyes, pain in and over the eyes that increases with eyeball movement, red and teary eyes, and a headache that intensifies with continued exposure to light. Prolonged exposure to these rays can result in permanent eye damage. To treat snow blindness, bandage your eyes until the symptoms disappear.

You can prevent snow blindness by wearing sunglasses. If you don't have sunglasses, improvise. Cut slits in a piece of cardboard, thin wood, tree bark, or other available material (Figure 15-3). Putting soot under your eyes will help reduce shine and glare.

Constipation

It is very important to relieve yourself when needed. Do not delay because of the cold condition. Delaying relieving yourself because of the cold, eating dehydrated foods, drinking too little liquid, and irregular eating habits can cause you to become constipated. Although not disabling, constipation can cause some discomfort. Increase your fluid intake to at least 2 liters above your normal 2 to 3 liters daily intake and, if available, eat fruit and other foods that will loosen the stool.

Insect Bites

Insect bites can become infected through constant scratching. Flies can carry various disease-producing germs. To prevent insect bites, use insect repellent, netting, and wear proper clothing. See Chapter 11 for information on insect bites and Chapter 4 for treatment.

SHELTERS

Your environment and the equipment you carry with you will determine the type of shelter you can build. You can build shelters in wooded areas, open country, and barren areas. Wooded areas usually provide the best location, while barren areas have only snow as building material. Wooded areas provide timber for shelter construction, wood for fire, concealment from observation, and protection from the wind.

Note: In extreme cold, do not use metal, such as an aircraft fuselage, for shelter. The metal will conduct away from the shelter what little heat you can generate.

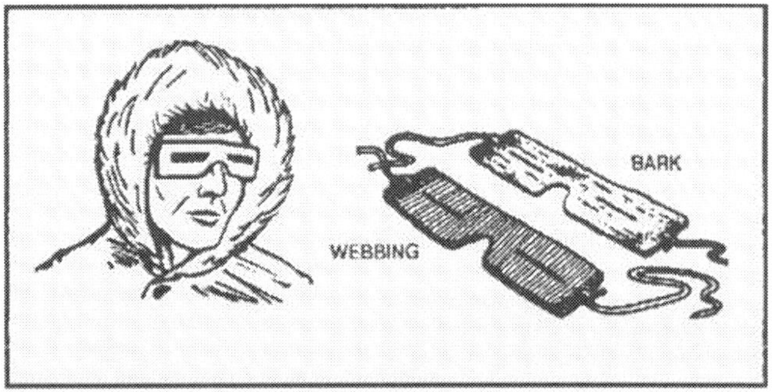

Figure 15-3. Improvised sunglasses.

Shelters made from ice or snow usually require tools such as ice axes or saws. You must also expend much time and energy to build such a shelter. Be sure to ventilate an enclosed shelter, especially if you intend to build a fire in it. Always block a shelter's entrance, if possible, to keep the heat in and the wind out. Use a rucksack or snow Dlock. Construct a shelter no larger than needed. This will reduce the amount of space to heat. A fatal error in cold weather shelter construction is making the shelter so large that it steals body heat rather than saving it. Keep shelter space small.

Never sleep directly on the ground. Lay down some pine boughs, grass, or other insulating material to keep the ground from absorbing your body heat.

Never fall asleep without turning out your stove or lamp. Carbon monoxide poisoning can result from a fire Durning in an unventilated shelter. Carbon monoxide is a great danger. It is colorless and odorless. Any time you have an open flame, it may generate carbon monoxide. Always check your ventilation. Even in a ventilated shelter, incomplete combustion can cause carbon monoxide poisoning. Usually, there are no symptoms. Unconsciousness and death can occur without warning. Sometimes, however, pressure at the temples, burning of the eyes, headache, pounding pulse, drowsiness, or nausea may occur. The one characteristic, visible sign of carbon monoxide poisoning is a cherry red coloring in the tissues of the lips, mouth, and mside of the eyelids. Get into fresh air at once if you have any of these symptoms.

There are several types of field-expedient shelters you can quickly build or employ. Many use snow for insulation.

Snow Cave Shelter

The snow cave shelter (Figure 15-4) is a most effective shelter because of the insulating qualities of snow. Remember that it takes time and energy to build and that you will get wet while building it. First, you need to nnd a drift about 3 meters deep into which you can dig. While building this shelter, keep the roof arched for strength and to allow melted snow to drain down the sides. Build the sleeping platform higher than the entrance. Separate the sleeping platform from the snow cave's walls or dig a small trench between the platform and the wall. This platform will prevent the melting snow from wetting you and your equipment. This construction is especially important if you have a goocTsource of

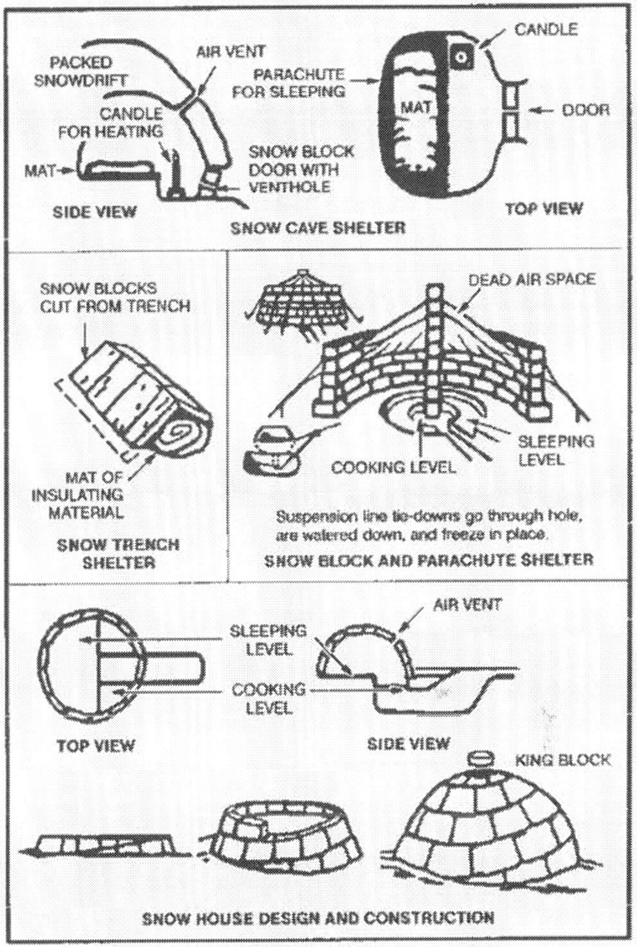

Figure 15-4. Snow houses.

heat in the snow cave. Ensure the roof is high enough so that you can sit up on the sleeping platform. Block the entrance with a snow block or other material and use the lower entrance area for cooking. The walls and ceiling should be at least 30 centimeters thick. Install a ventilation shaft. If you do not have a drift large enough to build a snow cave, you can make a variation of it by piling snow into a mound large enough to dig out.

Snow Trench Shelter

The idea behind this shelter (Figure 15-4) is to get you below the snow and wind level and use the snow's insulating qualities. If you are in an area of compacted snow, cut snow blocks and use them as overhead cover. If not, you can use a poncho or other material. Build only one entrance and use a snow block or rucksack as a door.

Snow Block and Parachute Shelter

Use snow blocks for the sides and parachute material for overhead cover (Figure 154). If snowfall is heavy, you will have to clear snow from the top at regular intervals to prevent the collapse of the parachute material.

Snow House or Igloo

In certain areas, the natives frequently use this type of shelter (Figure 15-4) as hunting and fishing shelters. They are efficient shelters but require some practice to make them properly. Also, you must be in an area that is suitable for cutting snow mocks and have the equipment to cut them (snow saw or knife).

Lean-To Shelter

Construct this shelter in the same manner as for other environments; however, pile snow around the sides for insulation (Figure 15-5).

Fallen Tree Shelter

To build this shelter, find a fallen tree and dig out the snow underneath it (Figure 15-6). The snow will not be deep under the tree. If you must remove branches from the inside, use them to line the floor.

Tree-Pit Shelter

Dig snow out from under a suitable large tree. It will not be as deep near the base of the tree. Use the cut branches to line the shelter. Use a ground sheet as overhead cover to prevent snow from falling off the tree into the shelter. If built properly, you can have 360-degree visibility (Figure 5-12, Chapter 5, page 5-17).

Figure 15-5. Lean-to made from natural shelter.

Figure 15-6. Fallen tree as shelter.

20-Man Life Raft

This raft is the standard overwater raft on U.S. Air Force aircraft. You can use it as a shelter. Do not let large amounts of snow build up on the overhead protection. If placed in an open area, it also serves as a good signal to overhead aircraft.

FIRE

Fire is especially important in cold weather. It not only provides a means to prepare food, but also to get warm and to melt snow or ice for water. It also provides you with a significant psychological boost by making you feel a Tittle more secure in your situation.

Use the techniques described in Chapter 7 to build and light your fire. If you are in enemy territory, remember that the smoke, smell, and light from your fire may reveal your location. Light reflects from surrounding trees or rocks, making even indirect light a source of danger. Smoke tends to go straight up in cold, calm weather, making it a beacon during the day, but helping to conceal the smell at night. In warmer weather, especially in a wooded area, smoke tends to hug the ground, making it less visible in the day, but making its odor spread.

If you are in enemy territory, cut low tree boughs rather than the entire tree for firewood. Fallen trees are easily seen from the air.

All wood will burn, but some types of wood create more smoke than others. For instance, coniferous trees that contain resin and tar create more and darker smoke than deciduous trees.

There are few materials to use for fuel in the high mountainous regions of the arctic. You may find some grasses and moss, but very little. The lower the elevation, the more fuelavailable. You may find some scrub willow and small, stunted spruce trees above the tree line. On sea ice, fuels are seemingly nonexistent. Driftwood or fats may be the only fuels available to a survivor on the barren coastlines in the arctic and subarctic regions.

Abundant fuels within the tree line are—

- Spruce trees are common in the interior regions. As a conifer, spruce makes a lot of smoke when burned m the spring and summer months. However, it burns almost smoke-free in late fall and winter.
- The tamarack tree is also a conifer. It is the only tree of the pine family that loses its needles in the fall. Without its needles, it looks like a dead spruce, but it has many knobby buds and cones on its bare branches. When burning, tamarack wood makes a lot of smoke and is excellent for signaling purposes.
- Birch trees are deciduous and the wood burns hot and fast, as if soaked with oil or kerosene. Most birches grow near streams and lakes, but occasionally you will find a few on higher ground and away from water.
- Willow and alder grow in arctic regions, normally in marsh areas or near lakes and streams. These woods burn hot and fast without much smoke.

Dried moss, grass, and scrub willow are other materials you can use for fuel. These are usually plentiful near streams in tundras (open, treeless plains). By bundling or twisting grasses or other scrub vegetation to form a large, solid mass, you will have a slower burning, more productive fuel.

If fuel or oil is available from a wrecked vehicle or downed aircraft, use it for fuel. Leave the fuel in the tank for storage, drawing on the supply only as you need it. Oil congeals in extremely cold temperatures, therefore, drain it from the vehicle or aircraft while still warm if there is no danger of explosion or fire. If you have no container, let the oil drain onto the snow or ice. Scoop up the fuel as you need it.

> **CAUTION**
>
> Do not expose flesh to petroleum, oil, and lubricants in extremely cold temperatures. The liquid state of these products is deceptive in that it can cause frostbite.

Some plastic products, such as MRE spoons, helmet visors, visor housings, aid foam rubber will ignite quickly from a burning match. They will also burn long enough to help start a fire. For example, a plastic spoon will burn for about 10 minutes.

In cold weather regions, there are some hazards in using fires, whether to keep warm or to cook. For example—

- Fires have been known to burn underground, resurfacing nearby. Therefore, do not build a fire too close to a shelter.

- In snow shelters, excessive heat will melt the insulating layer of snow that may also be your camouflage.
- A fire inside a shelter lacking adequate ventilation can result in carbon monoxide poisoning.
- A person trying to get warm or to dry clothes may become careless ancl burn or scorch his clothing and equipment.
- Melting overhead snow may get you wet, bury you and your equipment, and possibly extinguish your fire.

In general, a small fire and some type of stove is the best combination for cooking purposes. A hobo stove (Figure 15-7) is particularly suitable to the arctic. It is easy to make out of a tin can, and it conserves fuel. A bed of hot coals provides the best cooking heat. Coals from a crisscross fire will settle uniformly. Make this type of fire by crisscrossing the firewood. A simple crane propped on a forked stick will hold a cookmg container over a fire.

For heating purposes, a single candle provides enough heat to warm an enclosed shelter. A small fire about the size of a man's hand is ideal for use in enemy territory. It requires very little fuel, yet it generates considerable warmth and is hot enough to warm liquids.

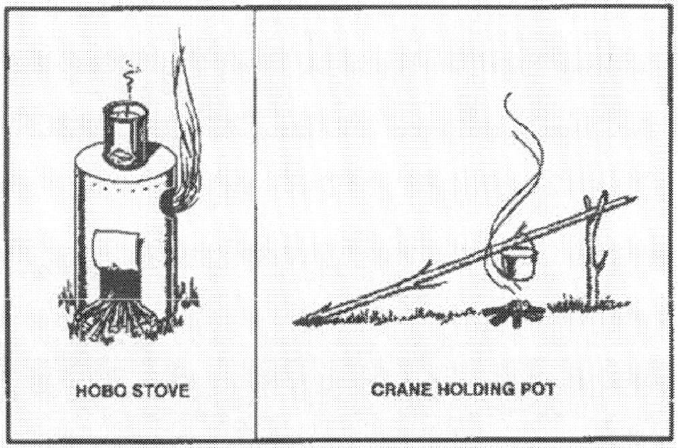

Figure 15-7. Cooking fire/stove.

WATER

There are many sources of water in the arctic and subarctic. Your location and the season of the year will determine where and how you obtain water.

Water sources in arctic and subarctic regions are more sanitary than in other regions due to the climatic and environmental conditions. However, *always purify* the water before drinking it. During the summer months, the best natural sources of water are freshwater lakes, streams, ponds, rivers, and springs. Water from ponds or lakes may be slightly stagnant, but still usable. Running water in streams, rivers, and bubbling sprmgs is usually fresh and suitable for drinking.

The brownish surface water found in a tundra during the summer is a good source of water. However, you may have to filter the water before purifying it.

You can melt freshwater ice and snow for water. Completely melt both before putting them in your mouth. Trying to melt ice or snow in your mouth takes away body heat and may cause internal cold injuries. If on or near pack ice m the sea, you can use old sea ice to melt for water. In time, sea ice loses its salinity. You can identify this ice by its rounded corners and bluish color.

You can use body heat to melt snow. Place the snow in a water bag and place the bag between your layers of clothing. This is a slow process, but you can use it on the move or when you have no fire.

Note: Do not waste fuel to melt ice or snow when drinkable water is available from other sources.

When ice is available, melt it, rather than snow. One cup of ice yields more water than one cup of snow. Ice also takes less time to melt. You can melt ice or snow in a water bag, MRE ration bag, tin can, or improvised container by placing the container near a fire. Begin with a small amount of ice or snow in the container and, as it turns to water, add more ice or snow.

Another way to melt ice or snow is by putting it in a bag made from porous material and suspending the bag near the fire. Place a container under the bag to catch the water.

During cold weather, avoid drinking a lot of liquid before going to bed. Crawling out of a warm sleeping bag at night to relieve yourself means less rest and more exposure to the cold.

Once you have water, keep it next to you to prevent refreezing. Also, do not fill your canteen completely. Allowing the water to slosh around will help keep it from freezing.

FOOD

There are several sources of food in the arctic and subarctic regions. The type of food—fish, animal, fowl, or plant —and the ease in obtaining it depend on the time of the year and your location.

Fish

During the summer months, you can easily get fish and other water life from coastal waters, streams, rivers, and lakes. Use the techniques described in Chapter 8 to catch fish.

The North Atlantic and North Pacific coastal waters are rich in seafood. You can easily find crawfish, snails, clams, oysters, and king crab. In areas where there is a great difference between the high and low tide water levels, you can easily find shellfish at low tide. Dig in the sand on the tidal flats. Look in tidal pools and on offshore reefs. In areas where there is a small difference between the high- and low-tide water levels, storm waves often wash shellfish onto the beaches.

The eggs of the spiny sea urchin that lives in the waters around the Aleutian Islands and southern Alaska are excellent food. Look for the sea urchins in tidal pools. Break the shell by placing it between two stones. The eggs are bright yellow in color.

Most northern fish and fish eggs are edible. Exceptions are the meat of the arctic shark and the eggs of the sculpins.

The bivalves, such as clams and mussels, are usually more palatable than spiral-shelled seafood, such as snails.

WARNING

The black mussel, a common mollusk of the far north, may be poisonous in any season. Toxins sometimes found in the mussel's tissue are as dangerous as strychnine.

The sea cucumber is another edible sea animal. Inside its body are five long white muscles that taste much like clam meat.

In early summer, smelt spawn in the beach surf. Sometimes you can scoop them up with your nands.

You can often find herring eggs on the seaweed in midsummer. Kelp, the long ribbonlike seaweed, and other smaller seaweed that grow among offshore rocks are also edible.

Sea Ice Animals

You find polar bears in practically all arctic coastal regions, but rarely inland. Avoid them if possible. They are the most dangerous of all bears. They are tireless, clever hunters with good sight and an extraordinary sense of smell, if you must kill one for food, approach it cautiously. Aim for the brain; a bullet elsewhere will rarely kill one. Always cook polar bear meat before eating it.

CAUTION

Do not eat polar bear liver as it contains a toxic concentration of vitamin A.

Earless seal meat is some of the best meat available. You need considerable skill, however, to get close enough to an earless seal to kill it. In spring, seals often bask on the ice beside their breathing holes. They raise their heads about every 30 seconds, however, to look for their enemy, the polar bear.

To approach a seal, do as the Eskimos do—stay downwind from it, cautiously moving closer while it sleeps. If it moves, stop and imitate its movements bylying flat on the ice, raising your head up and down, and wriggling your body slightly. Approach the seal with your body sideways to it and your arms close to your body so that you look as much like another seal as possible. The ice at the edge of the breathing hole is usually smooth and at an incline, so the least movement of the seal may cause it to slide into the water. Therefore, try to get within 22 to 45 meters of the seal and kill it instantly (aim for the bram). Try to reach the seal before it slips into the water. In winter, a dead seal will usually float, but it is difficult to retrieve from the water.

Keep the seal blubber and skin from coming into contact with any scratch or broken skin you may have. You could get "spekk-finger," that is, a reaction that causes the hands to become badly swollen.

Keep in mind that where there are seals, there are usually polar bears, and polar bears have stalked and killed seal hunters.

You can find porcupines in southern subarctic regions where there are trees. Porcupines feed on bark; if you find tree limbs stripped bare, you are likely to find porcupines in the area.

Ptarmigans, owls, Canadian jays, grouse, and ravens are the only birds that remain in the arctic during the winter. They are scarce north of the tree line. Ptarmigans and owls are as good for food as any game bird. Ravens are too thin to be worth the effort it takes to catch them. Ptarmigans, which change color to blend with their surroundings, are hard to spot. Rock ptarmigans travel in pairs and you can easily approach them. Willow ptarmigans live among willow clumps in bottom-lands. They gather in large flocks and you can easily snare them. During the summer months all arctic birds have a 2- to 3-week molting period during which they cannot fly and are easy to catch. Use one of the techniques described in Chapter 8 to catch, them.

Skin and butcher game (see Chapter 8) while it is still warm. If you do not have time to skin the game, at least remove its entrails, musk glands, and genitals before storing. If time allows, cut the meat into usable pieces and freeze each separately so that you can use the pieces as needed. Leave the fat on all animals except seals. During the winter, game freezes quickly if left in the open. During the summer, you can store it in underground ice holes.

Plants

Although tundras support a variety of plants during the warm months, all are small, however, when compared to plants in warmer climates. For instance, the arctic willow and birch are shrubs rather than trees. The following is a list of some plant foods found in arctic and subarctic regions (see Appendix B for descriptions).

ARCTIC FOOD PLANTS

- Arctic raspberry and blueberry
- Arctic willow
- Bearberry
- Cranberry
- Crowberry
- Dandelion
- Eskimo potato
- Fireweed
- Iceland moss
- Marsh marigold
- Reindeer moss
- Rock tripe
- Spatterdock

There are some plants growing in arctic and subarctic regions that are poisonous if eaten (see Appendix C). Use the plants that you know are edible. When in doubt, follow the Universal Edibility Test in Chapter 9, Figure 9-5, page 9-8.

TRAVEL

As a survivor or an evader in an arctic or subarctic region, you will face many obstacles. Your location and the time of the year will determine the types of obstacles and the inherent dangers. You should—

- Avoid traveling during a blizzard.
- Take care when crossing thin ice. Distribute your weight by lying flat and crawling.
- Cross streams when the water level is lowest. Normal freezing and thawing action may cause a stream level to vary as much as Zto 2.5 meters per day. This variance may occur any time during the day, depending on the distance from a glacier, the temperature, and the

terrain. Consider this variation in water level when selecting a campsite near a stream.

- Consider the clear arctic air. It makes estimating distance difficult. You more frequently underestimate than overestimate distances.
- Do not travel in "whiteout" conditions. The lack of contrasting colors makes it impossible to judge the nature of the terrain.
- Always cross a snow bridge at right angles to the obstacle it crosses. Find the strongest part of the bridge by poking ahead of you with a pole or ice axe. Distribute your weight Dy crawling or by wearing snowshoes or skis.
- Make camp early so that you have plenty of time to build a shelter.
- Consider frozen or unfrozen rivers as avenues of travel. However, some rivers that appear frozen may have soft, open areas that make travel very difficult or may not allow walking, skiing, or sledding.
- Use snowshoes if you are traveling over snow-covered terrain. Snow 30 or more centimeters deep makes traveling difficult. If you do not have snowshoes, make a pair using willow, strips of cloth, leather, or other suitable material.

It is almost impossible to travel in deep snow without snowshoes or skis. Traveling by foot leaves a well-marked trail for any pursuers to follow. If you must travel in deep snow, avoid snow-covered streams. The snow, which acts as an insulator, may have prevented ice from forming over the water. In hilly terrain, avoid areas where avalanches appear possible. Travel in the early morning in areas where there is danger of avalanches. On ridges, snow gathers on the lee side in overhanging piles called cornices. These often extend far out from the ridge and may break loose if stepped on.

WEATHER SIGNS

There are several good indicators of climatic changes.

Wind

You can determine wind direction by dropping a few leaves or grass or by watching the treetops. Once you determine the wind direction, you can predict the type of weather that is imminent. Rapidly shifting winds indicate an unsettled atmosphere and a likely change in the weather.

Clouds

Clouds come in a variety of shapes and patterns. A general knowledge of clouds and the atmospheric conditions they indicate can help you predict the weather. See Appendix G for details.

Smoke

Smoke rising in a thin vertical column indicates fair weather. Low rising or "flattened out" smoke indicates stormy weather.

Birds and Insects

Birds and insects fly lower to the ground than normal in heavy, moisture-laden air. Such flight indicates that rain is likely. Most insect activity increases before a storm, but bee activity increases before fair weather.

Low-Pressure Front

Slow-moving or imperceptible winds and heavy, humid air often indicate a low-pressure front. Such a front promises bad weather that will probably linger for several days. You can "smell" and "hear" this front. The sluggish, humid air makes wilderness odors more pronounced than during nigh-pressure conditions. In addition, sounds are sharper and carry farther in low-pressure than high-pressure conditions.

CHAPTER 16

SEA SURVIVAL

Perhaps the most difficult survival situation to be in is sea survival. Short- or long-term survival depends upon rations and equipment available and your ingenuity. You must be resourceful to survive.

Water covers about 75 percent of the earth's surface, with about 70 percent being oceans and seas. You can assume that you will sometime cross vast expanses of water. There is always the chance that the plane or ship you are on will become crippled by such hazards as storms, collision, fire, or war.

THE OPEN SEA

As a survivor on the open sea, you will face waves and wind. You may also face extreme heat or cold. To keep these environmental hazards from becoming serious problems, take precautionary measures as soon as possible. Use the available resources to protect yourself from the elements and from heat or extreme cold and humidity.

Protecting yourself from the elements meets only one of your basic needs. You must also be able to obtain water and food. Satisfying these three basic needs will help prevent serious physical and psychological problems. However, you must know how to treat health problems that may result from your situation.

Precautionary Measures

Your survival at sea depends upon—

- Your knowledge of and ability to use the available survival equipment.
- Your special skills and ability to apply them to cope with the hazards you face.
- Your will to live.

When you board a ship or aircraft, find out what survival equipment is on board, where it is stowed, and what it contains. For instance, how many life preservers and lifeboats or rafts are on board? Where are they located? What type of survival equipment do they have? How much food, water, and medicine do they contain? How many people are they designed to support?

If you are responsible for other personnel on board, make sure you know where they are and they know where you are.

Down at Sea

If you are in an aircraft that goes down at sea, take the following actions once you clear the aircraft. Whether you are in the water or in a raft —

- Get clear and upwind of the aircraft as soon as possible, but stay in the vicinity until the aircraft sinks.
- Get clear of fuel-covered water in case the fuel ignites.
- Try to find other survivors.

A search for survivors usually takes place around the entire area of and near the crash site. Missing personnel may be unconscious and floating low in the water. Figure 16-1 illustrates rescue procedures.

Figure 16-1. Rescue from water.

The best technique for rescuing personnel from the water is to throw them a life preserver attached to a line. Another is to send a swimmer (rescuer) from the raft with a line attached to a flotation device that will support the rescuer's weight. This device will help conserve a rescuer's energy while recovering the survivor. The least acceptable technique is to send an attached swimmer without flotation devices to retrieve a survivor. In all cases, the rescuer wears a life preserver. A rescuer should not underestimate the strength of a panic-stricken person in the water. A careful approach can prevent injury to the rescuer.

When the rescuer approaches a survivor in trouble from behind, there is little danger the survivor will kick, scratch, or grab him. The rescuer swims to a point directly behind the survivor ancf grasps the life preserver's backstop. The rescuer uses the sidestroke to drag the survivor to the raft.

If you are in the water, make your way to a raft. If no rafts are available, try to find a large piece of floating debris to cling to. Relax; a person who knows how to relax in ocean water is in very little danger of drowning. The body's natural buoyancy will keep at least the top of the head above water, but some movement is needed to keep the face above water.

Floating on your back takes the least energy. Lie on your back in the water, spread your arms and legs, and arch your back. By controlling your breathing in and out, your face will always be out of the water and you may even sleep in this position for short periods. Your head will be partially submerged, but your face will be above water. If you cannot float on your back or if the sea is too rough, float facedown in the water as shown in Figure 16-2.

The following are the best swimming strokes during a survival situation:

- *Dog paddle.* This stroke is excellent when clothed or wearing a life jacket. Although slow in speed, it requires very little energy.
- *Breaststroke.* Use this stroke to swim underwater, through oil or debris, or in rough seas. It is probably the best stroke for long-range swimming: it allows you to conserve your energy and maintain a reasonable speed.
- *Sidestroke.* It is a good relief stroke because you use only one arm to maintain momentum and buoyancy.

- *Backstroke.* This stroke is also an excellent relief stroke. It relieves the muscles that you use for other strokes. Use it if an underwater explosion is likely.

If you are in an area where surface oil is burning—

- Discard your shoes and buoyant life preserver.

  Note: If you have an uninflated life preserver, keep it.
- Cover your nose, mouth, and eyes and quickly go underwater.
- Swim underwater as far as possible before surfacing to breathe.
- Before surfacing to breathe and while still underwater, use your hands to push burning fluid away from the area where you wish to surface. Once an area is clear of burning liquid, you can surface and take a few breaths. Try to face downwind before inhaling.
- Submerge feet first and continue as above until clear of the flames.

If you are in oil-covered water that is free of fire, hold your head high to keep the oil out of your eyes. Attach your life preserver to your wrist and then use it as a raft.

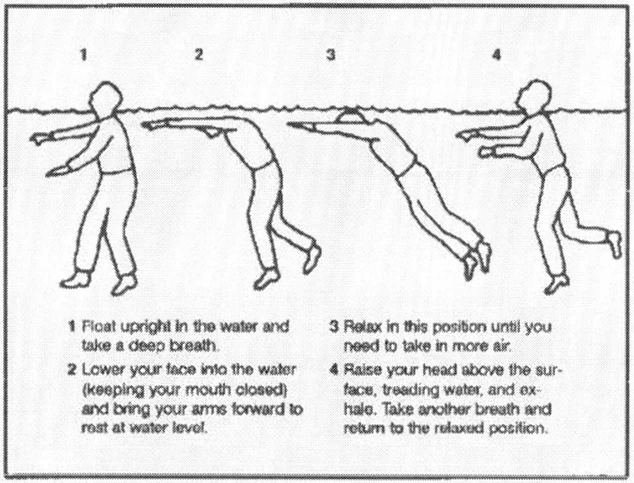

Figure 16-2. Floating position.

If you have a life preserver, you can stay afloat for an indefinite period. In this case, use the "HELP" body position: Heat Escaping Lessening Posture (HELP). Remain still and assume the fetal position to help you retain body heat. You lose about 50 percent of your body heat through your head. Therefore, keep your head out of the water. Other areas of high heat loss are the neck, the sides, and the groin. Figure 16-3 illustrates the HELP position.

If you are in a raft —

- Check the physical condition of all on board. Give first aid if necessary. Take seasickness pills if available. The best way to take these pills is to place them under the tongue and let them dissolve. There are also suppositories or injections against seasickness. Vomiting, whether from seasickness or other causes, increases the danger of dehydration.

Figure 16-3. HELP position.

- Try to salvage all floating equipment—rations; canteens, thermos jugs, and other containers; clothing; seat cushions; parachutes; and anything else that will be useful to you. Secure the salvaged items in or to your raft. Make sure the items have no sharp edges that can puncture the raft.
- If there are other rafts, lash the rafts together so they are about 7.5 meters apart. Be ready to draw them closer together if you see or hear an aircraft. It is easier for an aircrew to spot rafts that are close together rather than scattered.
- Remember, rescue at sea is a cooperative effort. Use all available visual or electronic signaling devices to signal and make contact with rescuers. For example, raise a flag or reflecting material on an oar as high as possible to attract attention.
- Locate the emergency radio and get it into operation. Operating instructions are on it. Use the emergency transceiver only when friendly aircraft are likely to be in the area.
- Have other signaling devices ready for instant use. If you are in enemy territory, avoid using a signaling device that will alert the enemy. However, if your situation is desperate, you may have to signal the enemy for rescue if you are to survive.
- Check the raft for inflation, leaks, and points of possible chafing. Make sure the main buoyancy chambers are firm (well rounded) but not overly tight (Figure 16-4). Check inflation regularly. Air expands with heat; therefore, on hot days, release some air and add air when the weather cools.
- Decontaminate the raft of all fuel. Petroleum will weaken its surfaces and break down its glued joints.
- Throw out the sea anchor, or improvise a drag from the raft's case, bailing bucket, or a roll of clothing. A sea anchor helps you stay close to your ditching site, making it easier for searchers to find you if you have relayed your location. Without a sea anchor, your raft may drift over 160 kilometers in a day, making it much harder to find you. You can adjust the sea anchor to act as a drag to slow down the rate of travel with the current, or as a means to travel with the current. You make this adjustment by opening or closing the sea anchor's apex. When open, the sea anchor (Figure 16-5) acts as a drag that keeps you in the general area. When closed, it forms a pocket for the current to strike and propels the raft in the current's direction.

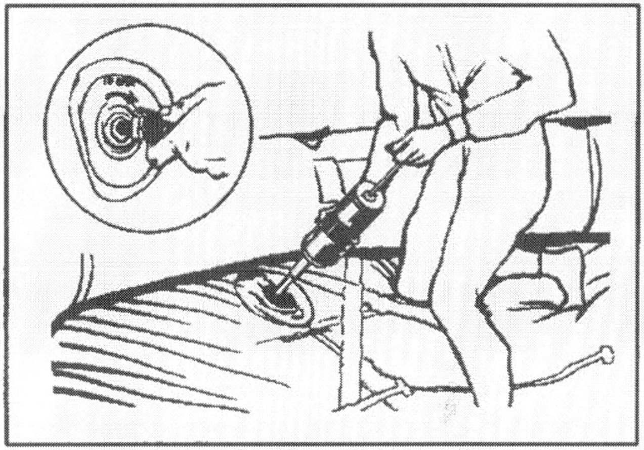

Figure 16-4. Inflating the 20-man raft.

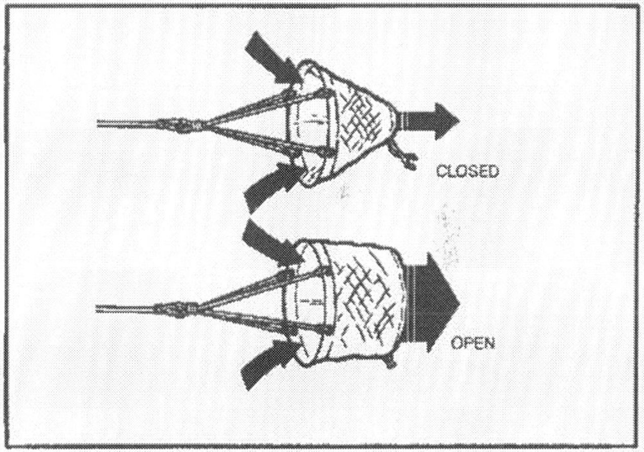

Figure 16-5. Sea anchor.

Additionally, adjust the sea anchor so that when the raft is on the wave's crest, the sea anchor is in the wave's trough (Figure 16-6).

- Wrap the sea anchor rope with cloth to prevent its chafing the raft. The anchor also helps to keep the raft headed into the wind and waves.
- In stormy water, rig the spray and windshield at once. In a 20-man raft, keep the canopy erected at all times. Keep your raft as dry as possible. Keep it properly balanced. All personnel should stay seated, the heaviest one m the center.
- Calmly consider all aspects of your situation and determine what you and your companions must do to survive. Inventory all equipment, food, and water. Waterproof items that salt water may affect. These include compasses, watches, sextant, matches, and lighters. Ration food and water.
- Assign a duty position to each person: for example, water collector, food collector, lookout, radio operator, signaler, and water bailers. Note *Lookout duty should not exceed 2 hours. Keep in mind and remind others that cooperation is one of the keys to survival.*
- Keep a log. Record the navigator's last fix, the time of ditching, the names and physical condition of personnel, and the ration schedule. Also record the winds, weather, direction of swells, times of sunrise and sunset, and other navigational data.

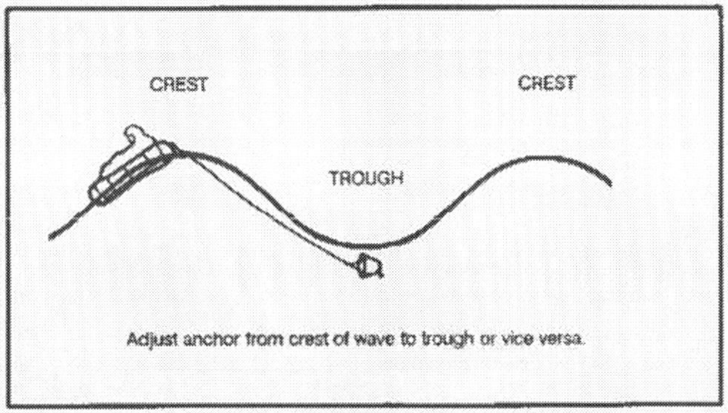

Figure 16-6. Deployment of the sea anchor.

- If you are down in unfriendly waters, take special security measures to avoid detection. Do not travel in the daytime. Throw out the sea anchor and wait for nightfall before paddling or hoisting sail. Keep low in the raft; stay covered with the blue side of the camouflage cloth up. Be sure a passing ship or aircraft is friendly or neutral before trying to attract its attention. If the enemy detects you and you are close to capture, destroy the log book, radio, navigation equipment, maps, signaling equipment, and firearms. Jump overboard and submerge if the enemy starts strafing.
- Decide whether to stay in position or to travel. Ask yourself, "How much information was signaled before the accident? Is your position known to rescuers? Do you know it yourself? Is the weather favorable for a search? Are other ships or aircraft likely to pass your present position? How many days supply of food and water do you have?"

Cold Weather Considerations

If you are in a cold climate—

- Put on an antiexposure suit. If unavailable, put on any extra clothing available. Keep clothes loose and comfortable.
- Take care not to snag the raft with shoes or sharp objects. Keep the repair kit where you can readily reach it.
- Rig a windbreak, spray shield, and canopy.
- Try to keep the floor of the raft dry. Cover it with canvas or cloth for insulation.
- Huddle with others to keep warm, moving enough to keep the blood circulating. Spread an extra tarpaulin, sail, or parachute over the group.
- Give extra rations, if available, to men suffering from exposure to cold.

The greatest problem you face when submerged in cold water is death due to hypothermia. When you are immersea in cold water, hypothermia occurs rapidly due to the decreased insulating quality of wet clothing and the result of water displacing the layer or still air that normally surrounds the body. The rate of heat exchange in water is about 25 times greater than it is in air of the same temperature. Figure 16-7 lists life expectancy times for immersion in water.

Your best protection against the effects of cold water is to get into the life raft, stay dry, and insulate your body from the cold surface of the bottom of the raft. If these actions are not possible, wearing an antiexposure suit will extend your life expectancy considerably. Remember, keep your head and neck out of the water and well insulated from the cold water's effects when the temperature is below 19 degrees C. Wearing life preservers increases the predicted survival time as Dody position in the water increases the chance of survival.

Hot Weather Considerations

If you are in a hot climate—

- Rig a sunshade or canopy. Leave enough space for ventilation.
- Cover your skin, where possible, to protect it from sunburn. Use sunburn cream, if available, on all exposed skin. Your eyelids, the back of your ears, and the skin under your chin sunburn easily.

Raft Procedures

Most of the rafts in the U.S. Army and Air Force inventories can satisfy the needs for personal protection, mode of travel, and evasion and camouflage.

Note: Before boarding any raft, remove and tether (attach) your life preserver to yourself or the raft. Ensure there are no other metallic or sharp objects on your clothing or equipment that could damage the raft. After boarding the raft, don your life preserver again.

One-Man Raft

The one-man raft has a main cell inflation. If the CO_2 bottle should malfunction or if the raft develops a leak, you can inflate it by mouth.

Water Temperature	Time
21.0-15.5 degrees C (70-60 degrees F)	12 hours
15.5-10.0 degrees C (60-50 degrees F)	6 hours
10.0-4.5 degrees C (50-40 degrees F)	1 hour
4.5 degrees C (40 degrees F) and below	less than 1 hour
Note: Wearing an antiexposure suit may increase these times up to a maximum of 24 hours.	

Figure 16-7. Life expectancy times for immersion in water.

The spray shield acts as a shelter from the cold, wind, and water. In some cases, this shield serves as insulation. The raft's insulated bottom limits the conduction of cold thereby protecting you from hypothermia (Figure 16-8).

You can travel more effectively by inflating or deflating the raft to take advantage of the wind or current. You can use the spray shield as a sail white the ballast buckets serve to increase drag in the water. You may use the sea anchor to control the raft's speed and direction.

There are rafts developed for use in tactical areas that are black. These rafts blend with the sea's background. You can further modify these rafts for evasion by partially deflating them to obtain a lower profile.

A lanyard connects the one-man raft to a parachutist (survivor) landing in the water. You (the survivor) inflate it upon landing. You do not swim to the raft, but pull it to you via the lanyard. The raft may hit the water upside down, but you can right it by approaching the side to which the bottle is attached and flipping the raft over. The spray shield must be in

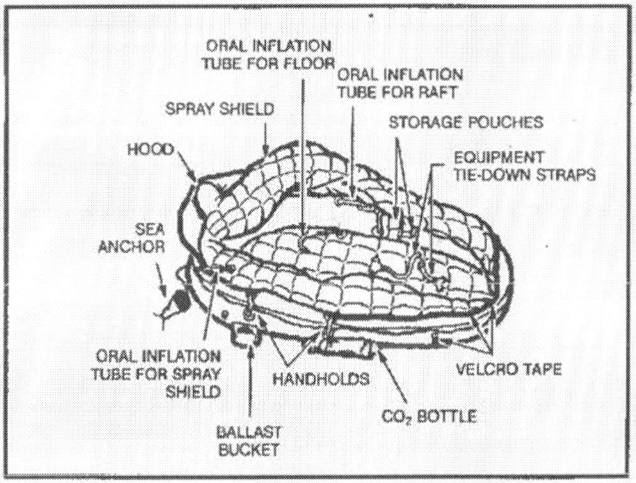

Figure 16-8. One-man raft with spray shield.

the raft to expose the boarding handles. Follow the steps outlined in the note under raft procedures above when boarding the raft (Figure 16-9).

If you have an arm injury, the best way to board is by turning your back to the small end of the raft, pushing the raft under your buttocks, and lying back. Another way to board the raft is to push down on its small end until one knee is inside and He forward (Figure 16-10).

In rough seas, it may be easier for you to grasp the small end of the raft and, in a prone position, to kick and pull yourself into the raft. When you are lymg face down in the raft, deploy and adjust the sea anchor. To sit upright, you may have to disconnect one side of the seat kit and roll to that side. Then you adjust the spray shield. There are two variations of the one-man raft; the improved model incorporates an inflatable spray shield and floor that provide additional insulation. The spray shield helps keep you dry and warm in cold oceans and protects you from the sun in the hot climates (Figure 16-11).

Figure 16-9. Boarding the one-man raft.

Figure 16-10. Boarding the one-man raft (other methods).

Seven-Man Raft

Some multiplace aircraft carry the seven-man raft. It is a component of the survival drop kit (Figure 16-12). This raft may inflate upside down and require you to right the raft before boarding. Always work from the bottle side to prevent injury if the raft turns over. Facing into the wind, the wind provides additional help in righting the raft. Use the handles on the inside bottom of the raft for boarding (Figure 16-13).

Figure 16-11. One-man raft with spray shield inflated.

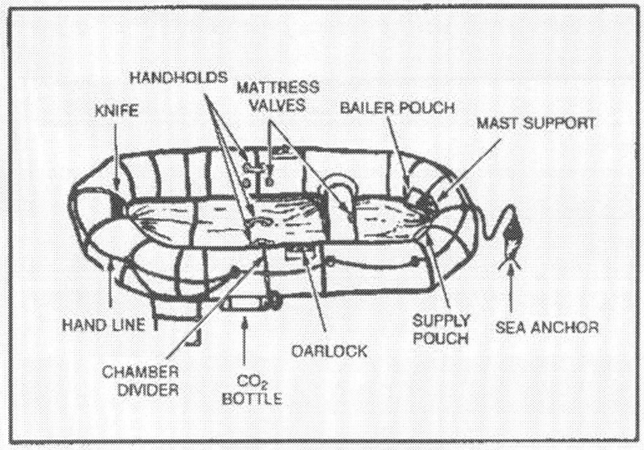

Figure 16-12. Seven-man raft.

Figure 16-13. Method of righting raft.

Use the boarding ramp if someone holds down the raft's opposite side. If you don't have help, again work from the bottle side with the wind at your back to help hold down the raft. Follow the steps outlined in the note under raft procedures above. Then grasp an oarlock and boarding handle, kick your legs to get your body prone on the water, and then kick and pull yourself into the raft. If you are weak or injured, you may partially deflate the raft to make boarding easier (Figure 16-14).

Use the hand pump to keep the buoyancy chambers and cross seat firm. Never overinflate the raft.

Twenty- or Twenty-Five-Man Rafts

You may find 20- or 25-man rafts in multiplace aircraft (Figures 16-15 and 16-16). You will find them in accessible areas of the fuselage or in raft compartments. Some may be automatically deployed from the cockpit, while others may need manual deployment. No matter how the raft lands in the water, it is ready for boarding. A lanyard connects the accessory kit to the raft and you retrieve the kit by hand. You must manually inflate the center chamber with the hand pump. Board the 20- or 25-man raft from the aircraft, if possible. If not, board in the following manner:

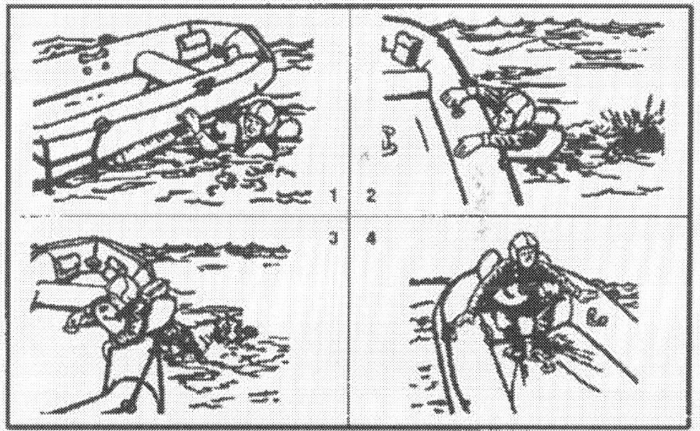

Figure 16-14. Method of boarding seven-man raft.

- Approach the lower boarding ramp.
- Remove your life preserver and tether it to yourself so that it trails behind you.
- Grasp the boarding handles and kick your legs to get your body into a prone position on the water's surface; then kick and pull until you are inside the raft.

An incompletely inflated raft will make boarding easier. Approach the intersection of fne raft and ramp grasp the upper boarding handle, and swing one leg onto the center of the ramp, as in mounting a horse (Figure 16-17).

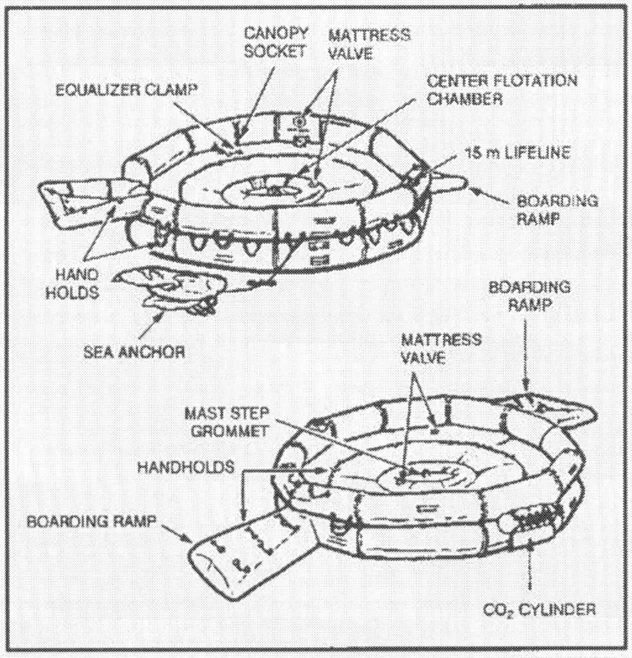

Figure 16-15. Twenty-man raft.

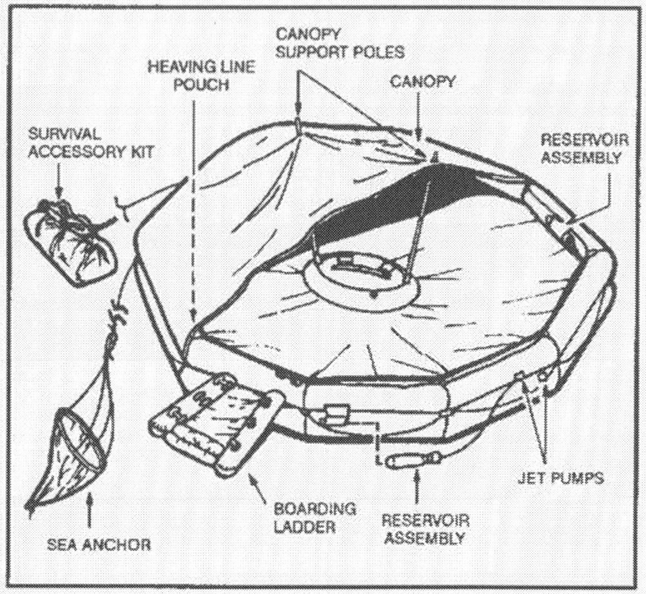

Figure 16-16. Twenty-five-man raft.

Figure 16-17. Boarding the 20-man raft.

Immediately tighten the equalizer clamp upon entering the raft to prevent deflating trie entire raft in case of a puncture (Figure 16-18).

Use the pump to keep these rafts' chambers and center ring firm. They should be well rounded but not overly tight.

Sailing Rafts

Rafts do not have keels, therefore, you can't sail them into the wind. However, anyone can sail a raft downwind. You can successfully sail multiplace (except 20- to 25-man) rafts 10 degrees off from the direction of the wind. Do not try to sail the raft unlessland is near. If you decide to sail and the wind is mowing toward a desired destination, fully inflate

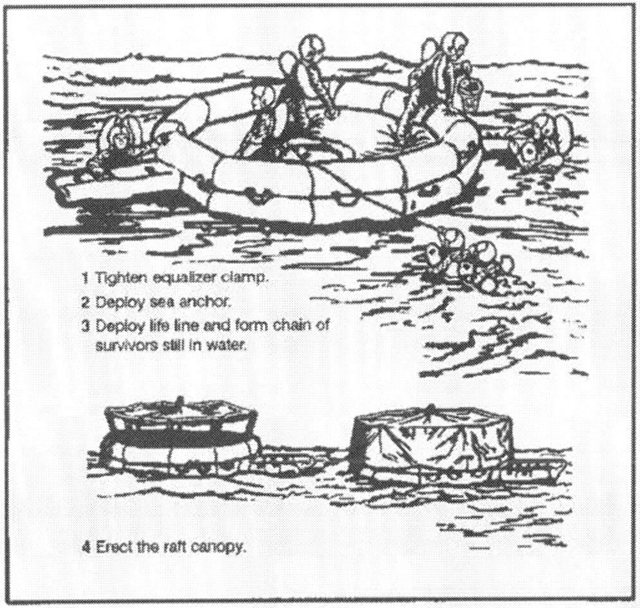

Figure 16-18. Immediate action—multiplace raft.

the raft, sit high, take in the sea anchor, rig a sail, and use an oar as a rudder.

In a multiplace (except 20- to 25-man) raft, erect a square sail in the bow using the oars and their extensions as the mast and crossbar (Figure 16-19). You may use a waterproof tarpaulin or parachute material for

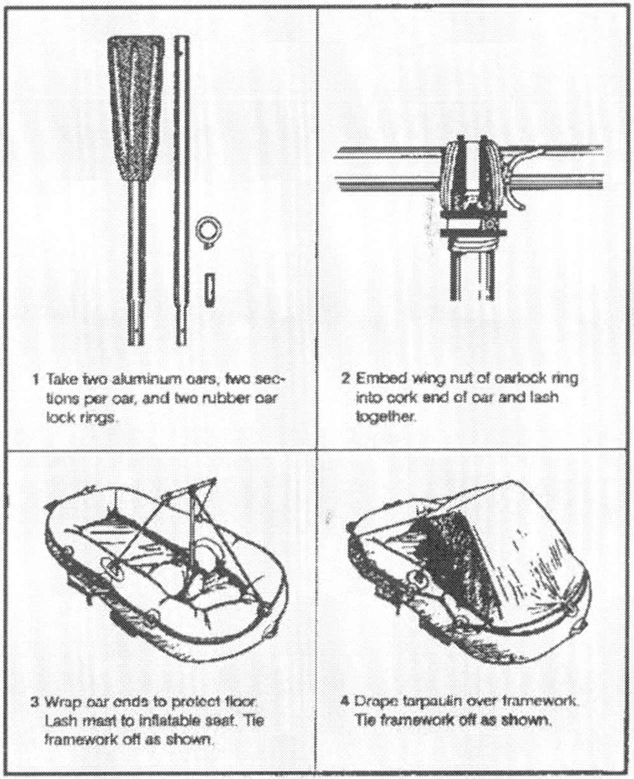

Figure 16-19. Sail construction.

the sail. If the raft has no regular mast socket and step, erect the mast by tying it securely to the front cross seat using braces. Pad the bottom or the mast to prevent it from chafing or punching a hole through the floor, whether or not there is a socket. The heel or a shoe, with the toe wedged under the seat, makes a good improvised mast step. Do not secure the comers of the lower edge of the sail. Hold the lines attached to the comers with your hands so that a gust of wind will not rip the sail, break the mast, or capsize the raft.

Take every precaution to prevent the raft from turning over. In rough weather, keep the sea anchor away from the bow. Have the passengers sit low in the raft, with their weight distributed to hold the upwind side down. To prevent falling out, they should also avoid sitting on the sides of the raft or standing up. Avoid sudden movements without warning the other passengers. When the sea anchor is not in use, tie it to the raft and stow it in such a manner that it will hold immediately if the raft capsizes.

Water

Water is your most important need. With it alone, you can live for ten days or longer, depending on your will to live. When drinking water, moisten your lips, tongue, and throat before swallowing.

Short Water Rations

When you have a limited water supply and you can't replace it by chemical or mechanical means, use the water efficiently. Protect freshwater supplies from seawater contamination. Keep your body well shaded, both from overhead sun and from reflection off the sea surface. Allow ventilation of air; dampen your clothes during the hottest part of the day. Do not exert yourself! Relax and sleep when possible. Fix your daily water ration after considering the amount of water you have, the output of solar stills and desalting kit, and the number and physical condition of your party.

If you don't have water, don't eat. If your water ration is two liters or more per day, eat any part of your ration or any additional food that you may catch, such as birds, fish, shrimp. The life raft's motion and anxiety may cause nausea. If you eat when nauseated, you may lose your food immediately. If nauseated, rest and relax as much as you can, and take only water.

To reduce your loss of water through perspiration, soak your clothes in the sea and wring them out before putting them on agam. Don't overdo this during hot days when no canopy or sun shield is available. This is a trade-off between cooling and saltwater boils and rashes that will result. Be careful not to get the oottom of the raft wet.

Watch the clouds and be ready for any chance of showers. Keep the tarpaulin handy for catching water. If it is encrusted with dried salt, wash it m seawater.Normally, a small amount of seawater mixed with rain will hardly be noticeable and will not cause any physical reaction. In rough seas you cannot get uncontaminated fresh water.

At night, secure the tarpaulin like a sunshade, and turn up its edges to collect dew. It is also possible to collect dew along the sides of the raft using a sponge or cloth. When it rains, drink as much as you can hold.

Solar Still

When solar stills are available, read the instructions and set them up immediately. Use as many stills as possible, depending on the number of men in the raft and the amount of sunlight available. Secure solar stills to the raft with care. This type of solar still only works on flat, calm seas.

Desalting Kits

When desalting kits are available in addition to solar stills, use them only for immediate water needs or during long overcast periods when you cannot use solar stills, hi any event, Keep desalting kits and emergency water stores for periods when you cannot use solar stills or catch rainwater.

Water From Fish

Drink the aqueous fluid found along the spine and in the eyes of large fish. Carefully cut the fish in half to get the fluid along the spine and suck the eye. If you are so short of water that you need to do this, then do not drink any of the other body fluids. These other fluids are rich in protein and fat and will use up more of your reserve water in digestion than they supply.

Sea Ice

In arctic waters, use old sea ice for water. This ice is bluish, has rounded comers, and splinters easily. It is nearly free of salt. New ice is gray, milky, hard, and salty. Water from icebergs is fresh, but icebergs are dangerous to approach. Use them as a source of water only in emergencies.

REMEMBER!
Do not drink seawater.
Do not drink urine.
Do not drink alcohol.
Do not smoke.
Do not eat, unless water is available.

Sleep and rest are the best ways of enduring periods of reduced water and food intake. However, make sure that you have enough shade when napping during the day. if the sea is rough, tie yourself to the raft, close any cover, and ride out the storm as best you can. Relax is the key word—at least try to relax.

Food Procurement

In the open sea, fish will be the main food source. There are some poisonous and dangerous ocean fish, but, in general, when out of sight of land, fish are safe to eat. Nearer the shore there are fish that are Doth dangerous and poisonous to eat. There are some fish, such as the red snapper and barracuda, that are normally edible but poisonous when taken from the waters of atolls and reefs. Flying fish will even jump into your raft!

Fish

When fishing, do not handle the fishing line with bare hands and never wrap it around your hands or tie it to a life raft. The salt that adheres to it can make it a sharp cutting edge, an edge dangerous both to the raft and your hands. Wear gloves, if they are available, or use a cloth to handle fish and to avoid injury from sharp fins and gill covers.

In warm regions, gut and bleed fish immediately after catching them. Cut fish that you do not eat immediately into thin, narrow strips and hang them to dry. A well-dried fish stays edible for several days. Fish not cleaned and dried may spoil in half a day. Fish with dark meat are very prone to decomposition. If you do not eat them all immediately, do not eat any of the leftovers. Use the leftovers for bait.

Never eat fish that have pale, shiny gills, sunken eyes, flabby skin and flesh, or an unpleasant odor. Good fish show the opposite characteristics. Sea fish have a saltwater or clean fishy odor. Do not confuse eels with sea snakes that have an obviously scaly body and strongly compressed, paddle-shaped tail. Both eels and sea snakes are edible, but you must handle the latter with care because of their poisonous bites. The heart, blood, intestinal wall, and liver of most fish are edible. Cook the intestines. Also edible are the partly digested smaller fish that you may find in the stomachs of large fish. In addition, sea turtles are edible.

Shark meat is a good source of food whether raw, dried, or cooked. Shark meat spoils very rapidly due to the high concentration of urea in the blood, therefore, bleed it immediately and soak it in several changes of water. People prefer some shark species over others. Consider them all edible except the Greenland shark whose flesh contains high quantities of vitamin A. Do not eat the livers, due to high vitamin A content.

Fishing Aids

You can use different materials to make fishing aids as described in the following paragraphs:

- *Fishing line.* Use pieces of tarpaulin or canvas. Unravel the threads and tie them together in short lengths in groups of three or more threads. Shoelaces and parachute suspension lme also work well.
- *Fish hook.* No survivor at sea should be without fishing equipment but if you are, improvise hooks as shown in Chapter 8.
- *Fish lures.* You can fashion lures by attaching a double hook to any shiny piece of metal.
- *Grapple.* Use grapples to hook seaweed. You may shake crabs, shrimp, or small fish out of the seaweed. These you may eat or use for bait. You may eat seaweed itself, but only when you have plenty of drinking water. Improvise grapples from wood. Use a heavy piece of wood as the main shaft, and fash three smaller pieces to the shaft as grapples.

- *Bait.* You can use small fish as bait for larger ones. Scoop the small fish up with a net. If you don't have a net, make one from cloth of some type. Hold the net under the water and scoop upward. Use all the guts from birds and fish for bait. When using bait, try to keep it moving in the water to give it the appearance otoeing alive.

Helpful Fishing Hints

Your fishing should be successful if you remember the following important hints:

- Be extremely careful with fish that have teeth and spines.
- Cut a large fish loose rather than risk capsizing the raft. Try to catch small rather than large fish.
- Do not puncture your raft with hooks or other sharp instruments.
- Do not fish when large sharks are in the area.
- Watch for schools of fish; try to move close to these schools.
- Fish at night using a light. The light attracts fish.
- In the daytime, shade attracts some fish. You may find them under your raft.
- Improvise a spear by tying a knife to an oar blade. This spear can help you catch larger fish, but you must get them into the raft quickly or they will slip off the blade. Also, tie the knife very securely or you may lose it.
- Always take care of your fishing equipment. Dry your fishing lines, clean and sharpen the hooks, and do not allow the hooks to stick into the fishing lines.

Birds

As stated in Chapter 8, all birds are edible. Eat any birds you can catch. Sometimes birds may land on your raft, but usually they are cautious. You may be able to attract some birds by towing a bright piece of metal behind the raft. This will bring the bird within shooting range, provided you have a firearm.

If a bird lands within your reach, you may be able to catch it. If the birds do not land close enough or land on the other end of the raft, you may be able to catch them with a bird noose. Bait the center of the noose and wait for the bird to land. When the bird's feet are in the center of the noose, pull it tight.

Use all parts of the bird. Use the feathers for insulation, the entrails and feet for bait, and so on. Use your imagination.

Medical Problems Associated With Sea Survival

At sea, you may become seasick, get saltwater sores, or face some of the same medical problems that occur on land, such as dehydration or sunburn. These problems can become critical if left untreated.

Seasickness

Seasickness is the nausea and vomiting caused by the motion of the raft. It can result in—

- Extreme fluid loss and exhaustion.
- Loss of the will to survive.
- Others becoming seasick.
- Attraction of sharks to the raft.
- Unclean conditions.

To treat seasickness—

- Wash both the patient and the raft to remove the sight and odor of vomit.
- Keep the patient from eating food until his nausea is gone.
- Have the patient lie down and rest.
- Give the patient seasickness pills if available. If the patient is unable to take the pills orally, insert them rectally for absorption by the body.

Note: Some survivors have said that erecting a canopy or using the horizon as a focal point helped overcome seasickness. Others have said that swimming alongside the raft for short periods helped, but extreme care must be taken if swimming.

Saltwater Sores

These sores result from a break in skin exposed to saltwater for an extended period. The sores may form scabs and pus. Do not open or drain. Flush the sores with fresh water, if available, and allow to dry. Apply an antiseptic, if available.

Immersion Rot, frostbite, and Hypothermia

These problems are similar to those encountered in cold weather environments. Symptoms and treatment are the same as covered in Chapter 15.

Blindness/Headache

If flame, smoke, or other contaminants get in the eyes, flush them immediately with salt water, then with fresh water, if available. Apply ointment, if available. Bandage both eyes 18 to 24 hours, or longer if damage is severe. If the glare from the sky and water causes your eyes to become bloodshot and inflamed, bandage them lightly. Try to prevent this problem by wearing sunglasses. Improvise sunglasses if necessary.

Constipation

This condition is a common problem on a raft. Do not take a laxative, as this will cause further dehydration. Exercise as much as possible and drink an adequate amount of water, if available.

Difficult Urination

This problem is not unusual and is due mainly to dehydration. It is best not to treat it, as it could cause further dehydration.

Sunburn

Sunburn is a serious problem in sea survival. Try to prevent sunburn by staying in shade and keeping your head and skm covered. Use cream or Chap Stick from your first aid kit. Remember, reflection from the water also causes sunburn.

Sharks

Whether you are in the water or in a boat or raft, you may see many Spes of sea life around you. Some may be more dangerous than others, Senerally, sharks are the greatest danger to you. Other animals such as whales, porpoises, and stingrays may look dangerous, but really pose little threat in the open sea.

Of the many hundreds of shark species, only about 20 species are known to attack man. The most dangerous are the great white shark, the hammerhead, the make, and the tiger shark. Other sharks known to attack man include the gray, blue, lemon, sand, nurse, bull, and oceanic white tip sharks. Consider any shark longer than 1 meter dangerous.

There are sharks in all oceans and seas of the world. While many live and feed in the depths of the sea, others hunt near the surface. The sharks living near the surface are the ones you will most likely see. Their dorsaffins frequently project above the water. Sharks in the tropical and subtropical seas are far more aggressive than those in temperate waters.

All sharks are basically eating machines. Their normal diet is live animals of any type, and they will strike at injured or helpless animals. Sight, smell, or sound may guide them to their prey. Sharks have an acute sense of smell and the smell of blood in the water excites them. They are also very sensitive to any abnormal vibrations in the water. The struggles of a wounded animal or swimmer, underwater explosions, or even a fish struggling on a fishline will attract a shark.

Sharks can bite from almost any position; they do not have to turn on their side to bite. The jaws of some of the larger sharks are so far forward that they can bite floating objects easily without twisting to the side.

Sharks may hunt alone, but most reports of attacks cite more than one shark present. The smaller sharks tend to travel in schools and attack in mass. Whenever one of the sharks finds a victim, the other sharks will quickly join it. Sharks will eat a wounded shark as quickly as their prey.

Sharks feed at all hours of the day and night. Most reported shark contacts and attacks were during daylight, and many of these have been in the late afternoon. Some of the measures that you can take to protect yourself against sharks when you are in the water are—

- *Stay with other swimmers.* A group can maintain a 360-degree watch. A group can either frighten or fight off sharks better than one man.
- *Always watch for sharks.* Keep all your clothing on, to include your shoes. Historically, sharks have attacked the unclothed men in groups first, mainly in the feet. Clothing also protects against abrasions should the shark brush against you.
- *Avoid urinating.* If you must, only do so in small amounts. Let it dissipate between discharges. If you must defecate, do so in small amounts and throw it as far away from you as possible. Do the same if you must vomit.

If a shark attack is imminent while you are in the water, splash and yell just enough to keep the shark at bay. Sometimes yelling underwater or slapping the water repeatedly will scare the shark away. Conserve your strength for fighting in case the shark attacks.

If attacked, kick and strike the shark. Hit the shark on the gills or eyes if possible. If you hit the shark on the nose, you may injure your hand if it glances off and hits its teeth.

When you are in a raft and see sharks—

- Do not fish. If you have hooked a fish, let it go. Do not clean fish in the water.
- Do not throw garbage overboard.
- Do not let your arms, legs, or equipment hang in the water.
- Keep quiet and do not move around.
- Bury all dead as soon as possible. If there are many sharks in the area, conduct the burial at night.

When you are in a raft and a shark attack is imminent, hit the shark with anything you have, except your hands. You will do more damage to your hands than the shark. If you strike with an oar, be careful not to lose or break it.

Detecting Land

You should watch carefully for any signs of land. There are many indicators that land is near.

A fixed cumulus cloud in a clear sky or in a sky where all other clouds are moving often hovers over or slightly downwind from an island.

In the tropics, the reflection of sunlight from shallow lagoons or shelves of coral reefs often causes a greenish tint in the sky.

In the arctic, light-colored reflections on clouds often indicate ice fields or snow-covered land. These reflections are quite different from the dark gray ones caused by open water.

Deep water is dark green or dark blue. Lighter color indicates shallow water, which may mean land is near.

At night, or in fog, mist, or rain, you may detect land by odors and sounds. The musty odor of mangrove swamps and mud flats carry a long way. You hear the roar of surf long before you see the surf. The continued cries of seabirds coming from one direction indicate their roosting place on nearby land.

There usually are more birds near land than over the open sea. The direction from which flocks fly at dawn and to which they fly at dusk may indicate the direction of land. During the day, birds are searching for food and the direction of flight has no significance.

Mirages occur at any latitude, but they are more likely in the tropics, especially during the middle of the day. Be careful not to mistake a mirage for nearby land. A mirage disappears or its appearance and elevation change when viewed from slightly different heights.

You may be able to detect land by the pattern of the waves (refracted) as they approach land (Figure 16-20). By traveling with the waves and parallel to the slightly turbulent area marked "a on the illustration, you should reachland.

Rafting or Beaching Techniques

Once you have found land, you must get ashore safely. To raft ashore, you can usually use the one-man raft without danger. However, going ashore in a strong surf is dangerous. Take your time. Select yourlanding

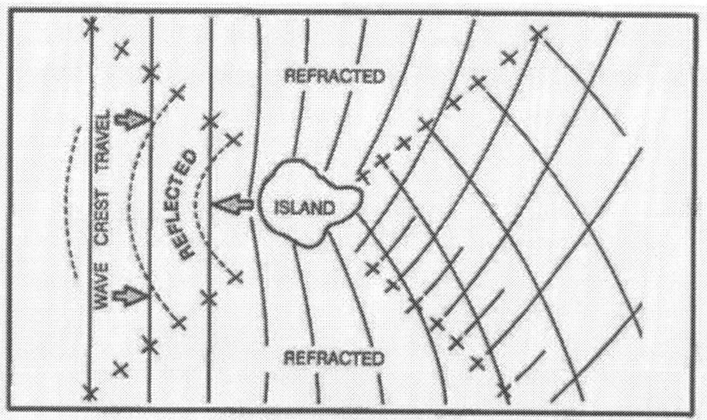

Figure 16-20. Wave patterns about an island.

point carefully. Try not to land when the sun is low and straight in front of you. Try to land on the lee side of an island or on a point of land jutting out into the water. Keep your eyes open for gaps in the surf line, and head for them. Avoid coral reefs and rocky cliffs.There are no coral reefs near the mouths of freshwater streams. Avoid rip currents or strong tidal currents that may carry you far out to sea. Either signal ashore for help or sail around and look for a sloping beach where the surf is gentle.

If you have to go through the surf to reach shore, take down the mast. Keep your clothes and shoes on to avoid severe cuts. Adjust and inflate your life vest. Trail the sea anchor over the stem using as much line as you have. Use the oars or paddles and constantly adjust the sea anchor to keep a strain on the anchor line. These actions will keep the raft pointed toward shore and prevent the sea from throwing the stern around and capsizing you.TJse the oars or paddles to help ride in on the seaward side of a large wave.

The surf may be irregular and velocity may vary, so modify your procedure as conditions demand. A good method of getting through the surf is to have half the men sit on one side of the raft, half on the other, facing away from each other. When a heavy sea bears down, half should row (pull) toward the sea until the crest passes; then the other half should row (pull) toward the shore until the next heavy sea comes along.

Against a strong wind and heavy surf, the raft must have all possible speed to pass rapidly through the oncoming crest to avoid being turned broadside or thrown end over end. If possible, avoid meeting alarge wave at the moment it breaks.

If in a medium surf with no wind or offshore wind, keep the raft from passing over a wave so rapidly that it drops suddenly after topping the crest, fi the raft turns over in the surf, try to grab hold of it and ride it in.

As the raft nears the beach, ride in on the crest of a large wave. Paddle or row hard and ride in to the beach as far as you can. Do not jump out of the raft until it has grounded, then quickly get out and beach it.

If you have a choice, do not land at night. If you have reason to believe that people live on the shore, lay away from the beach, signal, and wait for the inhabitants to come out and bring you in.

If you encounter sea ice, land only on large, stable floes. Avoid icebergs that may capsize and small floes or those obviously disintegrating. Use oars and hands to keep the raft from rubbing on the edge of the ice. Take the raft out of the water and store it well back from the floe's edge. You may be able to use it for shelter. Keep the raft inflated and ready for use. Any floe may break up without warning.

Swimming Ashore

If rafting ashore is not possible and you have to swim, wear your shoes and at least one thickness of clothing. Use the sidestroke or oreaststroke to conserve strength.

If the surf is moderate, ride in on the back of a small wave by swimming forward with it. Dive to a shallow depth to end the ride just before the wave breaks.

In high surf, swim toward shore in the trough between waves. When the seaward wave approaches, face it and submerge. After it passes, work toward shore in the next trough. If caught in the undertow of a large wave, push off the bottom or swim to the surface and proceed toward shore as above.

If you must land on a rocky shore, look for a place where the waves rush up onto the rocks. Avoid places where the waves explode with a high, white spray. Swim slowly when making your approach. You will need your strength to hold on to the rocks. You should be fully clothed and wear shoes to reduce injury.

After selecting your landing point, advance behind a large wave into the breakers. Face toward shore and take a sitting position with your feet in front, 60 to 90 centimeters (2 or 3 feet) lower than your head. This position will let your feet absorb the shock when you land or strike submerged boulders or reefs. If you do not reach shore behind the wave you picked, swim with your hands only. As the next wave approaches, take a sitting position with your feet forward. Repeat the procedure until you land.

Water is quieter in the lee of a heavy growth of seaweed. Take advantage of such growth. Do not swim through the seaweed; crawl over the top by graspmg the vegetation with overhand movements.

Cross a rocky or coral reef as you would land on a rocky shore. Keep your feet dose together and your knees slightly bent in a relaxed sitting posture to cushion the blows against the coral.

Pickup or Rescue

On sighting rescue craft approaching for pickup (boat, ship, conventional aircraft, or helicopter), quickly clear any lines (fishing lines, desalting kit lines) or other gear that could cause entanglement during rescue. Secure all loose items in the raft. Take down canopies and sails to ensure a safer pickup. After securing all items, put on your helmet, if available. Fully inflate your life preserver. Remain in the raft, unless otherwise instructed, and remove all equipment except the preservers. If possible, you will receive help from rescue personnellowered into the water. Remember, follow all mstructions given by the rescue personnel.

If the helicopter recovery is unassisted, do the following before pickup:

- Secure all the loose equipment in the raft, accessory bag, or in pockets.
- Deploy the sea anchor, stability bags, and accessory bag.
- Partially deflate the raft and fill it with water.
- Unsnap the survival kit container from the parachute harness.
- Grasp the raft handhold and roll out of the raft.
- Allow the recovery device or the cable to ground out on the water's surface.
- Maintain the handhold until the recovery device is in your other hand.
- Mount the recovery device, avoiding entanglement with the raft.
- Signal the hoist operator for pickup.

SEASHORES

Search planes or ships do not always spot a drifting raft or swimmer. You may have to land along the coast before being rescued. Surviving along the seashore is different from open sea survival. Food and water are more abundant and shelter is obviously easier to locate and construct.

If you are in friendly territory and decide to travel, it is better to move along the coast than to go inland. Do not leave the coast except to avoid obstacles (swamps and cliffs) or unless you find a trail that you know leads to human habitation.

In time of war, remember that the enemy patrols most coastlines. These patrols may cause problems for you if you land on a hostile shore. You will have extremely limited travel options in this situation. Avoid all contact with other humans, and make every effort to cover all tracks you leave on the shore.

Special Health Hazards

Coral, poisonous and aggressive fish, crocodiles, sea urchins, sea biscuits, sponges, anemones, and tides and undertow pose special health hazards.

Coral

Coral, dead or alive, can inflict painful cuts. There are hundreds of water hazards that can cause deep puncture wounds, severe bleeding, and the danger of infection. Clean all coral cuts thoroughly. Do not use iodine to disinfect any coral cuts. Some coral polyps feed on iodine and may grow inside your flesh if you use iodine.

Poisonous Fish

Many reef fish have toxic flesh. For some species, the flesh is always poisonous, for other species, only at certain times of the year. The poisons are present in all parts of the fish, but especially in the liver, intestines, and eggs.

Fish toxins are water soluble—no amount of cooking will neutralize them. They are tasteless, therefore the standard edibility tests are useless. Birds are least susceptible to the poisons. Therefore, do not think that because a bird can eat a fish, it is a safe species for you to eat.

The toxins will produce a numbness of the hps, tongue, toes, and tips of the fingers, severe itching, and a clear reversal of temperature sensations. Cold items appear hot and hot items cold. There will probably also be nausea, vomiting, loss of speech, dizziness, and a paralysis that eventually brings death.

In addition to fish with poisonous flesh, there are those that are dangerous to touch. Many stingrays have a poisonous barb in their tail. There are also species that can deliver an electric shock. Some reef fish, such as stonefish and toadfish, have venomous spines that can cause very painful although seldom fatal injuries. The venom from these spines causes a burning sensation or even an agonizing pain that is out of proportion to the apparent severity of the wound. Jellyfish, while not usually fatal, can inflict a very painful sting if it touches you with its tentacles. See Chapter 11 and Appendix F for details on particularly dangerous fish of the sea and seashore.

Aggressive Ksh

You should also avoid some ferocious fish. The bold and inquisitive barracuda has attacked men wearing shiny objects. It may charge lights or shiny objects at night. The sea bass, which can grow to 1.7 meters, is another fish to avoid. The moray eel, which has many sharp teeth and grows to 1.5 meters, can also be aggressive if disturbed.

Sea Snakes

Sea snakes are venomous and sometimes found in mid ocean. They are unlikely to bite unless provoked. Avoid them.

Crocodiles

Crocodiles inhabit tropical saltwater bays and mangrove-bordered estuaries and range up to 65 kilometers into the open sea. Few remain near inhabited areas. You commonly find crocodiles in the remote areas of the East Indies and Southeast Asia. Consider specimens over 1 meter long dangerous, especially females guarding their nests. Crocodile meat is an excellent source of food when available.

Sea Urchins, Sea Biscuits, Sponges, and Anemones

These animals can cause extreme, though seldom fatal, pain. Usually found in tropical shallow water near coral formations, sea urchins resemble small, round porcupines. If stepped on, they slip fine needles of lime or silica into the skin, where they Dreak off and fester. If possible, remove the spines and treat the injury for infection. The other animals mentioned inflict injury similarly.

Tides and Undertow

These are another hazard to contend with. If caught in a large wave's undertow, push off the bottom or swim to the surface and proceed shoreward m a trough between waves. Do not fight against the pull of the undertow. Swim with it or perpendicular to it until it loses strength, then swim for shore.

Food

Obtaining food along a seashore should not present a problem. There are many types of seaweed and other plants you can easily find and eat. See Chapter 9 and Appendix B for a discussion of these plants.

There is a great variety of animal life that can supply your need for food in this type of survival situation.

Mollusks

Mussels, limpets, clams, sea snails, octopuses, squids, and sea slugs are all edible. Shellfish will usually supply most of the protein eaten by coastal survivors. Avoid the blue-ringed octopus ana cone shells (described in Chapter 11 and Appendix F). Also beware of "red tides" that make mollusks poisonous. Apply the edibility test on each species before eating.

Worms

Coastal worms are generally edible, but it is better to use them for fish bait. Avoid bristle worms that look like fuzzy caterpillars. Also avoid tubeworms that have sharp-edged tubes. Arrowworms, alias amphioxus, are not true worms. You find them in the sand and are excellent either fresh or dried.

Crabs, Lobsters, and Barnacles

These animals are seldom dangerous to man and are an excellent food source. The pincers of larger crabs or lobsters can crush a man's finger. Many species have spines on their shells, making it preferable to wear gloves when catching them. Barnacles can cause scrapes or cuts and are difficult to detach from their anchor, but the larger species are an excellent food source.

Sea Urchins

These are common and can cause painful injuries when stepped on or touched. They are also a good source of food. Handle them with gloves, and remove all spines.

Sea Cucumbers

This animal is an important food source in the Indo-Pacific regions. Use them whole after evisceration or remove the five muscular strips that run the length of its body. Eat them smoked, pickled, or cooked.

CHAPTER 17

EXPEDIENT WATER CROSSINGS

In a survival situation, you may have to cross a water obstacle. It may be in the form of a river, a stream, a lake, a bog, quicksand, quagmire, or muskeg. Even in the desert, flash floods occur, making streams an obstacle. Whatever it is, you need to know how to cross it safely.

RIVERS AND STREAMS

You can apply almost every description to rivers and streams. They may be shallow or deep, slow or fast moving, narrow or wide. Before you try to cross a river or stream, develop a good plan.

Your first step is to look for a high place from which you can get a good view of the river or stream. From this place, you can look for a place to cross. If there is no high place, climb a tree. Good crossing locations include—

- A level stretch where it breaks into several channels. Two or three narrow channels are usually easier to cross than a wide river.
- A shallow bank or sandbar. If possible, select a point upstream from the bank or sandbar so that the current will carry you to it if you lose your footing.
- A course across the river that leads downstream so that you will cross the current at about a 45-degree angle.

The following areas possess potential hazards; avoid them, if possible:

- *Obstacles on the opposite side of the river that might hinder your travel* Try to select the spot from which travel will be the safest and easiest.
- *A ledge of rocks that crosses the river.* This often indicates dangerous rapids or canyons.
- *A deep or rapid waterfall or a deep channel.* Never try to ford a stream directly above or even close to such hazards.
- *Rocky places.* You may sustain serious injuries from slipping or falling on rocks. Usually, submerged rocks are very slick, making balance extremely difficult. An occasional rock that breaks the current, however, may help you.
- *An estuary of a river.* An estuary is normally wide, has strong currents, and is subject to tides. These tides can influence some rivers many kilometers from their mouths. Go back upstream to an easier crossing site.
- *Eddies.* An eddy can produce a powerful backward pull downstream of the obstruction causing the eddy and pull you under the surface.

The depth of a fordable river or stream is no deterrent if you can keep your footing. In fact, deep water sometimes runs more slowly and is therefore safer than fast-moving shallow water. You can always dry your clothes later, or if necessary, you can make a raft to carry your clothing and equipment across the river.

You must not try to swim or wade across a stream or river when the water is at very low temperatures. This swim could be fatal. Try to make a raft of some type. Wade across if you can get only your feet wet. Dry them vigorously as soon as you reach the other bank.

RAPIDS

If necessary, you can safely cross a deep, swift river or rapids. To swim across a deep, swift river, swim with the current, never fight it. Try to keep your body horizontal to the water. This will reduce the danger of being pulled under.

hi fast, shallow rapids, lie on your back, feet pointing downstream, finning your hands alongside your hips. This action wiliincrease buoyancy anahelp you steer away from obstacles. Keep your feet up to avoid getting them bruised or caught by rocks.

In deep rapids, he on your stomach, head downstream, angling toward the shore whenever you can. Watch for obstacles and be careful of backwater eddies and converging currents, as they often contain dangerous swirls. Converging currents occur where new watercourses enter the river or where water has been diverted around large obstacles such as small islands.

To ford a swift, treacherous stream, apply the following steps:

- Remove your pants and shirt to lessen the water's pull on you. Keep your footgear on to protect your feet and ankles from rocks. It will also provide you with firmer footing.
- Tie your pants and other articles to the top of your rucksack or in a bundle, if you have no pack. This way, if you have to release your equipment, all your articles will be together. It is easier to find one large pack than to find several small items.
- Carry your pack well up on your shoulders and be sure you can easily remove it, if necessary. Notbeing able to get a pack off quickly enough can drag even the strongest swimmers under.
- Find a strong pole about 7.5 centimeters in diameter and 2.1 to 2.4 meters long to help you ford the stream. Grasp the pole and plant it firmly on your upstream side to break the current. Plant your feet firmly with each step, and move the pole forward a little downstream from its previous position, but still upstream from you. With your next step, place your foot below the pole. Keep the pole well slanted so that the force of the current keeps the pole against your shoulder (Figure 17-1).
- Cross the stream so that you will cross the downstream current at a 45-degree angle.

Using this method, you can safely cross currents usually too strong for one person to stand against. Do not concern yourself about your pack's weight, as the weignt will help rather than hinder you in fording the stream.

If there are other people with you, cross the stream together. Ensure that everyone has prepared their pack and clothing as outlined above. Position the heaviest person on the downstream end of the pole and the lightest on the upstream end. In using this method, the upstream person breaks the current, and those below can move with relative ease in the eddy formed by the upstream person. If the upstream person gets temporarily swept off his feet, the others can hold steady while he regains his footing (Figure 17-2).

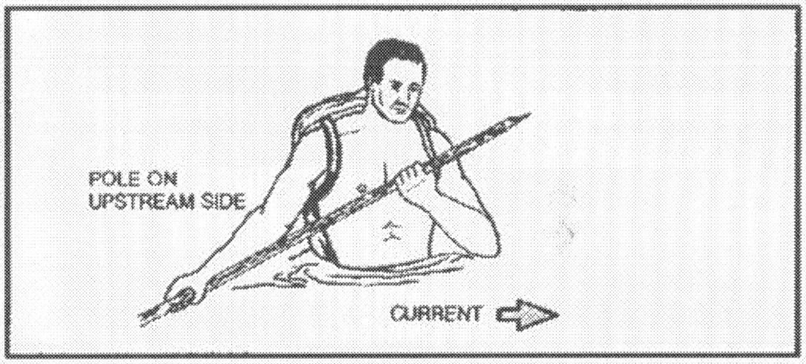

Figure 17-1. One man crossing swift stream.

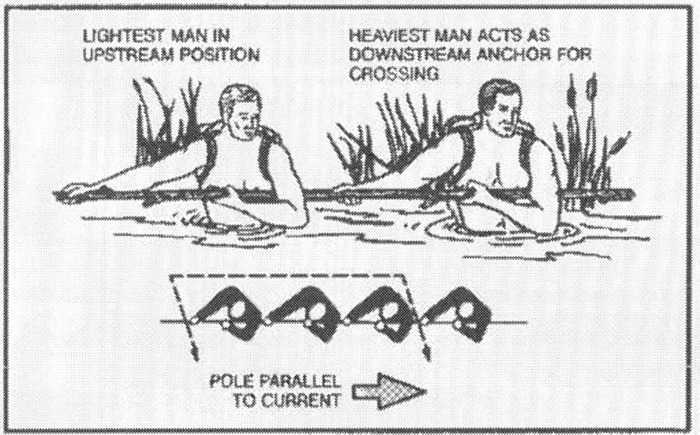

Figure 17-2. Several men crossing swift stream.

If you have three or more people and a rope available, you can use the technique shown in Figure 17-3 to cross the stream. The length of the rope must be three times the width of the stream.

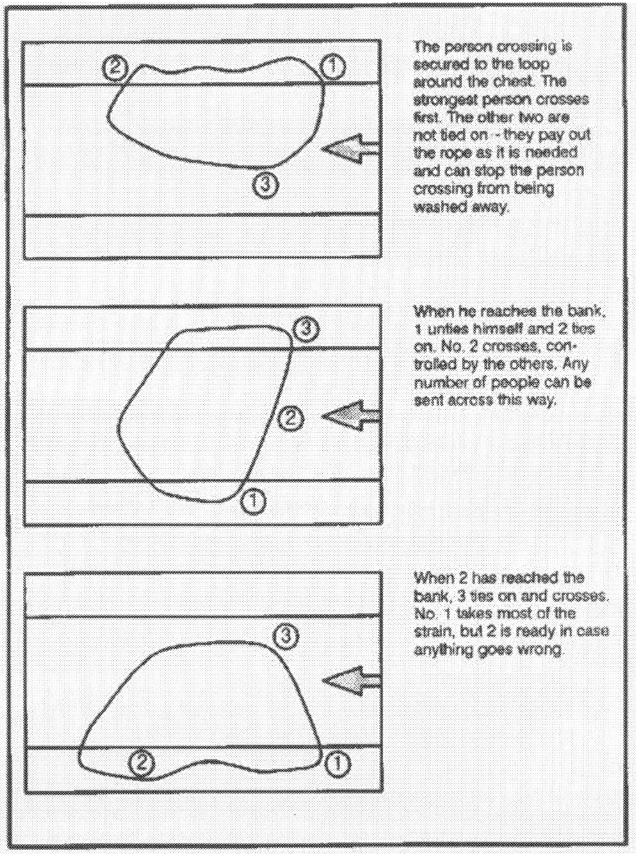

Figure 17-3. Individuals tied together to cross stream.

RAFTS

If you have two ponchos, you can construct a brush raft or an Australian poncho raft. With either of these rafts, you can safely float your equipment across a slow-moving stream or river.

Brush Raft

The brush raft, if properly constructed, will support about 115 kilograms. To construct it, use ponchos, fresh green brush, two small saplings, and rope or vine as follows (Figure 17-4)

- Push the hood of each poncho to the inner side and tightly tie off the necks using the drawstrings.
- Attach the ropes or vines at the corner and side grommets of each poncho. Make sure they are long enough to cross to and tie with the others attached at the opposite corner or side.
- Spread one poncho on the ground with the inner side up. Pile fresh, green brush (no thick branches) on the poncho until the brush stack is about 45 centimeters high. Pull the drawstring up through the center of the brush stack.

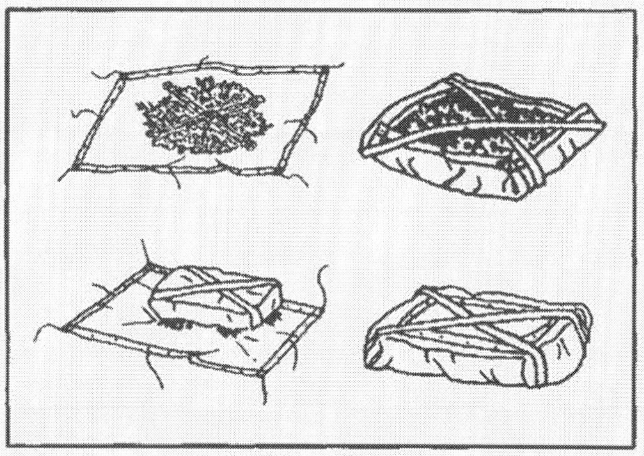

Figure 17-4. Brush raft.

- Make an X-frame from two small saplings and place it on top of the brush stack. Tie the X-frame securely in place with the poncho drawstring.
- Pile another 45 centimeters of brush on top of the X-frame, then compress the brush slightly.
- Pull the poncho sides up around the brush and, using the ropes or vines attached to the comer or side grommets, tie them diagonally from comer to corner and from side to side.
- Spread the second poncho, inner side up, next to the brush bundle.
- Roll the brush bundle onto the second poncho so that the tied side is down. Tie the second poncho around the brush bundle in the same manner as you tied the first poncho around the brush.
- Place it in the water with the tied side of the second poncho facing up.

Australian Poncho Raft

If you do not have time to gather brush for a brush raft, you can make an Australian poncho raft. This raft, although more waterproof than the poncho brush raft, will only float about 35 kilograms of equipment. To construct this raft, use two ponchos, two rucksacks, two 1.2-meter poles or branches, and ropes, vines, bootlaces, or comparable material as follows (Figure 17-5):

- Push the hood of each poncho to the inner side and tightly tie off the necks using the drawstrmgs.
- Spread one poncho on the ground with the inner side up. Place and center the two 1.2-meter poles on the poncho about 45 centimeters apart.
- Place your rucksacks or packs or other equipment between the poles. **Also** place other items that you want to keep dry between the poles. Snap the poncho sides together.
- Use your buddy's help to complete the raft. Hold the snapped portion of the poncho in the air and roll it tightly down to the equipment. Make sure you roll the full width of the poncho.
- Twist the ends of the roll to form pigtails in opposite directions. Fold the pigtails over the bundle and tie them securely in place using ropes, Dootlaces, or vines.
- Spread the second poncho on the ground, inner side up. If you need more buoyancy, place some fresh green brush on this poncho.
- Place the equipment bundle, tied side down, on the center of the second poncho. Wrap the second poncho around the equipment bundle following the same procedure you used for wrappmg the equipment in the first poncho.
- Tie ropes, bootlaces, vines, or other binding material around the raft about 30 centimeters from the end of each pigtail. Place and secure weapons on top of the raft.
- Tie one end of a rope to an empty canteen and the other end to the raft. This will help you to tow the raft.

Poncho Donut Raft

Another type of raft is the poncho donut raft. It takes more time to construct than the brush raft or Australian poncho raft, but it is effective. To construct it, use one poncho, small saplings, willow or vines, and rope, bootlaces, or other binding material (Figure 17-6) as follows:

- Make a framework circle byplacing several stakes in the ground that roughly outline an inner ana outer circle.
- Using young saplings, willow, or vines, construct a donut ring within the circles of stakes.
- Wrap several pieces of cordage around the donut ring about 30 to 60 centimeters apart and tie them securely.

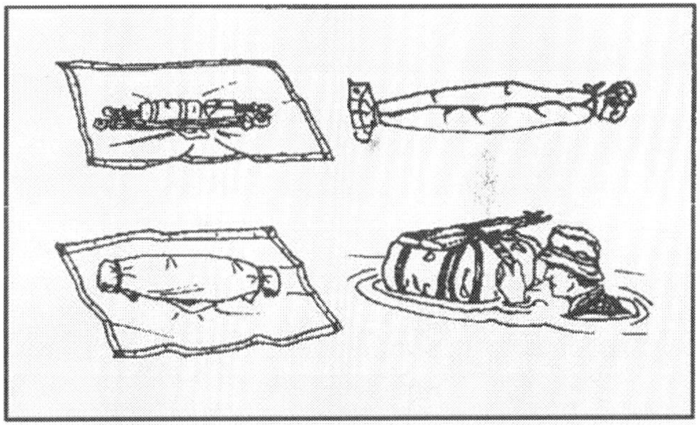

Figure 17-5. Australian poncho raft.

- Push the poncho's hood to the inner side and tightly tie off the neck using the drawstring. Place the poncho on the ground, inner side up.
- Place the donut ring on the center of the poncho. Wrap the poncho up and over the donut ring and tie off each grommet on the poncho to the ring.
- Tie one end of a rope to an empty canteen and the other end to the raft. This rope will help you to tow the raft.

When launching any of the above rafts, take care not to puncture or tear it by dragging it on the ground. Before you start to cross the river or stream, let the raft lay on the water a few minutes to ensure that it floats.

If the river is too deep to ford, push the raft in front of you while you are swimming. The design of the above rafts does not allow them to carry a person's full body weight. Use them as a float to get you and your equipment safely across the river or stream.

Be sure to check the water temperature before trying to cross a river or water obstacle. If the water is extremely cold and you are unable to find a shallow fording place in the river, do not try to ford it. Devise other means for crossing. For instance, you might improvise a bridge by felling a tree over the river. Or you might build a raft large enough to carry you and your equipment. For this, however, you will need an axe, a knife, a rope or vines, and time.

Log Raft

You can make a raft using any dry, dead, standing trees for logs. However, spruce trees found m polar and subpolar regions make the best

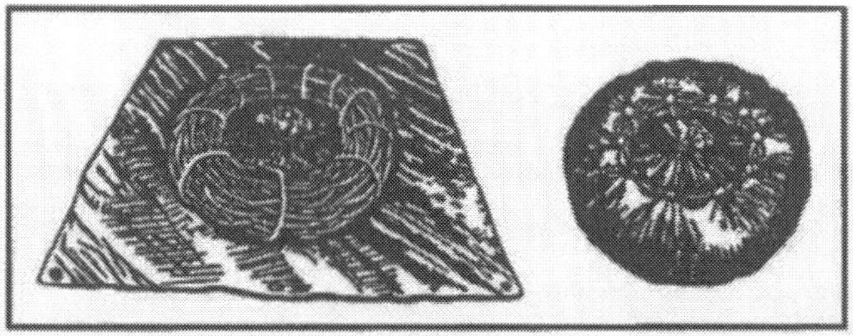

Figure 17-6. Poncho donut raft.

rafts. A simple method for making a raft is to use pressure bars lashed securely at each end of the raft to hold the logs together (Figure 17-7).

FLOTATION DEVICES

If the water is warm enough for swimming and you do not have the time or materials to construct one of the poncho-type rafts, you can use various flotation devices to negotiate the water obstacle. Some items you can use for flotation devices are—

- *Trousers.* Knot each trouser leg at the bottom and close the fly. With both hands, grasp the waistband at the sides and swing the trousers in the air to trap air in each leg. Quickly press the sides of the waistband together and hold it underwater so that the air will not escape. You now have water wings to keep you afloat as you cross the body of water.

  Note: Wet the trousers before inflating to trap the air better You may have to reinflate the trousers several times when crossing a large body of water.
- *Empty containers.* Lash together her empty gas cans, water jugs, ammo cans, boxes, or other items that will trap or hold air. Use them as water wings. Use this type of flotation device only in a slow-moving river or stream.
- *Plastic bags and ponchos.* Fill two or more plastic bags with air and secure them together at the opening. Use your poncho and roll green vegetation tightly inside it so that you have a roll at least 20 centimeters in diameter. Tie the ends of the roll securely. You can wear it

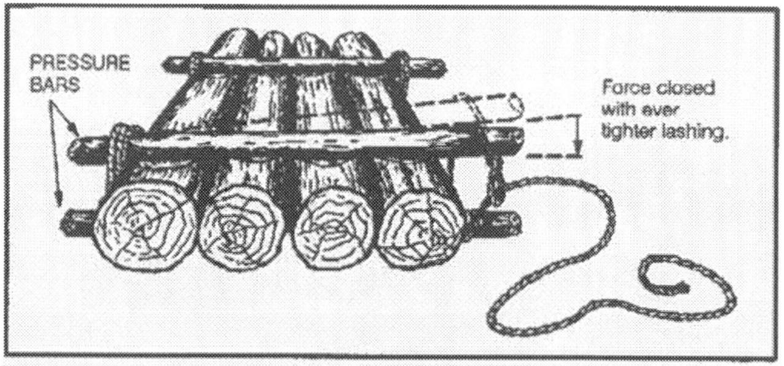

Figure 17-7. Use of pressure bars.

around your waist or across one shoulder and under the opposite arm.

- *Logs.* Use a stranded drift log if one is available, or find a log near the water to use as a float. Be sure to test the log before starting to cross. Some tree logs, palm for example, will sink even when the wood is dead. Another method is to tie two logs about 60 centimeters apart. Sit between the logs with your back against one and your legs over the other (Figure 17-8).
- *Cattails.* Gather stalks of cattails and tie them in a bundle 25 centimeters or more in diameter. The many air cells in each stalk cause a stalk to float until it rots. Test the cattail bundle to be sure it will support your weight before trying to cross a body of water.

There are many other flotation devices that you can devise by using some imagination. Just make sure to test the device before trying to use it.

OTHER WATER OBSTACLES

Other water obstacles that you may face are bogs, quagmire, muskeg, or quicksand. Do not try to walk across these. Trying to lift your feet while standing upright will make you sink deeper. Try to bypass these obstacles. If you are unable to bypass them, you may be able to bridge them using logs, branches, or foliage.

A way to cross a bog is to lie face down, with your arms and legs spread. Use a flotation device or form pockets of air in your clothing. Swim or pull your way across moving slowly and trying to keep your body horizontal.

Figure 17-8. Log flotation.

In swamps, the areas that have vegetation are usually firm enough to support your weight. However, vegetation will usually not be present in open mud or water areas. If you are an average swimmer, however, you should have no problem swimming, crawling, or pulling your way through miles or bog or swamp.

Quicksand is a mixture of sand and water that forms a shifting mass. It yields easily to pressure and sucks down and engulfs objects resting on its surface. It varies in depth and is usually localized. Quicksand commonly occurs on flat shores, in silt-choked rivers with shifting watercourses, and near the mouths of large rivers. If you are uncertain whether a sandy area is quicksand, toss a small stone on it. The stone will sink in quicksand. Although quicksand has more suction than mud or muck, you can cross it just as you would cross a bog. Lie face down, spread your arms and legs, and move slowly across.

VEGETATION OBSTACLES

Some water areas you must cross may have underwater and floating plants that will make swimming difficult. However, you can swim through relatively dense vegetation if you remain calm and do not thrash about. Stay as near the surface as possible and use the breaststroke with shallow leg and arm motion. Remove the plants around you as you would clothing. When you get tired, float or swim on your back until you have rested enough to continue with the breaststroke.

The mangrove swamp is another type of obstacle that occurs along tropical coastlines. Mangrove trees or shrubs throw out many prop roots that form dense masses. To get through a mangrove swamp, wait for low tide. If you are on the inland side, look for a narrow grove of trees and work your way seaward through these. You can also try to find the bed of a waterway or creek through the trees and follow it to the sea. If you are on the seaward side, work inland along streams or channels. Be on the lookout for crocodiles that you find along channels and in shallow water. If there are any near you, leave the water and scramble over the mangrove roots. While crossing a mangrove swamp, it is possible to gather food from tidal pools or tree roots.

To cross a large swamp area, construct some type of raft.

CHAPTER 18

FIELD-EXPEDIENT DIRECTION FINDING

In a survival situation, you will be extremely fortunate if you happen to have a map and compass. If you do have these two pieces of equipment, you will most likely be able to move toward help. If you are not proficient in using a map and compass, you must take the steps to gain this skill.

There are several methods by which you can determine direction by using the sun and the stars. These methods, however, will give you only a general direction. You can come up with a more nearly true direction if you know the terrain of the territory or country.

You must learn all you can about the terrain of the country or territory to which you or your unit may be sent, especially any prominent features or landmarks. This knowledge of the terrain together with using the methods explained below will let you come up with fairly true directions to help you navigate.

USING THE SUN AND SHADOWS

The earth's relationship to the sun can help you to determine direction on earth. The sun always rises in the east and sets in the west, but not exactly due east or due west. There is also some seasonal variation. In the northern hemisphere, the sun will be due south when at its highest point in the sky, or when an object casts no appreciable shadow. In the southern hemisphere, this same noonday sun will mark due north. In the northern hemisphere, shadows will move clockwise. Shadows will move counterclockwise in the southern hemisphere. With practice, you can use shadows to determine both direction and time of day. The shadow methods used for direction finding are the shadow-tip and watch methods.

Shadow-Tip Methods

In the first shadow-tip method, find a straight stick 1 meter long, and a level spot free of brush on which the stick will cast a definite shadow. This method is simple and accurate and consists of four steps:

Step 1. Place the stick or branch into the ground at a level spot where it will cast a distinctive shadow. Mark the shadow's tip with a stone, twig, or other means. This first shadow mark is always west —everywhere on earth.

Step 2. Wait 10 to 15 minutes until the shadow tip moves a few centimeters. Mark the shadow tip's new position in the same way as the first.

Step 3. Draw a straight line through the two marks to obtain an approximate east-west line.

Step 4. Stand with the first mark (west) to your left and the second mark to your right-you are now facing north. This fact is true everywhere on earth.

An alternate method is more accurate but requires more time. Set up your shadow stick and mark the first shadow m the morning. Use a piece of string to draw a clean arc through this mark and around the stick. At midday, the shadow will shrink and disappear. In the afternoon, it will lengthen again and at the point where it touches the arc, make a second mark. Draw a line through the two marks to get an accurate east-west line (see Figure 18-1).

Figure 18-1. Shadow-tip method.

The Watch Method

You can also determine direction using a common or analog watch—one that has hands. The direction will be accurate if you are using true local time, without any changes for daylight savings time. Remember, the further you are from the equator, the more accurate this method will be. If you only have a digital watch, you can overcome this obstacle. Quickly draw a watch on a circle of paper with the correct time on it and use it to determine your direction at that time.

In the northern hemisphere, hold the watch horizontal and point the hour hand at the sun.Bisect the angle between the hour hand and the 12 o'clock mark to get the north-south line (Figure 18-2). If there is any doubt as to which end of the line is north, remember that the sun rises in the east, sets in the west, and is due south at noon. The sun is in the east before noon and in the west after noon.

Note: If your watch is set on daylight savings time, use the midway point between the hour hand and 1 o clock to determine the north-south line.

In the southern hemisphere, point the watch's 12 o'clock mark toward the sun and a midpoint halfway between 12 and the hour hand will give you the north-south line (Figure 18-2).

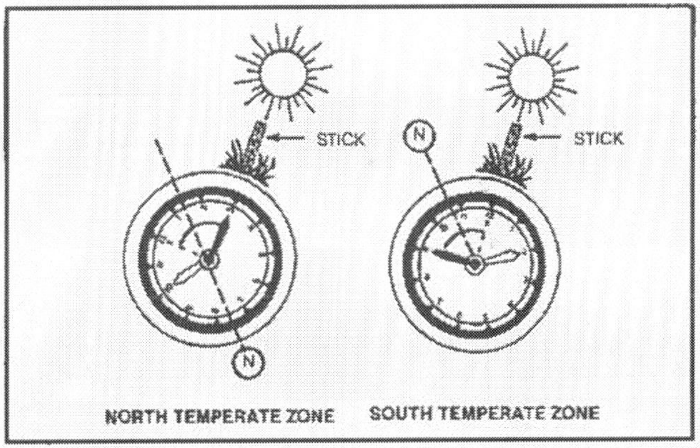

Figure 18-2. Watch method.

USING THE MOON

Because the moon has no light of its own, we can only see it when it reflects the sun's light. As it orbits the earth on its 28-day circuit, the shape of the reflected light varies according to its position. We say there is a new moon or no moon when it is on the opposite side of the earth from the sun. Then, as it moves away from the earth's shadow, it begins to reflect light from its right side and waxes to become a full moon before waning, or losing shape, to appear as a sliver on the left side. You can use this information to identify direction.

If the moon rises before the sun has set, the illuminated side will be the west. If the moon rises after midnight, the illuminated side will be the east. This obvious discovery provides us with a rough east-west reference during the night.

USING THE STARS

Your location in the Northern or Southern Hemisphere determines which constellation you use to determine your north or south direction.

The Northern Sky

The main constellations to learn are the Ursa Major, also known as the Big Dipper or the Plow, and Cassiopeia (Figure 18-3). Neither of these constellations ever sets. They are always visible on a clear night. Use them to locate Polaris, also known as the polestar or the North Star. The North Star forms part of the Little Dipper handle and can be confused with the Big Dipper. Prevent confusion by using both the Big Dipper and Cassiopeia together. The Big Dipper and Cassiopeia are always directly opposite each, other and rotate counterclockwise around Polaris, with Polaris in the center. The Big Dipper is a seven star constellation in the shape of a dipper. The two stars forming the outer lip of this dipper are the "pointer stars" because they point to the North Star. Mentally draw a fine from the outer bottom star to the outer top star of the Big Dipper's bucket. Extend this line about five times the distance between the pointer stars. You will find the North Star along this line.

Cassiopeia has five stars that form a shape like a "W" on its side. The North Star is straight out from Cassiopeia's center star.

After locating the North Star, locate the North Pole or true north by drawing an imaginary line directly to the earth.

The Southern Sky

Because there is no star bright enough to be easily recognized near the south celestial pole, a constellation known as the Southern Cross is used as a signpost to the South (Figure 18-4). The Southern Cross or Crux has five stars. Its four brightest stars form a cross that tilts to one side. The two stars that make up the cross's long axis are the pointer stars. To determine south, imagine a distance five times the distance between These stars and the point where this imaginary line ends is in the general direction of south. Look down to the horizon from this imaginary point and select a landmark to steer by. In a static survival situation, you can fix this location in daylight if you drive stakes in the ground at night to point the way.

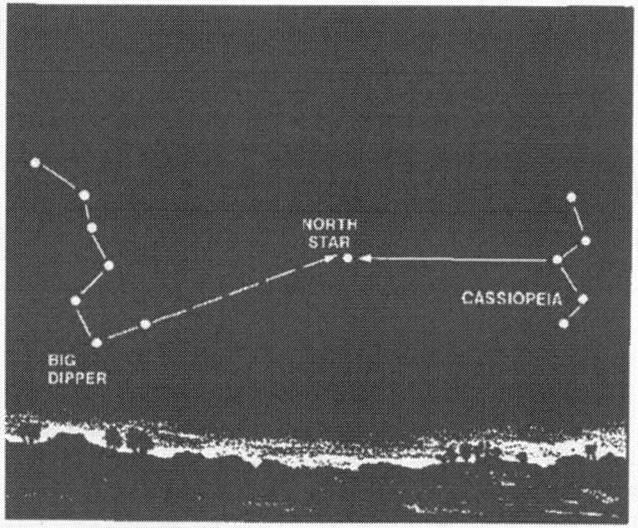

Figure 18-3. The Big Dipper and Cassiopeia.

MAKING IMPROVISED COMPASSES

You can construct improvised compasses using a piece of ferrous metal that can be needle shaped or a flat double-edged razor blade and a piece of nonmetallic string or long hair from which to suspend it. You can magnetize or polarize the metal by slowly stroking it in one direction on a piece of silk or carefully through your hair using deliberate strokes. You can also polarize metal Tby stroking it repeatedly at one end with a magnet. Always rub in one direction only. Ifyou have a battery and some electric wire, you can polarize the metal electrically. The wire should be insulated. If not insulated, wrap the metal object in a single, thin strip of paper to prevent contact. The battery must oe a minimum of 2 volts. Form a coil with the electric wire and touch its ends to the battery's terminals. Repeatedly insert one end of the metal object in and out of the coil. The needle will become an electromagnet. When suspended from a piece of nonmetallic string, or floated on a small piece of wood in water, it will align itself with a north-south line.

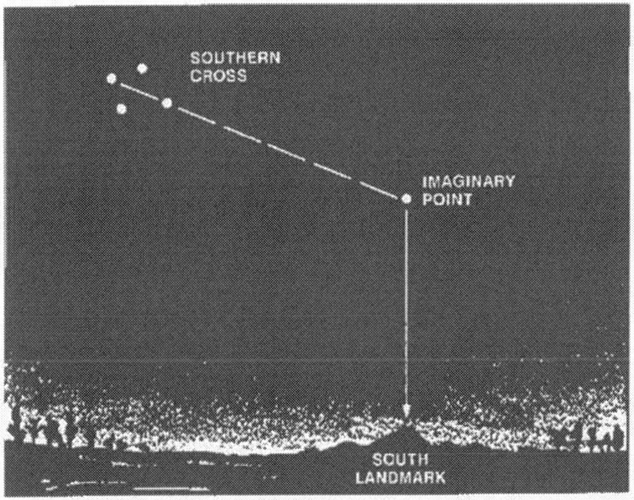

Figure 18-4. Southern Cross.

You can construct a more elaborate improvised compass using a sewing needle or thin metallic object, a nonmetallic container (for example, a plastic dip container), its lid with the center cut out and waterproofed, and the silver tip from a pen. To construct this compass, take an ordinary sewing needle and break in half. One half will form your direction pointer and the other will act as the pivot point. Push the portion used as the pivot point through the bottom center of your container; this portion should be flush on the bottom and not interfere with the lid. Attach the center of the other portion (the pointer) of the needle on the pen's silver tip using glue, tree sap, or melted plastic. Magnetize one end of the pointer ana rest it on the pivot point.

OTHER MEANS OF DETERMINING DIRECTION

The old saying about using moss on a tree to indicate north is not accurate because moss grows completely around some trees. Actually, growth is more lush on the side or the tree facing the south in the Northern Hemisphere and vice versa in the Southern Hemisphere. If there are several felled trees around for comparison, look at the stumps. Growth is more vigorous on the side toward the equator and the tree growth rings will be more widely spaced. On the other hand, the tree growth rings will be closer together on the side toward the poles.

Wind direction may be helpful in some instances where there are prevailing directions and you know what they are.

Recognizing the differences between vegetation and moisture patterns on north- and south-facing slopes can aid in determining direction. In the northern hemisphere, north-facing slopes receive less sun than south-facing slopes and are therefore cooler and damper. In the summer, north-facing slopes retain patches of snow. In the winter, the trees and open areas on south-facing slopes are the first to lose their snow, and ground snowpack is shallower.

CHAPTER 19

SIGNALING TECHNIQUES

One of your first concerns when you find yourself in a survival situation is to communicate with your friends or allies. Generally, communication is the giving and receiving of information. As a survivor, you must get your rescuer's attention first, and second, send a message your rescuer understands. Some attention-getters are man-made geometric patterns such as straight lines, circles, triangles, or X's displayed in uninhabited areas; a large fire or flash of light; a large, bright object moving slowly; or contrast, whether from color or shadows. The type of signal used will depend on your environment and the enemy situation.

APPLICATION

If in a noncombat situation, you need to find the largest available clear and flat area on the highest possible terrain. Use as obvious a signal as you can create. On the other hand, you will have to be more discreet in combat situations. You do not want to signal and attract the enemy. Pick an area that is visible from the air, but ensure there are hiding places nearby. Try to have a hill or other object between the signal site and the enemy to mask your signal from the enemy. Perform a thorough reconnaissance of the area to ensure there are no enemy forces nearby.

Whatever signaling technique or device you plan to use, know how to use it and be ready to put it into operation on short notice. If possible, avoid using signals or signaling techniques that can physically endanger you. Keep m mind that signals to your friends may alert the enemy of your presence and location. Before signaling, carefully weigh your rescue chances by friends against the danger of capture by the enemy.

A radio is probably the surest and quickest way to let others know where you are and to let you receive their messages. Become familiar with the radios in your unit. Learn how to operate them and how to send and receive messages.

You will find descriptions of other signaling techniques, devices, and articles you can use. Learn how to use them. Think of ways in which you can adapt or change them for different environments. Practice using these signaling techniques, devices, and articles before you need them. Planned, prearranged signaling techniques may improve your chance of rescue.

MEANS FOR SIGNALING

There are two main ways to get attention or to communicate— visual and audio. The means you use will depend on your situation and the material you have available. Whatever the means, always have visual and audio signals ready for use.

Visual Signals

These signals are materials or equipment you use to make your presence known to rescuers.

Ere

During darkness, fire is the most effective visual means for signaling. Build three fires in a triangle (the international distress signal) or in a straight line with about 25 meters between the fires. Build them as soon as time and the situation permit and protect them until you need them. If you are alone, maintaining three fires may be difficult. If so, maintain one signal fire.

When constructing signal fires, consider your geographic location. If in a jungle, find a natural clearing or the edge of a stream where you can build fires that the jungle foliage will not hide. You may even have to clear an area. If in a snow-covered area, you may have to clear the ground of snow or make a platform on which to build the fire so that melting snow will not extinguish it.

A burning tree (tree torch) is another way to attract attention (Figure 19-1). You can set pitch-bearing trees afire, even when green. You can get other types of trees to burn by placing dry wood in the lower branches and igniting it so that the flames flare up and ignite the foliage. Before the primary tree is consumed, cut and add more small green trees to the fire to produce more smoke. Always select an isolated tree so that you do not start a forest fire and endanger yourself.

Figure 19-1. Tree torch.

Smoke

During daylight, build a smoke generator and use smoke to gain attention (Figure 19-2). The international distress signal is three columns of smoke. Try to create a color of smoke that contrasts with the background; dark smoke against a light background and vice versa. If you practically smother a large fire with green leaves, moss, or a little water, the fire will produce white smoke. If you add rubber or oil-soaked rags to a fire, you will get black smoke.

Figure 19-2. Smoke generator—ground.

In a desert environment, smoke hangs close to the ground, but a pilot can spot it in open desert terrain.

Smoke signals are effective only on comparatively calm, clear days. High winds, ram, or snow disperse smoke, lessening its chances of being seen.

Smoke Grenades

If you have smoke grenades with you, use them in the same pattern as described for fires. Keep them dry so that they will work when you need them. Take care not to ignite the vegetation in the area when you use them.

Pen Flares

These flares are part of an aviator's survival vest. The device consists of a pen-shaped gun with a flare attached by a nylon cord. When fired, the pen flare sounds like a pistol shot and fires the flare about 150 meters high. It is about 3 centimeters in diameter.

To have the pen flare ready for immediate use, take it out of its wrapper, attach the flare, leave the gun uncocked, and wear it on a cord or chain around your neck. Be ready to fire it in front of search aircraft and be ready with a secondary signal. Also, be ready to take cover in case the pilot mistakes the flare for enemy fire.

Tracer Ammunition

You may use rifle or pistol tracer ammunition to signal search aircraft. Do not fire the ammunition in front of the aircraft. As with pen flares, be ready to take cover if the pilot mistakes your tracers for enemy fire.

Star Clusters

Red is the international distress color; therefore, use a red star cluster whenever possible. Any color, however, will let your rescuers know where you are. Star clusters reach a height of 200 to 215 meters, burn an average of 6 to 10 seconds, and descend at a rate of 14 meters per second.

Star Parachute Flares

These flares reach a height of 200 to 215 meters and descend at a rate of 2.1 meters per second. The M126 (red) burns about 50 seconds and the M127 (white) about 25 seconds. At night you can see these flares at 48 to 56 kilometers.

Mirrors or Shiny Objects

On a sunny day, a mirror is your best signaling device. If you don't have a mirror, polish your canteen cup, your celt buckle, or a similar object that will reflect tric sun's rays. Direct the flashes in one area so that they are secure from enemy observation. Practice using a mirror or shiny object for signaling now; do not wait until you need it.Tf you have an MK-3 signal mirror, follow the instructions on its back (Figure 19-3).

Wear the signal mirror on a cord or chain around your neck so that it is ready for immediate use. However, be sure the glass side is against your body so that it will not flash; the enemy can see the flash.

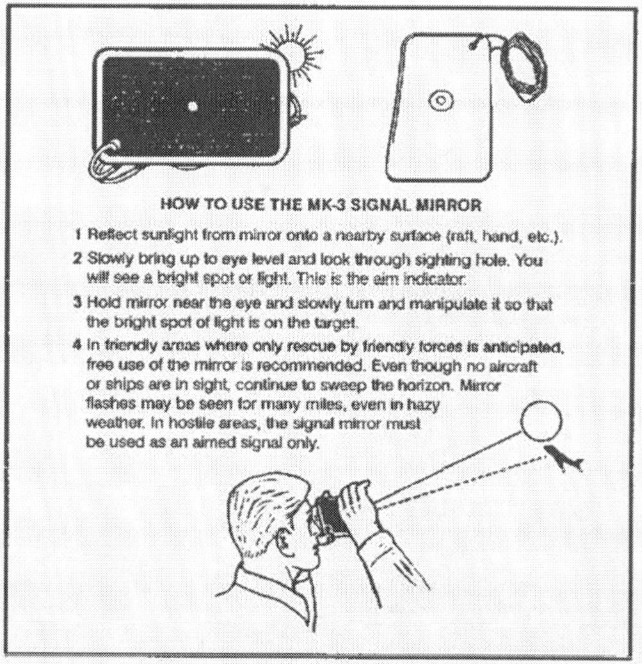

Figure 19-3. Signal mirror.

> **CAUTION**
> Do not flash a signal mirror rapidly because a pilot may mistake the flashes for enemy fire. Do not direct the beam in the aircraft's cockpit for more than a few seconds as it may blind the pilot.

Haze, ground fog, and mirages may make it hard for a pilot to spot signals from a flashing object. So, if possible, get to the highest point in your area when signaling. If you can't determine the aircraft's location, flash your signal in the direction of the aircraft noise.

Note *Pilots have reported seeing mirror flashes up to 160 kilometers away under ideal conditions.*

Figures 19-4 and 19-5 show methods of aiming a signal mirror for signaling.

Flashlight or Strobe Light

At night you can use a flashlight or a strobe light to send an SOS to an aircraft. When using a strobe light, take care to prevent the pilot from mistaking it for incoming ground fire. The strobe light flashes 60 times per minute. Some strobe lights have infrared covers and lenses. Blue flash collimators are also available for strobe lights.

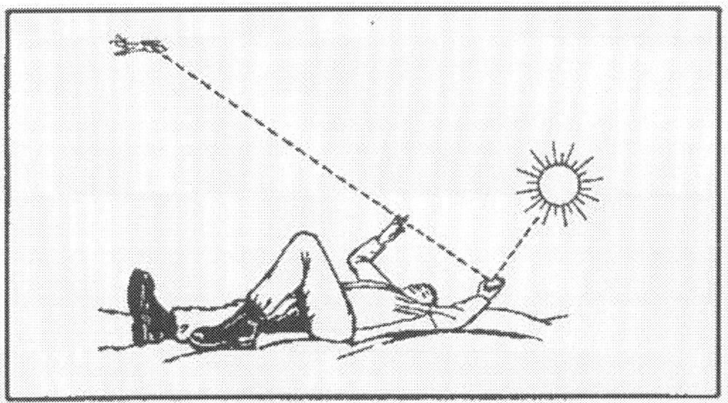

Figure 19-4. Aiming an improvised signal mirror.

VS-17 Panel

During daylight you can use a VS-17 panel to signal. Place the orange side up as it is easier to see from the air than the violet side. Flashing the panel will make it easier for the aircrew to spot. You can use any bright orange or violet cloth as a substitute for the VS-17.

Clothing

Spreading clothing on the ground or in the top of a tree is another way to signal. Select articles whose color will contrast with the natural surroundings. Arrange them in a large geometric pattern to make them more likely to attract attention.

Natural Material

If you lack other means, you can use natural materials to form a symbol or message that can be seen from the air. Build mounds that cast shadows; you can use brush, foliage of any type, rocks, or snow blocks.

In snow-covered areas, tramp the snow to form letters or symbols and fill the depression with contrasting material (twigs or branches). In sand, use poulders, vegetation, or seaweed to form a symbol or message. In brush-covered areas, cut out patterns in the vegetation or sear the ground. In tundra, dig trenches or turn the sod upside down.

In any terrain, use contrasting materials that will make the symbols visible to the aircrews.

Figure 19-5. Aiming an improvised signal mirror—stationary object.

Sea Dye Markers

All Army aircraft involved in operations near or over water will normally carry a water survival kit that contains sea dye markers. If you are in a water survival situation, use sea dye markers during daylight to indicate your location. These spots of dye stay conspicuous tor about 3 hours, except in very rough seas. Use them only if you are in a friendly area. Keep the markers wrapped until you are ready to use them. Use them only when you hear or sight an aircraft. Sea dye markers are also very effective on snow-covered ground; use them to write distress code letters.

Audio Signals

Radios, whistles, and gunshots are some of the methods you can use to signal your presence to rescuers.

Radio Equipment

The AN/PRC-90 survival radio is a part of the Army aviator's survival vest. The AN/PRC-112 will eventually replace the AN/PRC-90. Both radios can transmit either tone or voice. Any other type of Army radio can do the same. The ranges of the different radios vary depending on the altitude of the receiving aircraft, terrain, vegetation density, weather, battery strength, type of radio, and interference. To obtain maximum performance from radios, use the following procedures:

- Try to transmit only in clear, unobstructed terrain. Since radios are line-of-sight communications devices, any terrain between the radio and the receiver will block the signal.
- Keep the antenna at right angles to the rescuing aircraft. There is no signal from the tip of the antenna.
- If the radio has tone capability, place it upright on a flat, elevated surface so that you can perform other survival tasks.
- Never let the antenna touch your clothing, body, foliage, or the ground. Such contact greatly reduces the range of the signal.
- Conserve battery power. Turn the radio off when you are not using it. Do not transmit or receive constantly. In hostile territory, keep transmissions short to avoid enemy radio direction finding.
- In cold weather, keep the battery inside your clothing when not using the radio. Cold quickly drains the battery's power. Do not expose the battery to extreme heat such as desert sun. High heat may cause the battery to explode. Try to keep the radio and battery as dry as possible, as water may destroy the circuitry.

Whistles

Whistles provide an excellent way for close up signaling. In some documented cases, they have been heard up to 1.6 kilometers away. Manufactured whistles have more range than a human whistle.

Gunshots

In some situations you can use firearms for signaling. Three shots fired at distinct intervals usually indicate a distress signal. Do not use this technique in enemy territory. The enemy will surely come to investigate shots.

CODES AND SIGNALS

Now that you know how to let people know where you are, you need to know how to give them more information. It is easier to form one symbol than to spell out an entire message. Therefore, learn the codes and symbols that all aircraft pilots understand.

SOS

You can use lights or flags to send an SOS—three dots, three dashes, three dots. The SOS is the internationally recognized distress signal in radio Morse code. A dot is a short, sharp pulse; a dash is a longer pulse. Keep repeating the signal. When using flags, hold flags on the left side for dashes ana on the right side for dots.

Ground-to-Air Emergency Code

This code (Figure 19-6) is actually five definite, meaningful symbols. Make these symbols a minimum of 1 meter wide and 6 meters long. If you make them larger, keep the same 1:6 ratio. Ensure the signal contrasts greatly with the ground it is on. Place it in an open area easily spotted from the air.

Body Signals

When an aircraft is close enough for the pilot to see you clearly, use body movements or positions (Figure 19-7) to convey a message.

Panel Signals

If you have a life raft cover or sail, or a suitable substitute, use the symbols shown in Figure 19-8 to convey a message.

Number	Message	Code symbol
1	Require assistance.	V
2	Require medical assistance.	X
3	No or negative.	N
4	Yes or affirmative.	Y
5	Proceed in this direction.	↑

Figure 19-6. Ground-to-air emergency code (pattern signals).

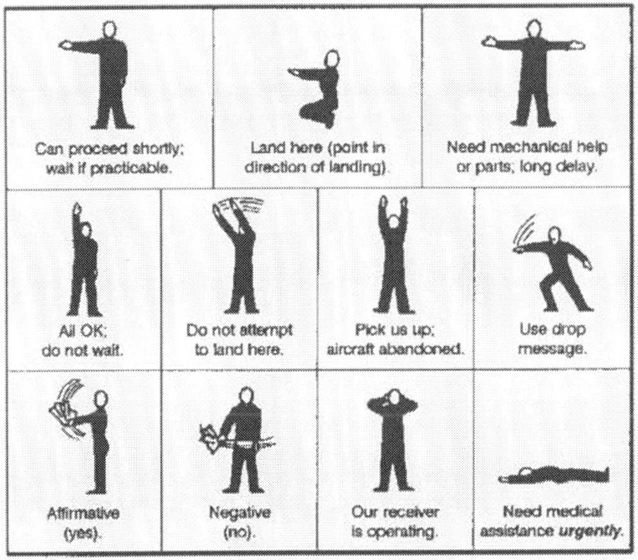

Figure 19-7. Body signals.

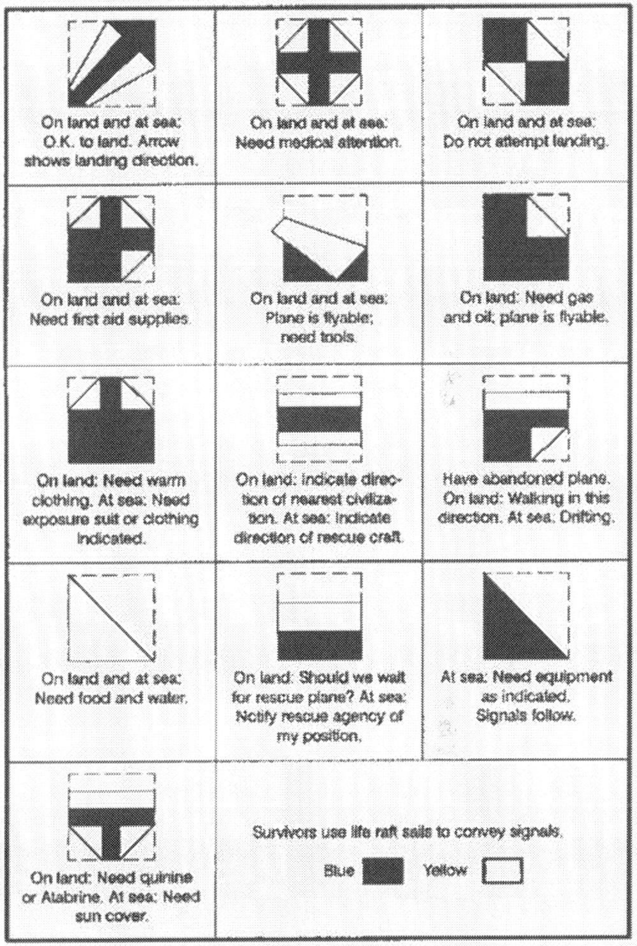

Figure 19-8. Panel signals.

Aircraft Acknowledgments

Once the pilot of a fixed-wing aircraft has sighted you, he will normally indicate he has seen you by flying low, moving the plane, and flashing lights as shown in Figure 19-9. Be ready to relay other messages to the pilot once he acknowledges that he received and understood your first message. Use a radio, if possible, to relay further messages. If no radio is available, use the codes covered in the previous paragraphs.

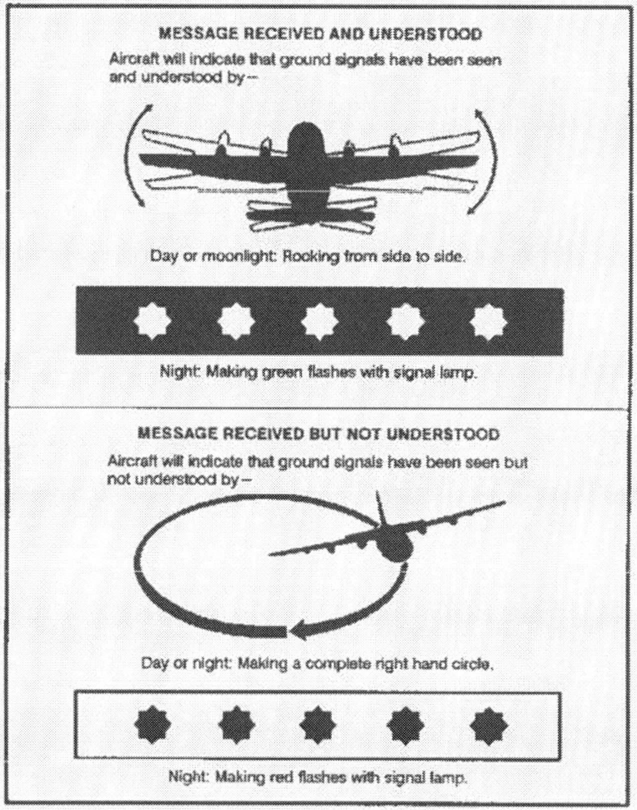

Figure 19-9. Aircraft acknowledgments.

AIRCRAFT VECTORING PROCEDURES

If you can contact a friendly aircraft with a radio, guide the pilot to your location. Use the following general format to guide the pilot:

- Mayday, Mayday.
- Call sign (if any).
- Name.
- Location.
- Number of survivors.
- Available landing sites.
- Any remarks such as medical aid or other specific types of help needed immediately.

Simply because you have made contact with rescuers does not mean you are safe. Follow instructions and continue to use sound survival and evasion techniques until you are actually rescued.

CHAPTER 20

SURVIVAL MOVEMENT IN HOSTILE AREAS

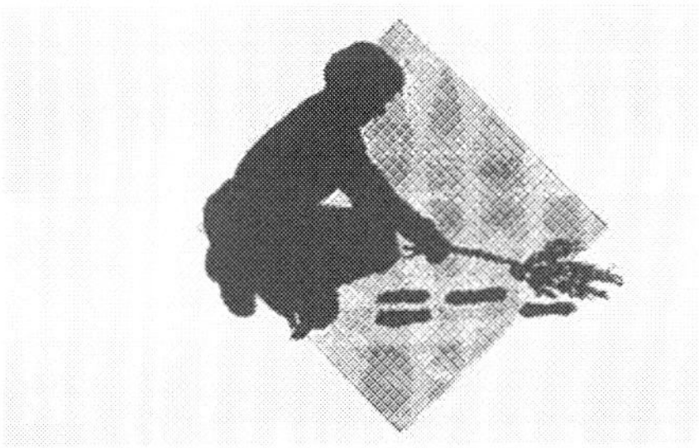

The "rescue at any cost" philosophy of previous conflicts is not likely to be possible in future conflicts. Our potential adversaries have made great progress in air defense measures and radio direction finding (RDF) techniques. We must assume that U.S. military forces trapped behind enemy lines in future conflicts may not experience quick recovery by friendly elements. Soldiers may have to move for extended times and distances to places less threatening to the recovery forces. The soldier will not likely know the type of recovery to expect. Each situation and the available resources determine the type of recovery possible. Since no one can be absolutely sure until the recovery effort begins, soldiers facing a potential cutoff from friendly forces should be familiar with all the possible types of recovery, their related problems, and their responsibilities to the recovery effort. Preparation and training can improve the chances of success.

PHASES OF PLANNING

Preparation is a requirement for all missions. When planningjvou must consider how to avoid capture and return to your unit. Contingency plans must be prepared m conjunction with unit standing operating procedures (SOPs). Courses of action you or your unit will take must also be considered.

Contingency Plan of Action (CPA)

Intelligence sections can help prepare personnel for contingency actions through information supplied m area studies, SERF (survival, evasion, resistance, and escape) contingency guides, threat briefings, current intelligence reports, and current contact and authentication procedures. Pre-mission preparation includes the completion of a CPA. The study and research needed to develop the CPA will make you aware of the current situation in your mission area. Your CPA will let recovery forces know your probable actions should you have to move to avoid capture.

Start preparing even before pre-mission planning. Many parts of the CPA are SOP for your unit. Include the CPA in your training. Planning starts in your daily training.

The CPA is jour entire plan for your return to friendly control. It consists of five paragraphs written in the operation order format. You can take most or paragraph 1, Situation, with you on the mission. Appendix H contains the CPA format. It also mdicates what portion of the CPA you can take with you.

A comprehensive CPA is a valuable asset to the soldier trapped behina enemy lines who must try to avoid capture. To complete paragraph 1, know your unit's assigned area or concentrate on potential mission areas or the world. Many open or closed sources contain the information you need to complete a CPA. Open sources may include newspapers, magazines, country or area handbooks, area studies, television, radio, persons familiar with the area, and libraries. Closed sources may include area studies, area assessments, SERE contingency guides, various classified field manuals, and intelligence reports.

Prepare your CPA in three phases. During your normal training, prepare paragraph 1, Situation. Prepare paragraphs 2,3,4, and 5 durmg your pre-mission planning. After deployment into an area, continually update your CPA based on mission changes and intelligence updates.

The CPA is a guide. You may add or delete certain portions based on the mission. The CPA may be a recovery force's only means of determining your location and intentions after you start to move. It is an essential tool for your survival and return to friendly control.

Standing Operating Procedures

Unit SOPs are valuable tools your unit has that will help your planning. When faced with a dangerous situation requiring immediate action, it is not the time to discuss options; it is the time to act. Many of the techniques used during small unit movement can be carried over to fit requirements for moving and returning to friendly control. Items from the SOP should include, but are not limited to—

- Movement team size (three to four persons per team).
- Team communications (technical and nontechnical).
- Essential equipment.
- Actions at danger areas.
- Signaling techniques.
- Immediate action drills.
- Linkup procedures.
- Helicopter recovery devices and procedures.
- Security procedures during movement and at hide sites.
- Rally points.

Rehearsals work effectively for reinforcing these SOP skills and also provide opportunities for evaluation and improvement.

Notification to Move and Avoid Capture

An isolated unit has several general courses of action it can take to avoid the capture of the group or individuals. These courses of action are not courses the commander can choose instead of his original mission. He cannot arbitrarily abandon the assigned mission. Rather, he may adopt these courses of action after completing his mission when Iris unit cannot complete its assigned mission (because of combat power losses) or when he receives orders to extract his unit from its current position. If such actions are not possible, the commander may decide to have the unit try to move to avoid capture and return to friendly control. In either case, as long as there is communication with higher headquarters, that headquarters will make the decision.

If the unit commander loses contact with higher headquarters, he must make the decision to move or wait. Heoases his decision on many factors, including the mission, rations and ammunition on hand, casualties, the chance of relief by friendly forces, and the tactical situation. The commander of an isolated unit faces other questions. What course of action will inflict maximum damage on the enemy? What course of action will assist in completing the higher headquarters' overall mission?

Movement teams conduct the execution portion of the plan when notified by higher headquarters or, if there is no contact with higher headquarters, when the highest ranking survivor decides that the situation requires the unit to try to escape capture or destruction. Movement team leaders receive their notification through prebriefed signals. Once the signal to try to avoid capture is given, it must be passed rapidly to all personnel. Notify higher headquarters, if possible. If unable to communicate with higher headquarters, leaders must recognize that organized resistance has ended, and that organizational control has ceased. Command and control is now at the movement team or individual level and is returned to higher organizational control only after reaching friendly lines.

EXECUTION

Upon notification to avoid capture, all movement team members will try to link up at the initial movement point. This point is where team members rally and actually begin their movement. Tentatively select the initial movement point during your planning phase through a map recon. Once on the ground, the team verifies this location or selects a better one. All team members must know its location. The initial movement point should be easy to locate and occupy for a minimum amount of time.

Once the team has rallied at the initial movement point, it must—

- Give first aid.
- inventory its equipment (decide what to abandon, destroy, or take along).
- Apply camouflage.
- Make sure everyone knows the tentative hide locations.
- Ensure everyone knows the primary and alternate routes and rally points en route to the hide locations.
- Always maintain security.
- Split the team into smaller elements. The ideal element should have two to three members; however, it could include more depending on team equipment and experience.

The movement portion of returning to friendly control is the most dangerous as you are now most vulnerable. It is usually better to move at night because of the concealment darkness offers. Exceptions to such movement would be when moving through hazardous terrain or dense vegetation (for example, jungle or mountainous terrain). When moving, avoid the following even if it takes more time and energy to bypass:

- Obstacles and barriers.
- Roads and trails.
- Inhabited areas.
- Waterways and bridges.
- Natural lines of drift.
- Man-made structures.
- All civilian and military personnel.

Movement in enemy-held territory is a very slow and deliberate process. The slower you move and the more careful you are, the better. Your best security will be using your senses. Use your eyes and ears to detect people before they detect you. Make frequent listening halts. In daylight, olbserve a section of your route before you move along it. The distance you travel before you hide will depend on the enemy situation, your health, the terram, the availability of cover and concealment for hiding, and the amount of darkness left.

Once you have moved into the area in which you want to hide (hide area), select a hide site. Keep the following formula in mind when selecting a hide site: BLISS.

B - Blends in with the surroundings.

L - Low in silhouette.

I - Irregular in shape.

S - Small in size.

S - Secluded.

Avoid the use of existing buildings or shelters. Usually, your best option will be to crawl into the thickest vegetation you can find. Construct any type of shelter within the hiac area only in cold weather and desert environments. If you build a shelter, follow the BLISS formula.

Hide Site Activities

After you have located your hide site, do not move straight into it. Use a button hook or other deceptive technique to move to a position outside of the hide site. Conduct a listening halt before moving individually into the hide site. Be careful not to disturb or cut any vegetation. Once you have occupied the hide site, limit your activities to maintaining security, resting, camouflaging, and planning your next moves.

Maintain your security through visual scanning and listening. Upon detection of the enemy, the security personnel alert all personnel, even if the team's plan is to stay hidden and not move upon sighting the enemy. Take this action so that everyone is aware of the danger and ready to react.

If any team member leaves the team, give him a five-point contingency plan. Take such steps especially when a recon team or a work party is out of the hole-up or hide site.

It is extremely important to stay healthy and alert when trying to avoid capture. Take every opportunity to rest, but do not sacrifice security. Rotate security so that all members of your movement team can rest. Treat all injuries, no matter how minor. Loss of your health will mean loss of your ability to continue to avoid capture.

Camouflage is an important aspect of both moving and securing a hide site. Always use a buddy system to ensure that camouflage is complete. Ensure that team members blend with the hide site. Use natural or man-made materials. If you add any additional camouflage material to the hide site, do not cut vegetation in the immediate area.

Plan your next actions while at the hide site. Start your planning process immediately upon occupying the hide site. Inform all team members of their current location and designate an alternate hide site location. Once this is done, start plannmg for the team's next movement.

Planning the team's movement begins with a map recon. Choose the next hide area first. Then choose a primary and an alternate route to the hide area. In choosing the routes, do not use straight lines. Use one or two radical changes in direction. Pick the routes that offer the best cover and concealment, the fewest obstacles, and the least likelihood of contact with humans. There should be locations along the route where the team can get water. To aid team navigation, use azimuths, distances, checkpoints or steering marks, and corridors. Plan rally points and rendezvous points at intervals along the route.

Other planning considerations may fall under what the team already has in the team SOP. Examples are immediate action drills, actions on sighting the enemy, and hand-and-arm signals.

Once planning is complete, ensure everyone knows and memorizes the entire plan. The team members should know the distances and azimuths for the entire route to the next hide area. They should study the map and know the various terrain they will be moving across so that they can move without using the map.

Do not occupy a hide site for more than 24 hours. In most situations, hide during the day and move at night. Limit your actions in the hide site to those discussed above. Once in the hide site, restrict all movement to less than 45 centimeters above the ground. Do not build fires or prepare food. Smoke and food odors will reveal your location. Before leaving the hide site, sterilize it to prevent tracking.

Hole-Up Areas

After moving and hiding for several days, usually three or fomyyou or the movement team will have to move into a hole-up area. This is an area where you can rest, recuperate, and get and prepare food. Choose an area near a water source. You then have a place to get water, to place fishing devices, and to trap game. Since waterways are a line of communication, locate your hide site well away from the water.

The hole-up area should offer plenty of cover and concealment for movement m and around the area. Always maintain security while in the hole-up area. Always man the hole-up area. Actions in the hole-up area are the same as in hide site, except that you can move away from the hole-up area to get and prepare food. Actions in the hole-up area include—

- Selecting and occupying the next hide site (remember you are still in a dangerous situation; this is not a friendly area).
- Reconnoitering the area for resources and potential concealed movement routes to the alternate hide site.
- Gathering food (nuts, berries, vegetables). When moving around the area for food, maintain security and avoid leaving tracks or other signs. When setting traps and snares, keep them well-camouflaged and in areas where people are not likely to discover them. Remember, the local population sometimes heavily travels trails near water sources.
- Getting water from sources within the hide area. Be careful not to leave tracks of signs along the banks of water sources when getting water. Moving on hard rocks or logs along the banks to get water will reduce the signs you leave.
- Setting clandestine fishing devices, such as stakeouts, below the surface of the water to avoid detection.
- Locating a fire site well away from the hide site. Use this site to prepare food or boil water. Camouflage and sterilize the fire site

after each use. Be careful that smoke and light from the fire does not compromise the hole-up area.

While in the hole-up area, security is still your primary concern. Designate team members to perform specific tasks. To limit movement around the area, you may have a two-man team perform more than one task. For example, the team getting water could also set the fishing devices. Do not occupy the hole-up area longer than 72 hours.

RETURN TO FRIENDLY CONTROL

Establishing contact with friendly lines or patrols is the most crucial part of movement and return to friendly control. All your patience, planning, and hardships will be in vain if you do not exercise caution when contacting friendly frontline forces. Friendly patrols have killed personnel operating behind enemy lines because they did not make contact properly. Most of the casualties could have been avoided if caution had been exercised and a few simple procedures followed. The normal tendency is to throw caution to the winds when in sight of friendly forces. You must overcome this tendency and understand that linkup is a very sensitive situation.

Border Crossings

If you have made your way to a friendly or neutral country, use the following procedures to cross the border and link up with friendly forces on the other side:

- Occupy a hide site on the near side of the border and send a team out to reconnoiter the potential crossing site.
- Surveil the crossing site for at least 24 hours, depending on the enemy situation.
- Make a sketch of the site, taking note of terrain, obstacles, guard routines and rotations, and any sensor devices or trip wires. Once the recon is complete, the team moves to the hide site, briefs the rest of the team, and plans to cross the border at night.
- After crossing the border, set up a hide site on the far side of the border and try to locate friendly positions. Do not reveal your presence.
- Depending on the size of your movement team, have two men surveil the potential linkup site with friendly forces until satisfied that the personnel are indeed friendly.
- Make contact with the friendly forces during daylight. Personnel chosen to make contact should be unarmed, have no equipment, and have positive identification readily available. The person who actually makes the linkup should be someone who LOOKS least like the enemy.
- During the actual contact, have only one person make the contact. The other person provides the security and observes the linkup area from a safe distance. The observer should be far enough away so that he can warn the rest of the movement team if something goes wrong.
- Wait until the party he is contacting looks in his direction so that he does not surprise the contact. He stands up from behind cover, with hands overhead and states that he is an American. After this, he follows any instructions given him. He avoids answering any tactical questions and does not give any indication that there are other team members.
- Reveal that there are other personnel with him only after verifying his identity and satisfying himself he has made contact with friendly forces.

Language problems or difficulties confirming identities may arise. The movement team should maintain security, be patient, and have a contingency plan.

Note *If you are moving to a neutral country, you are surrendering to that power and become a detained person.*

Linkup at the FEBA/FLOT

If caught between friendly and enemy forces and there is heavy fighting in the area, you may choose to hide and let the friendly lines pass over you. If overrun by friendly forces, you may try to link up from their rear during daylight hours. If overrun by enemy forces, you may move further to the enemy rear, try to move to the forward edge of the battle area (FEBA)/forward line of own troops (FLOT) durmg a lull in the fighting, or move to another area along the front.

The actual linkup will be done as for linkup during a border crossing. The only difference is that you must be more careful on the initial contact. Frontline personnel are more likely to shoot first and ask questions later, especially in areas of heavy fighting. You should be near or behind cover before trying to make contact.

Linkup With Friendly Patrols

If friendly lines are a circular perimeter or an isolated camp, for example, any direction you approach from will be considered enemy territory. You do not have the option of moving behind the lines and trying to link up. This move makes the linkup extremely dangerous. One option you have is to place the perimeter under observation and wait for a friendly patrol to move out in your direction, providing a chance for a linkup. You may also occupy a position outside of the perimeter and call out to get the attention of the friendly forces. Ideally, display anything that is white while making contact. If nothing else is available, use any article of clothing. The idea is to draw attention while staying behmd cover. Once you have drawn attention to your signal and called out, follow instructions given to you.

Be constantly on the alert for friendly patrols because these provide a means for return to friendly control. Tind a concealed position that allows you maximum visual coverage of the area. Try to memorize every terrain feature so that, if necessary, you can infiltrate to friendly positions under the cover of darkness. Remember, trying to infiltrate m darkness is extremely dangerous.

Because of the missions of combat and recon patrols and where they are operating, making contact can be dangerous. If you decide not to make contact, you can observe their route and approach friendly lines at about the same location. Such observation will enable you to avoid mines and booby traps.

Once you have spotted a patrol, remain in position and, if possible, allow the patrol to move toward you. When the patrol is 25 to 50 meters from your position, signal them and call out a greeting that is clearly and unmistakably of American origin.

If you have nothing white, an article of clothing will suffice to draw attention. If the distance is greater than 50 meters, a recon patrol may avoid contact and bypass your position. If the distance is less than 25 meters, a patrol member may react instantly by firing a fatal shot.

It is crucial, at the time of contact, that there is enough light for the patrol to identify you as an American.

Whatever linkup technique you decide to use, use extreme caution. From the perspective of the friendly patrol or friendly personnel occupying a perimeter, you are hostile until they make positive identification.

CHAPTER 21

CAMOUFLAGE

In a survival situation, especially in a hostile environment, you may find it necessary to camouflage yourself, your equipment, and your movement. It may mean the difference between survival and capture by the enemy. Camouflage and movement techniques, such as stalking, will also help you get animals or game for food using primitive weapons and skills.

PERSONAL CAMOUFLAGE

When camouflaging yourself, consider that certain shapes are particular to humans. The enemy will look for these shapes. The shape of a hat, helmet, or black boots can give you away. Even animals know and run from the shape of a human silhouette. Break up your outline by placing small amounts of vegetation from the surrounding area in your uniform, equipment, and headgear. Try to reduce any shine from skin or equipment. Blend in with the surrounding colors and simulate the texture of your surroundings.

Shape and Outline

Change the outline of weapons and equipment by tying vegetation or strips of cloth onto them. Make sure the added camouflage does not hinder the equipment's operation. When hiding, cover yourself and your equipment with leaves, grass, or other local debris. Conceal any signaling devices you have prepared, but keep them ready for use.

Color and Texture

Each area of the world and each climatic condition (arctic/winter, temperate/jungle, or swamp/desert) has color patterns and textures that are natural for that area. While color is self-explanatory, texture defines the surface characteristics of something when looking at it. For example, surface textures may be smooth, rough, rocky, leafy, or many other possible combinations. Use color and texture together to camouflage yourself effectively. It makes little sense to cover yourself with dead, brown vegetation in the middle of a large grassy field. Similarly, it would be useless to camouflage yourself with green grass in the middle of a desert or rocky area.

To hide and camouflage movement in any specific area of the world, you must take on the color and texture of the immediate surroundings. Use natural or man-made materials to camouflage yourself. Camouflage paint, charcoal from burned paper or wood, mud, grass, leaves, strips of cloth or burlap, pine bougns, and camouflaged uniforms are a few examples.

Cover all areas of exposed skin, including face, hands, neck, and ears. Use camouflage paint, charcoal, or mud to camouflage yourself. Cover with a darker color areas that stick out more and catch more light (forehead, nose, cheekbones, chin, and ears). Cover other areas, particularly recessed or shaded areas (around the eyes and under the chin), with lighter colors. Be sure to use an irregular pattern. Attach vegetation from the area or strips of cloth of the proper color to clothing and equipment. If you use vegetation, replace it as it wilts. As you move through an area, be alert to the color changes and modify your camouflage colors as necessary.

Figure 21-1 gives a general idea of how to apply camouflage for various areas and climates. Use appropriate colors for your surroundings. The blotches or slashes will help to simulate texture.

Shine

As skin gets oily, it becomes shiny. Equipment with worn off paint is also shiny. Even painted objects, if smooth, may shine. Glass objects such as mirrors, glasses, binoculars, and telescopes shine. You must cover these glass objects when not in use. Anything; that shines automatically attracts attention and will give away yourlocation.

Whenever possible, wash oily skin and reapply camouflage. Skin oil will wash off camouflage, so reapply it frequently. If you must wear glasses, camouflage them by applying a thin layer of dust to the outside of the lenses. This layer of dust will reduce the reflection of light. Cover shiny spots on equipment by painting, covering with mud, or wrapping with cloth or tape. Pay particular attention to covering boot eyelets, buckles on equipment, watches and jewelry, zippers, and uniform insignia. Carry a signar mirror in its designed pouch or in a pocket with the mirror portion facing your body.

Shadow

When hiding or traveling, stay in the deepest part of the shadows. The outer edges of the shadows are lighter and the deeper parts are darker. Remember, if you are in an area where there is plenty of vegetation, keep as much vegetation between you and a potential enemy as possible.

Area	Method
Temperate deciduous forest	Blotches
Coniferous forest	Broad slash
Jungle	Broad slash
Desert	Slash
Arctic	Blotches
Grass or open area	Slash

Figure 21-1. Camouflage methods for specific areas.

This action will make it very hard for the enemy to see you as the vegetation will partially mask you from his view. Forcing an enemy to look through many layers of masking vegetation will fatigue his eyes very quickly.

When traveling, especially in built-up areas at night, be aware of where you cast your shadow. It may extend out around the comer of a building and give away your position. Also, if you are in a dark shadow and there is a light source to one side, an enemy on the other side can see your silhouette against the light.

Movement

Movement, especially fast movement, attracts attention. If at all possible, avoid movement in the presence of an enemy. If capture appears imminent in your present location and you must move, move away slowly, making as little noise as possible. By moving slowly in a survival situation, you decrease the chance of detection and conserve energy that you may need for long-term survival or long-distance evasion.

When moving past obstacles, avoid going over them. If you must climb over an obstacle, keep your body level with its top to avoid silhouetting yourself. Do not silhouette yourself against the skyline when crossing hills or ridges. When you are moving, you will have difficulty detecting the movement of others. Stop frequently, listen, and look around slowly to detect signs of hostile movement.

Noise

Noise attracts attention, especially if there is a sequence of loud noises such as several snapping twigs. If possible, avoid making any noise at all. Slow down your pace as much as necessary to avoid making noise when moving around or away from possible threats.

Use background noises to cover the noise of your movement. Sounds of aircraft, trucks, generators, strong winds, and people talking will cover some or all the sounds producecfby your movement. Rain will mask a lot of movement noise, but it also reduces your ability to detect potential enemy noise.

Scent

Whether hunting animals or avoiding the enemy, it is always wise to camouflage the scent associated with humans. Start by washing yourself and your clothes without using soap. This washing method removes soap and body odors. Avoiding strong smelling foods, such as garlic, helps reduce body odors. Do not use tobacco products, candy, gum, or cosmetics.

You can use aromatic herbs or plants to wash yourself and your clothing, to rub on your body and clothing, or to chew on to camouflage your breath. Pine needles, mint, or any similar aromatic plant wilTnelp camouflage your scent from both animals and humans. Standing in smoke from a fire can help mask your scent from animals. While animals are afraid of fresh smoke from a fire, older smoke scents are normal smells after forest fires and do not scare them.

While traveling, use your sense of smell to help you find or avoid humans. Pay attention to smells associated with numans, such as fire, cigarettes, gasoline, oil, soap, and food. Such smells may alert you to their presence long before you can see or hear them, depending on wind speed and direction. Note the wind's direction and, when possible, approach from or skirt around on the downwind side when nearing humans or animals.

METHODS OF STALKING

Sometimes you need to move, undetected, to or from a location. You need more than just camouflage to make these moves successfully. The ability to stalk or move without making any sudden quick movement or loud noise is essential to avoiding detection.

You must practice stalking if it is to be effective. Use the following techniques when practicing.

Upright Stalking

Take steps about half your normal stride when stalking in the upright position. Such strides help you to maintain your balance. You should be able to stop at any pomt in that movement and hold that position as long as necessary. Curl the toes up out of the way when stepping down so the outside edge of the ball of the foot touches the ground. Feel for sticks and twigs that may snap when you place your weight on them. If you start to step on one, lift your foot and move it. After making contact with the outside edge of the ball of your foot, roll to the inside ball of your foot, place your heel down, followed by your toes. Then gradually shift your weight forward to the front foot. Lift the back foot to about knee height and start the process over again.

Keep your hands and arms close to your body and avoid waving them about or hitting vegetation. When moving in a crouch, you gain extra support by placing your hands on your knees. One step usually takes 1 minute to complete, but the time it takes will depend on the situation.

Crawling

Crawl on your hands and knees when the vegetation is too low to allow you to walk upright without being seen. Move one limb at a time and be sure to set it down softly, feeling for anything that may snap and make noise. Be careful that your toes and heels do not catch on vegetation.

Prone Staking

To stalk in the prone position, you do a low, modified push-up on your hands and toes, moving yourself forward slightly, and then lowering yourself again slowly. Avoid dragging and scraping along the ground as this makes excessive noise and leaves large trails for trackers to follow.

Animal Stalking

Before stalking an animal, select the best route. If the animal is moving, you will need an intercepting route. Pick a route that puts objects between you and the animal to conceal your movement from it. By positioning yourself in this way, you will be able to move faster, until you pass that object. Some objects, such as large rocks and trees, may totally conceal you, and others, such as small bushes and grass, may only partially conceal you. Pick the route that offers the best concealment and requires the least amount of effort.

Keep your eyes on the animal and stop when it looks your way or turns its ears your way, especially if it suspects your presence. As you get close, squint your eyes slightly to conceal both the light-dark contrast of the whites of the eyes and any shine from your eyes. Keep your mouth closed so that the animal does not see the whiteness or shine of your teeth.

CHAPTER 22

CONTACT WITH PEOPLE

Some of the best and most frequently given advice, when dealing with local peoples, is for the survivor to accept, respect, and adapt to their ways. Thus, "when in Rome, do as the Romans do." This is excellent advice, but there are several considerations involved in putting this advice into practice.

CONTACT WITH LOCAL PEOPLE

You must give serious consideration to dealing with the local people. Do they have a primitive culture? Are they farmers, fishermen, friendly people, or enemy? As a survivor, "cross-cultural communication" can vary radically from area to area and from people to people. It may mean interaction with people of an extremely primitive culture or contact with people who have a relatively modem culture. A culture is identified by standards of behavior that its members consider proper and acceptable but may or may not conform to your idea of what is proper. No matter who these people are, you can expect they will have laws, social and economic values, and political and religious beliefs that may be radically different from yours. Before deploying into your area of operations, study these different cultural aspects. Prior study and preparation will help you make or avoid contact if you have to deal with the local population.

People will be friendly, unfriendly, or they will choose to ignore you. Their attitude may be unknown. If the people are known to be friendly, try to keep them friendly through your courtesy and respect for their religion, politics, social customs, habits, and all other aspects of their culture. E the people are known to be enemies or are unknowns, make every effort to avoid any contact and leave no sign of your presence. A basic knowledge of the daily habits of the local people will be essential in this attempt If after careful observation you determine that an unknown people are friendly, you may contact them if you absolutely need their help.

Usually, you have little to fear and much to gain from cautious and respectful contact with local people of friendly or neutral countries. If you become familiar with the local customs, display common decency, and most important, show respect for their customs, you should be able to avoid trouble and possibly gain needed help. To make contact, wait until only one person is near and, if possible, let that person make the initial approach. Most people will be willing to help a survivor who appears to be m need. However, local political attitudes, instruction, or propaganda efforts may change the attitudes of otherwise friendly people. Conversely, in unfriendly countries, many people, especially in remote areas, may feel animosity toward their politicians and may be more friendly toward a survivor.

The key to successful contact with local peoples is to be friendly, courteous, and patient. Displaying fear, showmg weapons, and making sudden or threatening movements can cause a local person to fear you. Such actions can prompt a hostile response. When attempting a contact, smile as often as you can. Many local peoples are shy and seem unapproachable, or they may ignore you. Approach them slowly and do not rush your contact.

THE SURVIVOR'S BEHAVIOR

Use salt, tobacco, silver money, and similar items discreetly when trading with local people. Paper money is well-known worldwide. Do not overpay; it may lead to embarrassment and even danger. Always treat people with respect. Do not bully them or laugh at them.

Using sign language or acting out needs or questions can be very effective. Many people are used to such language and communicate using nonverbal sign language. Try to learn a few words and phrases of the local language in and around your potential area of operations. Trying to speak someone's language is one of the best ways to show respect for his culture. Since English is widely used, some of the local people may understand a few words of English.

Some areas may be taboo. They range from religious or sacred places to diseased or danger areas. In some areas, certain animals must not be killed. Learn the rules and follow them. Watch and learn as much as possible. Such actions will help to strengthen relations and provide new knowledge and skills that may be very important later. Seek advice on local hazards and find out from friendly people where the hostile people are. Always remember that people frequently insist that other peoples are hostile, simply because they do not understand different cultures and distant peoples. The people they can usually trust are their immediate neighbors—much the same as in our own neighborhood.

Frequently, local people, like ourselves, will suffer from contagious diseases. Build a separate shelter, if possible, and avoid physical contact without giving the impression of doing so. Personally prepare your food and drink, if you can do so without giving offense. Frequently, the local people will accept the use of "personal or religious custom" as an explanation for isolationist behavior.

Barter, or trading, is common in more primitive societies. Hard coin is usually good, whether for its exchange value or as jewelry or trinkets. In isolated areas, matches, tobacco, salt, razor blades, empty containers, or cloth may be worth more than any form of money.

Be very cautious when touching people. Many people consider "touching" taboo and such actions maybe dangerous. Avoid sexual contact.

Hospitality among some people is such a strong cultural trait that they may seriously reduce their own supplies to feed a stranger. Accept what they offer and share it equally with all present. Eat in the same way they eat and, most important, try to eat all they offer.

If you make any promises, keep them. Respect personal property and local customs and manners, even if they seem odd. Make some Kind of payment for food, supplies, and so forth. Respect privacy. Do not enter a house unless invited.

CHANGES TO POLITICAL ALLEGIANCE

In today's world of fast-paced international politics, political attitudes and commitments within nations are subject to rapid change. The population of many countries, especially politically hostile countries, must not be considered friendly just because they do not demonstrate open hostility. Unless briefed to the contrary; avoid all contact with such people.

CHAPTER 23

SURVIVAL IN MAN-MADE HAZARDS

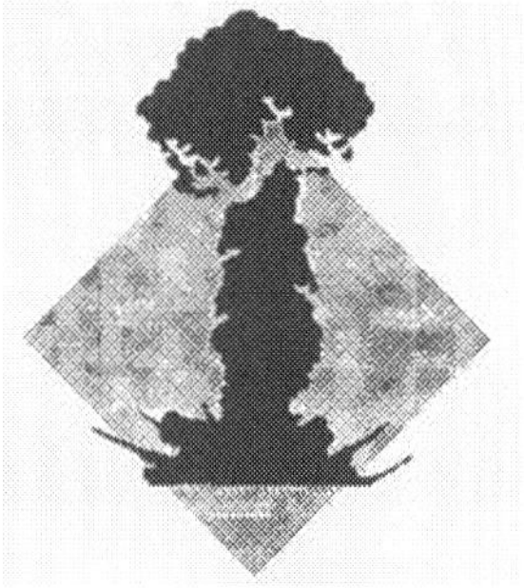

Nuclear, chemical, and biological weapons have become potential realities on any modern battlefield. Recent experience in Afghanistan, Cambodia, and other areas of conflict has proved the use of chemical and biological weapons (such as mycotoxins). The warfighting doctrine of the NATO and Warsaw Pact nations addresses the use of both nuclear and chemical weapons. The potential use of these weapons intensifies the problems of survival because of the serious dangers posed by either radioactive fallout or contamination produced by persistent biological or chemical agents.

You must use special precautions if you expect to survive in these man-made hazards. If you are subjected to any of the effects of nuclear, chemical, or biological warfare, the survival procedures recommended in this chapter may save your life. This chapter presents some background information on each type of hazard so that you may better understand the true nature of the hazard. Awareness of the hazards, knowledge of this chapter, and application of common sense should keep you alive.

THE NUCLEAR ENVIRONMENT

Prepare yourself to survive in a nuclear environment. Know how to react to a nuclear hazard.

Effects of Nuclear Weapons

The effects of nuclear weapons are classified as either initial or residual. Initial effects occur in the immediate area of the explosion and are hazardous In the first minute after the explosion. Residual effects can last for days or years and cause death. The principal initial effects are blast and radiation.

Blast

Defined as the brief and rapid movement of air away from the explosion's center and the pressure accompanying this movement. Strong winds accompany the blast. Blast hurls debris and personnel, collapses lungs, ruptures eardrums, collapses structures and positions, and causes immediate death or injury with its crushing effect.

Thermal Radiation

The heat and light radiation a nuclear explosion's fireball emits. Light radiation consists of both visible light and ultraviolet and infrared light. Thermal radiation produces extensive fires, skin burns, and flash bfiidness.

Nuclear Radiation

Nuclear radiation breaks down into two categories-initial radiation and residual radiation.

Initial nuclear radiation consists of intense gamma rays and neutrons produced during the first minute after the explosion. This radiation causes extensive damage to cells throughout the body. Radiation damage may cause headaches, nausea, vomiting, diarrhea, and even death, depending on the radiation dose received. The major problem in protecting yourself against the initial radiation's effects is that you may have received a lethal or incapacitating dose before taking any protective action. Personnel exposed to lethal amounts of initial radiation may well have been killea or fatally injured by blast or thermal radiation.

Residual radiation consists of all radiation produced after one minute from the explosion. It has more effect on you than initial radiation. A discussion of residual radiation takes place in a subsequent paragraph.

Types of Nuclear Bursts

There are three types of nuclear bursts—airburst, surface burst, and subsurface burst. The type of burst directly affects your chances of survival. A subsurface burst occurs completely underground or underwater. Its effects remain beneath the surface or in the immediate area where the surface collapses into a crater over the burst's location. Subsurface bursts cause you little or no radioactive hazard unless you enter the immediate area of the crater. No further discussion of this type of burst will take place.

An airburst occurs in the air above its intended target. The airburst provides the maximum radiation effect on the target and is, therefore, most dangerous to you in terms of *immediate nuclear* effects.

A surface burst occurs on the ground or water surface. Large amounts of fallout result, with serious long-term effects for you. This type of burst is your *greatest* nuclear hazard.

Nuclear Injuries

Most injuries in the nuclear environment result from the initial nuclear effects of the detonation. These injuries are classed as blast, thermal, or radiation injuries. Further radiation injuries may occur if you do not take proper precautions against fallout. Individuals in the area near a nuclear explosion will probably suffer a combination of all three types of injuries.

Blast Injuries

Blast injuries produced by nuclear weapons are similar to those caused by conventional high-explosive weapons. Blast overpressure can produce collapsed lungs and ruptured internal organs. Projectile wounds occur as the explosion s force hurls debris at you/Large pieces of debris striking you will cause fractured limbs or massive internal injuries. Blast overpressure may throw you long distances, and you will suffer severe injury upon impact with the ground or other objects. Substantial cover and distance from the explosion are the best protection against blast injury. Cover blast injury wounds as soon as possible to prevent the entry of radioactive dust particles.

Thermal Injuries

The heat and light the nuclear fireball emits causes thermal injuries. First-, second-, or third-degree burns may result. Flash blindness also occurs. This blindness may be permanent or temporary depending on the degree of exposure of the eyes. Substantial cover and distance from the explosion can prevent thermal injuries. Clothing will provide significant protection against thermal injuries. Cover as much exposed skin as possible before a nuclear explosion. First aid for thermal injuries is the same as first aid for burns. Cover open burns (second- or third-degree) to prevent the entry of radioactive particles. Wash all burns before covering.

Radiation Injuries

Neutrons, gamma radiation, alpha radiation, and beta radiation cause radiation injuries. Neutrons are high-speed, extremely penetrating particles that actually smash cells within your body. Gamma radiation is similar to X rays and is also a highly penetrating radiation. During the initial fireball stage of a nuclear detonation, initial gamma radiation and neutrons are the most serious threat. Beta and alpha radiation are radioactive particles normally associated with radioactive dust from fallout. They are short-range particles and you can easily protect yourself against them if you take precautions. See *Bodily Reactions to Radiation*, below, for the symptoms of radiation injuries.

Residual Radiation

Residual radiation is all radiation emitted after 1 minute from the instant of the nuclear explosion. Residual radiation consists of induced radiation and fallout.

Induced Radiation

It describes a relatively small, mtensely radioactive area directly underneath the nuclear weapon's fireball. The irradiated earth in this area will remain highly radioactive for an extremely long time. You should not travel into an area of induced radiation.

Fallout

Fallout consists of radioactive soil and water particles, as well as weapon fragments. During a surface detonation, or if an airburst's nuclear fireball touches the ground, large amounts of soil and water are vaporized along with the bomb's fragments, and forced upward to altitudes of 25,000 meters or more. When these vaporized contents cool, they can form more than 200 different radioactive products. The vaporized bomb contents condense into tiny radioactive particles that the wind carries and they fall back to earth as radioactive dust. Fallout particles emit alpha, beta, and gamma radiation. Alpha and beta radiation are relatively easy to counteract, and residual gamma radiation is much less intense than the gamma radiation emitted during the first minute after the explosion. Fallout is your most significant radiation hazard, provided you have not received a lethal radiation dose from the initial radiation.

Bodily Reactions to Radiation

The effects of radiation on the human body can be broadly classed as either chronic or acute. Chronic effects are those that occur some years after exposure to radiation. Examples are cancer and genetic defects. Chronic effects are of minor concern insofar as they affect your immediate survival in a radioactive environment. On the other hand, acute effects are of primary importance to your survival. Some acute effects occur within hours after exposure to radiation. These effects result from the radiation's direct physical damage to tissue. Radiation sickness and beta burns are examples of acute effects. Radiation sickness symptoms include nausea, diarrhea, vomiting, fatigue, weakness, and loss or hair. Penetrating beta rays cause radiation burns; the wounds are similar to fire burns.

Recovery Capability

The extent of body damage depends mainly on the part of the body exposed to radiation and now long it was exposed, as well as its ability to recover. The brain and kidneys have little recovery capability. Other parts (skin and bone marrow) have a great ability to recover from damage. Usually, a dose of 600 centigrams (cgys) to the entire body will result in almost certain death. If only your hands received this same dose, your overall health would not suffer much, although your hands would suffer severe damage.

External and Internal Hazards

An external or an internal hazard can cause body damage. Highly penetrating gamma radiation or the less penetrating beta radiation that causes burns can cause external damage. The entry of alpha or beta radiation-emitting particles into the body can cause internal damage. The external hazard produces overall irradiation and beta burns. The internal hazard results in irradiation of critical organs such as the gastrointestinal tract, thyroid gland, and bone. A very small amount of radioactive material can cause extreme damage to these and other internal organs. The internal hazard can enter the body either through consumption of contaminated water or food or by absorption through cuts or abrasions. Material that enters the body through breathing^ presents only a minor hazard. You can greatly reduce the internal radiation hazard by using good personal hygiene and carefully decontaminating your food and water.

Symptoms

The symptoms of radiation injuries include nausea, diarrhea, and vomiting. The severity of these symptoms is due to the extreme sensitivity of the gastrointestinal tract to radiation. The severity of the symptoms and the speed of onset after exposure are good indicators of the degree of radiation damage. The gastrointestinal damage can come from either the external or the internal radiation hazard.

Countermeasures Against Penetrating External Radiation

Knowledge of the radiation hazards discussed earlier is extremely important in surviving in a fallout area. It is also critical to know how to protect yourself from the most dangerous form of residual radiation-penetrating external radiation.

The means you can use to protect yourself from penetrating external radiation are time, distance, and shielding. You can reduce the level of radiation and help increase your chance of survival by controlling the duration of exposure. You can also get as far away from the radiation source as possible. Finally you can place some radiation-absorbing or shielding material between you and the radiation.

Time

Time is important to you, as the survivor, in two ways. First, radiation dosages are cumulative. The longer you are exposed to a radioactive source, the greater the dose you will receive. Obviously, spend as little time in a radioactive area as possible. Second, radioactivity decreases or decays over time. This concept is known as radioactive *half-life*. Thus, a radioactive element decays or loses half of its radioactivity within a certain time. The rule of thumb for radioactivity decay is that it decreases in intensity by a factor of ten for every sevenfold increase in time following the peak radiation level. For example, if a nuclear fallout area had a maximum radiation rate of 200 cgys per hour when fallout is complete, this rate would fall to 20 cgys per hour after 7 hours; it would fall still further to 2 cgys pernour after 49 hours. Even an untrained observer can see that the greatest hazard from fallout occurs immediately after detonation, and that the hazard decreases quickly over a relatively short time. As a survivor, try to avoid fallout areas until the radioactivity decays to safe levels. If you can avoid fallout areas long enough for most of the radioactivity to decay, you enhance your chance of survival.

Distance

Distance provides very effective protection against penetrating gamma radiation Because radiation intensity decreases by the square of the distance from the source. For example, if exposed to 1,000 cgys of radiation standing 30 centimeters from the source, at 60 centimeters, you would only receive 250 cgys. Thus, when you double the distance, radiation decreases to (0.5f or 0.25 the amount. While this formula is valid for concentrated sources of radiation in small areas, it becomes more complicated for large areas of radiation such as fallout areas.

Shielding

Shielding is the most important method of protection from penetrating radiation. Of the three countermeasures against penetrating radiation, shielding provides the greatest protection and is the easiest to use under survival conditions. Therefore, it is the most desirable method. If shielding is not possible, use the other two methods to the maximum extent practical.

Shielding actually works by absorbing or weakening the penetrating, radiation, thereby reducing the amount of radiation reaching your boay. The denser the material, the better the shielding effect. Lead, iron, concrete, and water are good examples of shielding materials.

Special Medical Aspects

The presence of fallout material in your area requires slight changes in first aid procedures. You must cover all wounds to prevent contamination and the entry of radioactive particles. You must first wash burns of beta radiation, then treat them as ordinary burns. Take extra measures to prevent infection. Your body will be extremely sensitive to infections due to changes in your blood chemistry. Pay close attention to the prevention of colds or respiratory infections. Rigorously practice personal hygiene to prevent infections. Cover your eyes with improvised goggles to prevent the entry of particles.

Shelter

As stated earlier, the shielding material's effectiveness depends on its thickness and density. An ample thickness of shielding material will reduce the level of radiation to negligible amounts.

The primary reason for finding and building a shelter is to get protection against the nigh-intensity radiation levels of early gamma fallout as fast as possible. Five minutes to locate the shelter is a good guide. Speed in finding shelter is absolutely essential. Without shelter, the dosage received in the first few hours will exceed that received during the rest of a week in a contaminated area. The dosage received in this first week will exceed the dosage accumulated during the rest of a lifetime spent in the same contaminated area.

Shielding Materials

The thickness required to weaken gamma radiation from fallout is far less than that needed to shield agamst initial gamma radiation. Fallout radiation has less energy than a nuclear detonation's initial radiation. For fallout radiation, a relatively small amount of shielding material can provide adequate protection. Figure 23-1 gives an idea of trie thickness of various materials needed to reduce residual gamma radiation transmission by 50 percent.

The principle of *half-value layer thickness'is* useful in understanding the absorption of gamma radiation by various materials. According to this principle, if 5 centimeters of brick reduce the gamma radiation level by one-half, adding another 5 centimeters of brick (another half-value layer) will reduce the intensity by another half, namely, to one-fourth the original amount. Fifteen centimeters will reduce gamma radiation fallout levels to one-eighth its original amount, 20 centimeters to one-sixteenth, and so on. Thus, a shelter protected by 1 meter of dirt would reduce a radiation intensity of 1,000 cgys per hour on the outside to about 0.5 cgy per hour inside the shelter.

Natural Shelters

Terrain that provides natural shielding and easy shelter construction is the ideal location for an emergency shelter. Good examples are ditches, ravines, rocky outcropping, hills, and river banks. In level areas without natural protection, dig a fighting position or slit trench.

Trenches

When digging a trench, work from inside the trench as soon as it is large enough to cover part of your body thereby not exposing all your body to radiation. In open country, try to dig the trench from a prone position, stacking the dirt carefully and evenly around the trench. On level

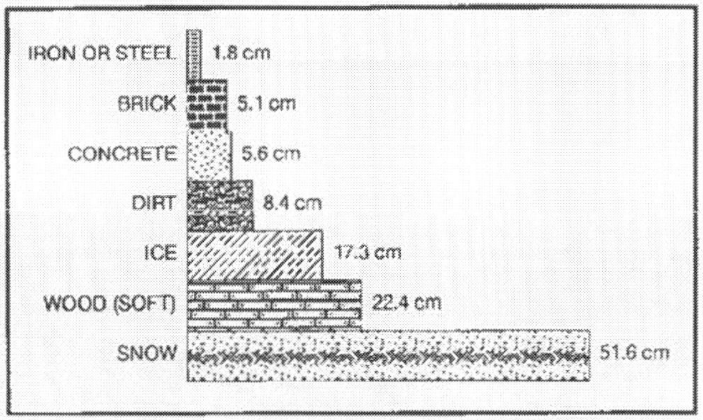

Figure 23-1. Thickness of materials to reduce gamma radiation.

ground, pile the dirt around your body for additional shielding. Depending upon soil conditions, shelter construction time will vary from a few minutes to a few hours. If you dig as quickly as possible, you will reduce the dosage you receive.

Other Shelters

While an underground shelter covered by 1 meter or more of earth provides the best protection against fallout radiation, the following unoccupied structures (in order listed) offer the next best protection:

- Caves and tunnels covered by more than 1 meter of earth.
- Storm or storage cellars.
- Culverts.
- Basements or cellars of abandoned buildings.
- Abandoned buildings made of stone or mud.

Roofs

It is not mandatory that you build a roof on your shelter. Build one only if the materials are readily available with only a brief exposure to outside contamination. If building a roof would require extended exposure to penetrating radiation, it would be wiser to leave the shelter roofless. A roof's sole function is to reduce radiation from the fallout source to your body. Unless you use a thick roof, a roof provides very little shielding.

You can construct a simple roof from a poncho anchored down with dirt, rocks, or other refuse from your shelter. You can remove large particles of dirt and debris from the top of the poncho by beating it off from the inside at frequent intervals. This cover will not offer shielding from the radioactive particles deposited on the surface, but it will increase the distance from the fallout source and keep the shelter area from further contamination.

Shelter Site Selection and Preparation

To reduce your exposure time and thereby reduce the dosage received, remember the following factors when selecting and setting up a shelter:

- Where possible, seek a crude, existing shelter that you can improve. If none is available, dig a trench.
- Dig the shelter deep enough to get good protection, then enlarge it as required for comfort.

- Cover the top of the fighting position or trench with any readily available material and a thick layer of earth, if you can do so without leaving the shelter. While a roof and camouflage are both desirable, it is probably safer to do without them than to expose yourself to radiation outside your fighting position.
- While building your shelter, keep all parts of your body covered with clothing to protect it against beta bums.
- Clean the shelter site of any surface deposit using a branch or other object that you can discard. Do this cleaning to remove contaminated materials from the area you will occupy. The cleaned area should extend at least 1.5 meters beyond your shelter's area.
- Decontaminate any materials you bring into the shelter. These materials include grass or foliage that you use as insulation or bedding, and your outer clothing (especially footgear). If the weather permits and you have heavily contaminated outer clothing, you may want to remove it and bury it under a foot of earth at the end of your shelter. You may retrieve it later (after the radioactivity decays) when leaving the shelter. If the clothing is dry, you may decontaminate it by beating or shaking it outside the shelter's entrance to remove the radioactive dust. You may use any body of water, even though contaminated, to rid materials of excess fallout particles. Simply dip the material into the water and shake it to get rid of the excess water. Do not wring it out, this action will trap the particles.
- If at all possible and without leaving the shelter, wash your body thoroughly with soap and water, even if the water on hand may be contaminated. This washing will remove most of the harmful radioactive particles that are likely to cause beta burns or other damage. If water is not available, wipe your face and any other exposed skin surface to remove contaminanted dust and dirt. You may wipe your face with a clean piece of cloth or a handful of uncontaminated dirt. You get this uncontaminated dirt by scraping off the top few inches of soil and using the "clean" dirt.
- Upon completing the shelter, lie down, keep warm, and sleep and rest as much as possible while in the shelter.
- When not resting, keep busy by planning future actions, studying your maps, or making the shelter more comfortable and effective.
- Don't panic if you experience nausea and symptoms of radiation sickness. Your mam danger from radiation sickness is infection. There is no first aid for this sickness. Resting, drinking fluids, taking any medicine that prevents vomiting, maintaining your food intake, and preventing additional exposure will help avoid infection and aid recovery. Even small doses of radiation can cause these symptoms which may disappear in a short time.

Exposure Timetable

The following timetable provides you with the information needed to avoid receiving serious dosage and still let you cope with survival problems:

- Complete isolation from 4 to 6 days following delivery of the last weapon.
- A very brief exposure to procure water on the third day is permissible, but exposure should not exceed 30 minutes.
- One exposure of not more than 30 minutes on the seventh day.
- One exposure of not more than 1 hour on the eighth day.
- Exposure of 2 to 4 hours from the ninth day through the twelfth day.
- Normal operation, followed by rest in a protected shelter, from the thirteenth day on.
- In all instances, make your exposures as brief as possible. Consider only mandatory requirements as valid reasons for exposure. Decontaminate at every stop.

The times given above are conservative. If forced to move after the first or second day, you may do so, Make sure that the exposure is no longer than absolutely necessary.

Water Procurement

In a fallout-contaminated area, available water sources may be contaminated. If you wait at least 48 hours before drinking any water to allow for radioactive decay to take place and select the safest possible water source, you will greatly reduce the danger of ingesting harmful amounts of radioactivity.

Although many factors (wind direction, rainfall, sediment) will influence your choice in selecting water sources, consider the following guidelines.

Safest Water Sources

Water from springs, wells, or other underground sources that undergo natural filtration will be your safest source. Any water found in the pipes or containers of abandoned houses or stores will also be free from radioactive particles. This water will be safe to drink, although you will have to take precautions against bacteria in the water.

Snow taken from 15 or more centimeters below the surface during the fallout is also a safe source of water.

Streams and Rivers

Water from streams and rivers will be relatively free from fallout within several days after the last nuclear explosion because of dilution. If at all possible, filter such water before drinking to get rid of radioactive particles. The best filtration method is to dig sediment holes or seepage basins along the side of a water source. The water will seep laterally into the hole through the intervening soil that acts as a filtering agent and removes the contaminated fallout particles that settled on the original body of water. This method can remove up to 99 percent of the radioactivity in water. You must cover the hole In some way in order to prevent further contamination. See Figure 6-9 on page 6-14 for an example of a water filter.

Standing Water

Water from lakes, pools, ponds, and other standing sources is likely to be heavily contaminated, though most of the heavier, long-lived radioactive isotopes will settle to the bottom. Use the settling technique to purify this water. First, fill a bucket or other deep contamer three-fourths full with contaminated water. Then take dirt from a depth of 10 or more centimeters below the ground surface and stir it into the water. Use about 2.5 centimeters of dirt for every 10 centimeters of water. Stir the water until you see most dirt particles suspended in the water. Let the mixture settle for at least 6 hours. The settling dirt particles will carry most of the suspended fallout particles to the bottom and cover them. You can then dip out the clear water. Purify this water using a filtration device.

Additional Precautions

As an additional precaution against disease, treat all water with water purification tablets from your survival kit or boil it.

Food Procurement

Although it is a serious problem to obtain edible food in a radiation-contaminated area, it is not impossible to solve. You need to follow a few special procedures in selecting and preparing rations and local foods for use. Since secure packaging protects your combat rations, they will be perfectly safe for use. Supplement your rations with any food you can find on trips outside your shelter. Most processed foods you may find in abandoned buildings are safe for use after decontaminating them. These include canned and packaged foods after removing the containers or wrappers or washing them free of fallout particles. These processed foods also include food stored in any closed container and food stored in protected areas (such as cellars), if you wash them before eating. Wash all food containers or wrappers before handling them to prevent further contamination.

If little or no processed food is available in your area, you may have to supplement your diet with local food sources. Local food sources are ammals and plants.

Animals as a Food Source

Assume that all animals, regardless of their habitat or living conditions, were exposed to radiation. The effects of radiation on animals are similar to those on humans. Thus, most of the wild animals living in a fallout area are likely to become sick or die from radiation during the first month after the nuclear explosion. Even though animals may not be free from harmful radioactive materials, you can and must use them in survival conditions as a food source if other foods are not available. With careful preparation and by following several important principles, animals can be safe food sources.

First, do not eat an animal that appears to be sick. It may have developed a bacterial infection as a result of radiation poisonmg. Contaminated meat, even if thoroughly cooked, could cause severe illness or death if eaten.

Carefully skin all animals to prevent any radioactive particles on the skin or fur from entering the body. Do not eat meat close to the bones and joints as an animal's skeleton contains over 90 percent of the radioactivity. The remaining animal muscle tissue, however, will be safe to eat. Before cooking it, cut the meat away from the bone, leaving at least a 3-millimeter thickness of meat on the bone. Discard all internal organs (heart, liver, and kidneys) since they tend to concentrate beta and gamma radioactivity.

Cook all meat until it is very well done. To be sure the meat is well done, cut it into less than 13-millrneter-thick pieces before cooking. Such cuts will also reduce cooking time and save fuel.

The extent of contamination in fish and aquatic animals will be much greater than that of land animals. This is also true for water plants, especially in coastal areas. Use aquatic food sources only in conditions of extreme emergency.

All eggs, even if laid during the period of fallout, will be safe to eat. Completely avoid milk from any animals in a fallout area because animals absorb large amounts of radioactivity from the plants they eat.

Plants as a Food Source

Plant contamination occurs by the accumulation of fallout on their outer surfaces or by absorption of radioactive elements through their roots. Your first choice of plant food should be vegetables such as potatoes, turnips, carrots, and other plants whose edible portion grows underground. These are the safest to eat once you scrub them and remove their skins.

Second in order of preference are those plants with edible parts that you can decontaminate by washing and peelmg their outer surfaces. Examples are bananas, apples, tomatoes, prickly pears, and other such fruits and vegetables.

Any smooth-skinned vegetable, fruit, or plant that you cannot easily peel or effectively decontaminate by washing will be your third choice of emergency food.

The effectiveness of decontamination by scrubbing is inversely proportional to the roughness of the fruit's surface. Smooth-surfaced fruits have lost 90 percent of their contamination after washing, while washing rough-surfaced plants removes only about 50 percent of the contamination.

You eat rough-surfaced plants (such as lettuce) only as a last resort because you cannot effectively decontaminate them by peeling or washing.

Other difficult foods to decontaminate by washing with water include dried fruits (figs, prunes, peaches, apricots, pears) and soya beans.

In general, you can use any plant food that is ready for harvest if you can effectively decontaminate it. Growing plants, however, can absorb some radioactive materials through their leaves as well as from the soil, especially if rains have occurred during or after the fallout period. Avoid using these plants for food except in an emergency.

BIOLOGICAL ENVIRONMENTS

The use of biological agents is real. Prepare yourself for survival by being proficient m the tasks identified m your Soldier's Manuals of Common Tasks (SMCTs). Know what to do to protect yourself against these agents.

Biological Agents and Effects

Biological agents are microorganisms that can cause disease among personnel, animals, or plants. They can also cause the deterioration of material. These agents fall into two broad categories-pathogens (usually called germs) and toxins. Pathogens are livmg microorganisms that cause lethal or incapacitating diseases. Bacteria, rickettsiae, fungi, and viruses are includea in the pathogens. Toxins are poisons that plants, animals, or microorganisms produce naturally. Possible biological warfare toxins include a variety of neurotoxic (affecting the central nervous system) and cytotoxic (causing cell death) compounds.

Germs

Germs are living organisms. Some nations have used them in the past as weapons. Only a few germs can start an infection, especially if inhaled into the lungs. Because germs are so small and weigh so little, the wind can spread them over great distances; they can also enter unf iltered or nonairtight places. Buildings and bunkers can trap them thus causing a higher concentration. Germs do not affect the body immediately. They must multiply inside the body and overcome the body's defenses—a process called the incubation period. Incubation periods vary from several hours to several months, depending on the germ. Most germs must live within another living organism (host), such as your body, to survive and grow. Weather conditions such as wind, rain, cold, and sunlight rapidly kill germs.

Some germs can form protective shells, or spores, to allow survival outside the host. Spore-producing agents are a long-term hazard you must neutralize by decontaminating infected areas or personnel. Fortunately, most live agents are not spore-producing. These agents must find a host within roughly a day of their delivery or they die. Germs have three basic routes of entry into your body: through the respiratory tract, through a break in the skin, and through the digestive tract. Symptoms of infection vary according to the disease.

Toxins

Toxins are substances that plants, animals, or germs produce naturally. These toxins are what actually harm man, not bacteria. Botulin, which produces botulism, is an example. Modern science has allowed large-scale production of these toxins without the use of the germ that produces the toxin. Toxins may produce effects similar to those of chemical agents. Toxic victims may not, however, respond to first aid measures used against chemical agents. Toxins enter the body in the same manner as germs. However, some toxins, unlike germs, can penetrate unbroken skm. Symptoms appear almost immediately, since there is no incubation period. Many toxins are extremely lethal, even in very small doses. Symptoms may mclude any of the following:

- Dizziness.
- Mental confusion.
- Blurred or double vision.
- Numbness or tingling of skin.
- Paralysis.
- Convulsions.
- Rashes or blisters.
- Coughing.
- Fever.
- Aching muscles.
- Tiredness.
- Nausea, vomiting, and/or diarrhea.
- Bleeding from body openings.
- Blood in urine, stool, or saliva.
- Shock.
- Death.

Detection of Biological Agents

Biological agents are, by nature, difficult to detect. You cannot detect them oy any of the five physical senses. Often, the first sign of a biological agent will be symptoms of the victims exposed to the agent. Your best chance of detecting biological agents before they can affect you is to recognize their means of delivery. The three main means of delivery are—

- *Bursting-type munitions.* These may be bombs or projectiles whose burst causes very little damage. The burst will produce a small cloud of liquid or powder in the immediate impact area. This cloud will disperse eventually; the rate of dispersion depends on terrain and weather conditions.
- *Spray tanks or generators.* Aircraft or vehicle spray tanks or ground-level aerosol generators produce an aerosol cloud of biological agents.
- *Vectors.* Insects such as mosquitoes, fleas, lice, and ticks deliver pathogens. Large infestations of these insects may indicate the use of biological agents.

Another sign of a possible biological attack is the presence of unusual substances on the ground or on vegetation, or sick-looking plants, crops, or animals.

Influence of Weather and Terrain

Your knowledge of how weather and terrain affect the agents can help you avoid contamination by biological agents. Major weather factors that affect biological agents are sunlight, wind, and precipitation. Aerosol sprays will tend to concentrate in low areas of terrain, similar to early morning mist.

Sunlight contains visible and ultraviolet solar radiation that rapidly kills most germs used as biological agents. However, natural or man-made cover may protect some agents from sunlight. Other man-made mutant strains of germs may be resistant to sunlight.

High wind speeds increase the dispersion of biological agents, dilute their concentration, and dehydrate them. The further downwind the agent travels, the less effective it becomes due to dilution and death of the pathogens. However, the downwind hazard area of the biological agent is significant and you cannot ignore it.

Precipitation in the form of moderate to heavy rain tends to wash biological agents out of the air, reducing downwmd hazard areas. However, the agents may still be very effective where they were deposited on the ground.

Protection Against Biological Agents

While you must maintain a healthy respect for biological agents, there is no reason for you to panic. You can reduce your susceptibility to biological agents by maintaining current immunizations, avoiding contaminated areas, and controlling rodents and pests. You must also use proper first aid measures in the treatment of wounds and only safe or properly decontaminated sources of food and water. You must ensure that you get enough sleep to prevent a run-down condition. You must always use proper field sanitation procedures.

Assuming you do not have a protective mask, always try to keep your face covered with some type of cloth to protect yourself against biological agent aerosols. Dust may contain biological agents; wear some type of mask when dust is in the air.

Your uniform and gloves will protect you against bites from vectors (mosquitoes and ticks) that carry diseases. Completely button your clothing and ruck your trousers tightly into your boots. Wear a chemical protective overgarment, if available, as it provides better protection than normal clothing. Covering your skin will also reduce the chance of the agent entering your body through cuts or scratches. Always practice high standards of personal hygiene and sanitation to help prevent the spread of vectors.

Bathe with soap and water whenever possible. Use germicidal soap, if available. Wash your hair and body thoroughly, and clean under your fingernails. Clean teeth, gums, tongue, and the roof of your mouth frequently. Wash your clothing in hot, soapy water if you can. If you cannot wash your clothing, lay it out in an area of bright sunlight and allow the light to kill the microorganisms. After a toxin attack, decontaminate yourself as if for a chemical attack using the M258A2 kit (if available) or by washing with soap and water.

Shelter

You can build expedient shelters under biological contamination conditions using the same techniques described in Chapter 5. However, you must make slight changes to reduce the chance of biological contamination. Do not build your shelter in depressions in the ground. Aerosol sprays tend to concentrate in these depressions. Avoid building, your shelter in areas of vegetation, as vegetation provides shade ana some degree of protection to biological agents. Avoid using vegetation in constructing your shelter. Place your shelter's entrance at a 90-degree angle to the prevailing winds. Such placement will limit the entry of airborne agents and prevent air stagnation in your shelter. Always keep your shelter clean.

Water Procurement

Water procurement under biological conditions is difficult but not impossible. Whenever possible, try to use water that has been in a sealed container. You can assume that the water inside the sealed container is not contaminated. Wash the water container thoroughly with soap and water or boil it for at least 10 minutes before breaking the seal.

If water in sealed containers is not available, your next choice, *only under emergency conditions,is* water from springs. Again, boil the water for at least 10 minutes before drinking. Keep the water covered while boiling to prevent contamination by airborne pathogens. Your *last choice, only in an extreme emergency,* is to use standing water. Vectors and germs can survive easily m stagnant water. Boil this water as long as practical to kill all orgamsms. Filter this water through a cloth to remove the dead vectors. Use water purification tablets in all cases.

Food Procurement

Food procurement, like water procurement, is not impossible, but you must take special precautions. Your combat rations are sealed, and you can assume they are not contaminated. You can also assume that sealed containers or packages of processed food are safe. To ensure safety, decontaminate all food containers by washing with soap and water or by boiling the container in water for 10 minutes.

You consider supplementing your rations with local plants or animals only in extreme emergencies. No matter what you do to prepare the food, there is no guarantee that cooking will kill ail the biological agents. Use local food only in life or death situations. Remember, you can survive for a long time without food, especially if the food you eat may kill you!

If you must use local food, select only healthy-looking plants and animals. Do not select known carriers of vectors such as rats or other vermin. Select and prepare plants as you would in radioactive areas. Prepare animals as you do plants. Always use gloves and protective clothing when handling animals or plants. Cook all plant and animal food by boiling only. Boil all food for at least 10 minutes to kill all pathogens. Do not try to fry, bake, or roast local food. There is no guarantee that all infected portions have reached the required temperature to kill all pathogens. Do not eat raw food.

CHEMICAL ENVIRONMENTS

Chemical agent warfare is real. It can create extreme problems in a survival situation, but you can overcome the problems with the proper equipment, knowledge, and training. As a survivor, your first line or defense against chemical agents is your proficiency in individual nuclear, biologicaf and chemical *(NBC)* teaming, to include donning and wearing the protective mask and overgarment, personal decontamination, recognition of chemical agent symptoms, and individual first aid for chemical agent contamination. The SMCTs cover these subjects. If you are not proficient in these skills, you will have little chance of surviving a chemical environment.

The subject matter covered below is not a substitute for any of the individual tasks in which you must be proficient. The SMCTs address the various chemical agents, their effects, and first aid for these agents. The following information is provided under the assumption that you are proficient in the use of chemical protective equipment and know the symptoms of various chemical agents.

Detection of Chemical Agents

The best method for detecting chemical agents is the use of a chemical agent detector. If you have one, use it. However, in a survival situation, you will most likely have to rely solely on the use of all of your physical senses. You must be alert and able to detect any clues indicating the use of chemical warfare. General indicators of the presence of chemical agents are tears, difficult breathing, choking, itching, coughing, and dizziness. With agents that are very hard to detect, you must watch for symptoms in fellow survivors. Your surroundings will provide valuable clues to the presence of chemical agents; for example, dead animals, sick people, or people and animals displaying abnormafbehavior.

Your sense of smell may alert you to some chemical agents, but most will be odorless. The odor of newly cut grass or hay may indicate the presence of choking agents. A smell of almonds may indicate blood agents.

Sight will help you detect chemical agents. Most chemical agents in the solid or liquid state have some color.Tn the vapor state, you can see some chemical agents as a mist or thin fog immediately after the bomb or shell bursts. By observing for symptoms in others and by observing delivery means, you may be able to have some warning of chemical agents. Mustard gas in the liquid state will appear as oily patches on leaves or on buildings.

The sound of enemy munitions will give some clue to the presence of chemical weapons. Muffled shell or bomb detonations are a good indicator.

Irritation in the nose or eyes or on the skin is an urgent warning to protect your body from chemical agents. Additionally, a strange taste in food, water, or cigarettes may serve as a warning that they have been contaminated.

Protection Against Chemical Agents

As a survivor, always use the following general steps, in the order listed, to protect yourself from a chemical attack:

- Use protective equipment.
- Give quick and correct self-aid when contaminated.
- Avoid areas where chemical agents exist.
- Decontaminate your equipment and body as soon as possible.

Your protective mask and overgarment are the key to your survival. Without these, you stand very Bttle chance of survival. You must take care of these items and protect them from damage. You must practice and know correct self-aid procedures before exposure to chemical agents. The detection of chemical agents and the avoidance of contaminated areas is extremely important to your survival. Use whatever detection kits may be available to help in detection. Since you are in a survival situation, avoid contaminated areas at all costs. You can expect no help should you become contaminated. If you do become contaminated, decontaminate yourself as soon as possible using proper procedures.

Shelter

If you find yourself in a contaminated area, try to move out of the area as fast as possible. Travel crosswind or upwind to reduce the time spent in the downwind hazard area. If you cannot leave the area immediately and have to build a shelter, use normal shelter construction techniques, with a few changes. Build the shelter in a clearing, away from all vegetation. Remove all topsoil in the area of the shelter to decontaminate the area. Keep the shelter's entrance closed and oriented at a 90-degree angle to the prevailing wind. Do not build a fire using contaminated wood—the smoke will be toxic. Use extreme caution when entering your shelter so that you will not bring contamination inside.

Water Procurement

As with biological and nuclear environments, getting water in a chemical environment is difficult. Obviously, water in sealed containers is your best and safest source. You must protect this water as much as possible. Be sure to decontaminate the containers before opening.

If you cannot get water in sealed containers, try to get it from a closed source such as underground water pipes. You may use rainwater or snow if there is no evidence of contamination. Use water from slow-moving streams, if necessary, but always check first for signs of contamination, and always filter the water as described under nuclear conditions. Signs of water source contamination are foreign odors such as garlic, mustard, geranium, or bitter almonds; oily spots on the surface of the water or nearby; and the presence of dead fish or animals. If these signs are present, do not use the water. Always boil or purify the water to prevent bacteriological infection.

Food Procurement

It is extremely difficult to eat while in a contaminated area. You will have to break the seal on your protective mask to eat. If you eat, find an area in which you can safely unmask.

The safest source of food is your sealed combat rations. Food in sealed cans or bottles will also be safe. Decontaminate all sealed food containers before opening, otherwise you will contaminate the food.

If you must supplement your combat rations with local plants or animals, *do not* use plants from contaminated areas or animals that appear to be sick. When handling plants or animals, always use protective gloves and clothing.

APPENDIX A

SURVIVAL KITS

The Army has several basic survival kits, primarily for issue to aviators. There are kits for cold climates, hot climates, and overwater. There is also an individual survival kit with general packet and medical packet. The cold climate, hot climate, and overwater kits are in canvas carrying bags. These kits are normally stowed in the helicopter's cargo/ passenger area.

An aviator's survival vest (SRU-21P), worn by helicopter crews, also contains survival items.

U.S. Army aviators flying fixed-wing aircraft equipped with ejection seats use the SRFU-31/P survival vest. The individual survival kits are stowed in the seat pan. Like all other kits, the rigid seat survival kit (RSSK) used depends on the environment.

Items contained in the kits may be ordered separately through supply channels. All survival kits and vests are CTA 50-900 items and can be ordered by authorized units.

- Food packets
- Snare wire
- Smoke, illumination signals
- Waterproof match box
- Saw/knife blade
- Wood matches
- First aid kit
- MC-1 magnetic compass
- Pocket knife
- Saw-knife-shovel handle
- Frying pan
- Illuminating candles
- Compressed trioxane fuel
- Signaling mirror
- Survival fishing kit
- Plastic spoon
- Survival Manual (AFM 64-5)
- Poncho
- Insect headnet
- Ejector snap
- Attaching strap
- Kit, outer case
- Kit, inner case
- Shovel
- Water bag
- Packing list
- Sleeping bag

Cold Climate Kit.

- Canned drinking water
- Waterproof matchbox
- Plastic whistle
- Smoke, illumination signals
- Pocket knife
- Signaling mirror
- Plastic water bag
- First aid kit
- Sunburn-preventive cream
- Plastic spoon
- Food packets
- Compressed trioxane fuel
- Fishing tackle kit
- MC-1 magnetic compass
- Snare wire
- Frying pan
- Wood matches
- Insect headnet
- Reversible sun hat
- Tool kit
- Kit, packing list
- Tarpaulin
- Survival manual (AFM 64-5)
- Kit, inner case
- Kit, outer case
- Attaching strap
- Ejector snap

Hot Climate Kit.

- Kit, packing list
- Raft boat paddle
- Survival manual (AFM 64-5)
- Insect headnet
- Reversible sun hat
- Water storage bag
- MC-1 magnetic compass
- Boat bailer
- Sponge
- Sunburn-preventive cream
- Wood matches
- First aid kit
- Plastic spoon
- Pocket knife
- Food packets
- Fluorescent sea marker
- Frying pan
- Seawater desálter kit
- Compressed trioxane fuel
- Smoke, illumination signals
- Signaling mirror
- Fishing tackle kit
- Waterproof match box
- Raft repair kit

Overwater Kit.

NSN	DESCRIPTION	QTY/UI
1680-00-205-0474	SURVIVAL KIT, INDIVIDUAL SURVIVAL VEST (OV-1), large, SC 1680-97-CL-A07	
1680-00-187-5716	SURVIVAL KIT, INDIVIDUAL SURVIVAL VEST (OV-1), small, SC 1680-97-CL-A07	
	Consisting of the following components:	
7340-00-098-4327	KNIFE, HUNTING: 5 in. lg blade, leather handle, w/sheath	1 ea
5110-00-526-8740	KNIFE, POCKET: one 3-1/16 in. lg cutting blade, & one 1-25/32 in. lg hook blade, w/safety lock & clevis	1 ea
4220-00-850-8655	LIFE PRESERVER, UNDERARM: gas or orally inflated, w/gas cyl, adult size, 10 in. h, orange color, shoulder & chest type harness w/quick release buckle & clip	1 ea
6230-00-938-1778	LIGHT, MARKER, DISTRESS: plastic body, rd, 1 in. w, accom 1 flashtube; one 5.4 v dry battery required	1 ea
6350-00-105-1252	MIRROR, EMERGENCY SIGNALING: glass, circular clear window in center or mirror for sighting, 3 in. lg, 2 in. w, 1/8 in. thk, w/o case, w/lanyard	1 ea
1370-00-490-7362	SIGNAL KIT, PERSONNEL DISTRESS: w/7 rocket cartridges & launcher	1 ea
6546-00-478-6504	SURVIVAL KIT, INDIVIDUAL *consisting of*	1 ea
4240-00-152-1578	GENERAL PACKET, INDIVIDUAL SURVIVAL KIT: w/ mandatory pack bag; 1 pkg ea of coffee & fruit flavored candy; 3 pkg chewing gum; 1 water storage container; 2 flash guards, w/infrared & blue filters; 1 mosquito headnet & pr mittens; 1 instruction card; 1 emergency signaling mirror; 1 fire starter & tinder; 5 safety	1 ea

Individual survival kit with general and medical packets.

NSN	DESCRIPTION	QTY/UI
	pins; 1 small straight-type surgical razor; 1 rescue/signal/medical instruction panel; 1 tweezer, & 1 wrist compass, strap & lanyard	
6545-00-231-9421	MEDICAL PACKET, INDIVIDUAL SURVIVAL KIT: w/carrying bag; 1 tube insect repellent & sun screen ointment; 1 medical instruction card; 1 waterproof receptacle, 1 bar soap & following items:	1 ea
6510-00-926-8881	ADHESIVE TAPE, SURGICAL: white rubber coating, 1/2 in. w, 360 in. lg, porous woven	1 ea
6505-00-118-1948	ASPIRIN TABLETS, USP: 0.324 gm, individually sealed in roll strip container	10 ea
6510-00-913-7909	BANDAGE, ADHESIVE: flesh, plastic coated, 3/4 in. w, 3 in. lg	1 ea
6510-00-913-7906	BANDAGE, GAUZE, ELASTIC: white, sterile, 2 in. w, 180 in. lg	1 ea
6505-00-118-1914	DIPHENOXYLATE HYDROCHLORIDE AND ATROPINE SULFATE TABLETS, USP: 0.025 mg atropine sulfate & 2.500 mg diphenoxylate hydrochloride active ingredients, individually sealed, roll strip container	10 ea
6505-00-183-9419	SULFACETAMIDE SODIUM OPHTHALMIC OINTMENT, USP: 10 percent	3.5 gm
6850-00-985-7166	WATER PURIFICATION TABLET, IODINE: 8 mg	50 ea
	VEST, SURVIVAL: nylon duck	1 ea
	8415-00-201-9098 large size	1 ea
	8415-00-201-9097 small size	1 ea
8465-00-254-8803	WHISTLE, BALL: plastic, olive drab w/lanyard	1 ea

Individual survival kit with general and medical packets (continued).

NSN	DESCRIPTION
8465-00-177-4819	Survival Vest
6515-00-383-0565	Tourniquet
5820-00-782-5308	AN/PRC-90 Survival Radio
1305-00-301-1692	.38 caliber tracer ammunition
1305-00-322-6391	.38 caliber ball ammunition
1005-00-835-9773	Revolver, .38 caliber
9920-00-999-6753	Lighter, butane
6350-00-105-1252	Mirror, signaling
6545-00-782-6412	Survival kit, individual tropical
1370-00-490-7362	Signal kit, foliage penetrating
6230-00-938-1778	Light, distress marker, SDU-5/E
8465-00-634-4499	Bag, storage, drinking water
5110-00-162-2205	Knife, pocket
4240-00-300-2138	Net, gill, fishing
6605-00-151-5337	Compass, magnetic, lensatic

SRU-21P Aviator's Survival Vest.

NSN	DESCRIPTION
1680-00-148-9233	Survival kit, cold climate (RSSK OV-1)
1680-00-148-9234	Survival kit, hot climate (RSSK OV-1)
1680-00-965-4702	Survival kit, overwater (RSSK OV-1)

OV-1 Rigid Seat Survival Kits.

APPENDIX B

EDIBLE AND MEDICINAL PLANTS

In a survival situation, plants can provide food and medicine. Their safe usage requires absolutely positive identification, knowing how to prepare them for eating, and knowing any dangerous properties they might have. Familiarity with botanical structures of plants and information on where they grow will make them easier to locate and identify.

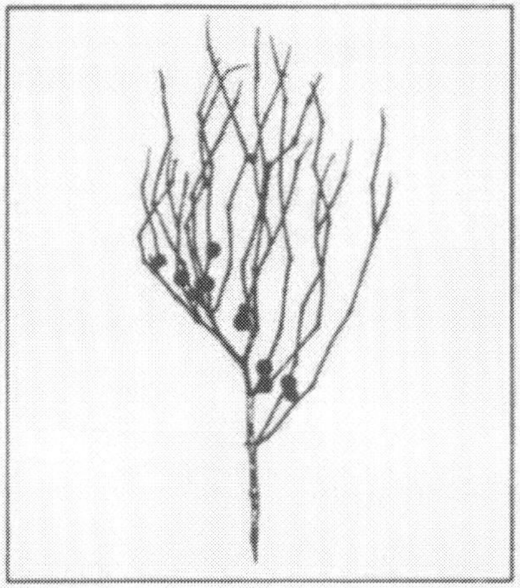

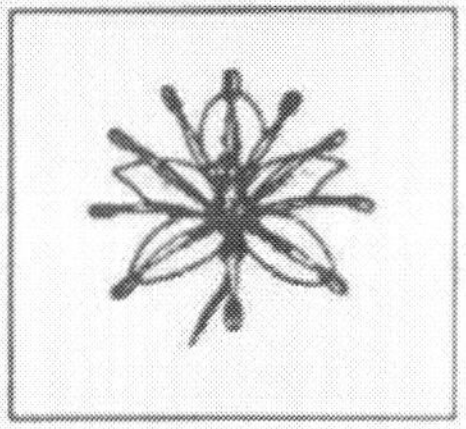

Abal

Calligonum comosum

Description: The abal is one of the few shrubby plants that exists in the shady deserts. This plant grows to about 1.2 meters, and its branches look like wisps from a broom. The stiff, green branches produce an abundance of flowers in the early spring months (March, April).

Habitat and Distribution: This plant is found in desert scrub and waste in any climatic zone. It inhabits much of the North African desert. It may also be found on the desert sands of the Middle East and as far eastward as the Rajputana desert of westen India.

Edible Parts: This plant's general appearance would not indicate its usefulness to the survivor, but while this plant is flowering in the spring, its fresh flowers can be eaten. This plant is common in the areas where it is found. An analysis of the food value of this plant has shown it to be high in sugar and nitrogenous components.

Acacia

Acacia farnesiana

Description: Acacia is a spreading, usually short tree with spines and alternate compound leaves. Its individual leaflets are small. Its flowers are ball-shaped, bright yellow, and very fragrant. Its bark is a whitish-gray color. Its fruits are dark brown and podlike.

Habitat and Distribution: Acacia grows in open, sunny areas. It is found throughout all tropical regions.

Note: There are about 500 species of acacia. These plants are especially prevalent in Africa, southern Asia, and Australia, but many species are found in the warmer and drier parts of America.

Edible Parts: Its young leaves, flowers, and pods are edible raw or cooked.

Agave

Agave species

Description: These plants have large clusters of thick, fleshy leaves borne close to the ground and surrounding a central stalk. The plants flower only once, then die. They produce a massive flower stalk.

Habitat and Distribution: Agaves prefer dry, open areas. They are found throughout Central America, the Caribbean, and parts of the western deserts of the United States and Mexico.

Edible Parts: Its flowers and flower buds are edible. Boil them before eating.

> **CAUTION**
>
> The juice of some species causes dermatitis in some individuals.

Other Uses: Cut the huge flower stalk and collect the juice for drinking. Some species have very fibrous leaves. Pound the leaves and remove the fibers for weaving and making ropes. Most species have thick, sharp needles at the tips of the leaves. Use them for sewing or making hacks. The sap of some species contains a chemical that makes the sap suitable for use as a soap.

Almond

Prunus amygdalus

Description: The almond tree, which sometimes grows to 12.2 meters, looks like a peach tree. The fresh almond fruit resembles a gnarled, unripe peach and grows in clusters. The stone (the almond itself) is covered with a thick, dry, woolly skin.

Habitat and Distribution: Almonds are found in the scrub and thorn forests of the tropics, the evergreen scrub forests of temperate areas, and in desert scrub and waste in all climatic zones. The almond tree is also found in the semidesert areas of the Old World in southern Europe, the eastern Mediterranean, Iran, the Middle East, China, Madeira, the Azores, and the Canary Islands.

Edible Parts: The mature almond fruit splits open lengthwise down the side, exposing the ripe almond nut. You can easily get the dry kernel by simply cracking open the stone. Almond meats are rich in food value, like all nuts. Gather them in large quantities and shell them for further use as survival food. You could live solely on almonds for rather long periods. When you boil them, the kernel's outer covering comes off and only the white meat remains.

Amaranth
Amaranthus species

Description: These plants, which grow 90 centimeters to 150 centimeters tall, are abundant weeds in many parts of the world. All amaranth have alternate simple leaves. They may have some red color present on the stems. They bear minute, greenish flowers in dense clusters at the top of the plants. Their seeds may be brown or black in weedy species and light-colored in domestic species.

Habitat and Distribution: Look for amaranth along roadsides, in disturbed waste areas, or as weeds in crops throughout the world. Some amaranth species have been grown as a grain crop and a garden vegetable in various parts of the world, especially in South America.

Edible Parts: All parts are edible, but some may have sharp spines you should remove before eating. The young plants or the growing tips of alder plants are an excellent vegetable. Simply boil the young plants or eat them raw. Their seeds are very nutritious. Shake the tops of alder plants to get the seeds. Eat the seeds raw, boiled, ground into flour, or popped like popcorn.

Arctic willow
Salix arctica

Description: The arctic willow is a shrub that never exceeds more than 60 centimeters in height and grows in clumps that form dense mats on the tundra.

Habitat and Distribution: The arctic willow is common on tundras in North America, Europe, and Asia. You can also find it in some mountainous areas in temperate regions.

Edible Parts: You can collect the succulent, tender young shoots of the arctic willow in early spring. Strip off the outer bark of the new shoots and eat the inner portion raw. You can also peel and eat raw the young underground shoots of any of the various kinds of arctic willow. Young willow leaves are one of the richest sources of vitamin C, containing 7 to 10 times more than an orange.

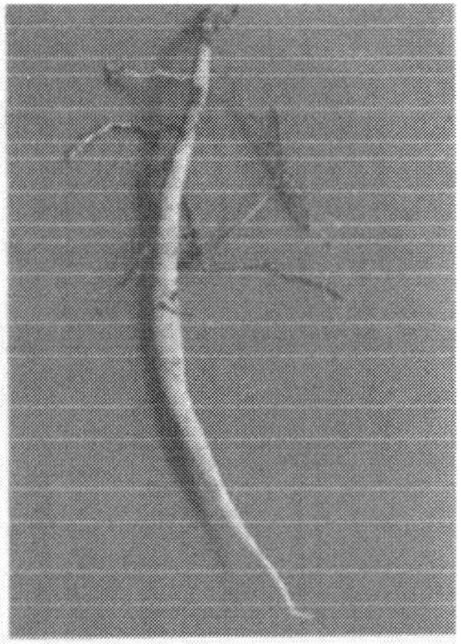

Arrowroot

Maranta and Sagittaria species

Description: The arrowroot is an aquatic plant with arrow-shaped leaves and potatolike tubers in the mud.

Habitat and Distribution: Aroowroot is found worldwide in temperate zones and the tropics. It is found in moist to wet habitats.

Edible Parts: The rootstock is a rich source of high quality starch. Boil the rootstock and eat it as a vegetable.

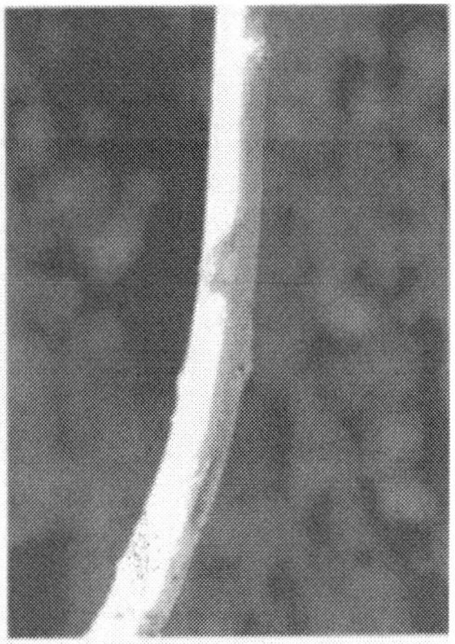

Asparagus

Asparagus officinalis

Description: The spring growth of this plant resembles a cluster of green fingers. The mature plant has fernlike, wispy foliage and red berries. Its flowers are small and greenish in color. Several species have sharp, thornlike structures.

Habitat and Distribution: Asparagus is found worldwide in temperate areas. Look for it in fields, old homesites, and fencerows.

Edible Parts: Eat the young stems before leaves form. Steam or boil them for 10 to 15 minutes before eating. Raw asparagus may cause nausea or diarrhea. The fleshy roots are a good source of starch.

WARNING

Do not eat the fruits of any species since some are toxic.

Bael fruit
Aegle marmelos

Description: This is a tree that grows from 2.4 to 4.6 meters tall, with a dense spiny growth. The fruit is 5 to 10 centimeters in diameter, gray or yellowish, and full of seeds.

Habitat and Distribution: Bael fruit is found in rain forests and semievergreen seasonal forests of the tropics. It grows wild in India and Burma.

Edible Parts: The fruit, which ripens in December, is at its best when just turning ripe. The juice of the ripe fruit, diluted with water and mixed with a small amount of tamarind and sugar or honey, is sour but refreshing. Like other citrus fruits, it is rich in vitamin C.

Bamboo
Various species including *Bambusa, Dendrocalamus, Phyllostachys*

Description: Bamboos are woody grasses that grow up to 15 meters tall. The leaves are grasslike and the stems are the familiar bamboo used in furniture and fishing poles.

Habitat and Distribution: Look for bamboo in warm, moist regions in open or jungle country, in lowland, or on mountains. Bamboos are native to the Far East (Temperate and Tropical zones) but have bean widely planted around the world.

Edible Parts: The young shoots of almost all species are edible raw or cooked. Raw shoots have a slightly bitter taste that is removed by boiling. To prepare, remove the tough protective sheath that is coated with tawny or red hairs. The seed grain of the flowering bamboo is also edible. Boil the seeds like rice or pulverize them, mix with water, and make into cakes.

Other Uses: Use the mature bamboo to build structures or to make containers, ladles, spoons, and various other cooking utensils. Also use bamboo to make tools and weapons. You can make a strong bow by splitting the bamboo and putting several pieces together.

CAUTION

Green bamboo may explode in a fire. Green bamboo has an internal membrane you must remove before using it as a food or water container.

Banana and plantain

Musa species

Description: These are treelike plants with several large leaves at the top. Their flowers are borne in dense hanging clusters.

Habitat and Distribution: Look for bananas and plantains in open fields or margins of forests where they are grown as a crop. They grow in the humid tropics.

Edible Parts: Their fruits are edible raw or cooked. They may be boiled or baked. You can boil their flowers and eat them like a vegetable. You can cook and eat the rootstocks and leaf sheaths of many species. The center or "heart" or the plant is edible year-round, cooked or raw.

Other Uses: You can use the layers of the lower third of the plants to cover coals to roast food. You can also use their stumps to get water (see Chapter 6). You can use their leaves to wrap other foods for cooking or storage.

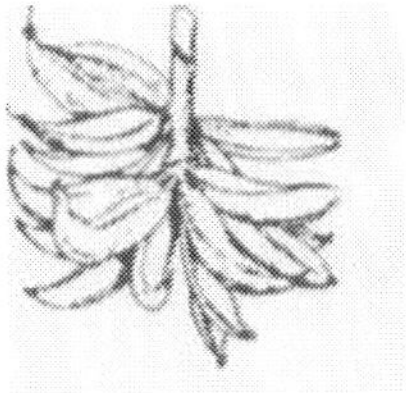

Baobab

Adansonia digitata

Description: The baobab tree may grow as high as 18 meters and may have a trunk 9 meters in diameter. The tree has short, stubby branches and a gray, thick bark. Its leaves are compound and their segments are arranged like the palm of a hand. Its flowers, which are white and several centimeters across, hang from the higher branches. Its fruit is shaped like a football, measures up to 45 centimeters long, and is covered with short dense hair.

Habitat and Distribution: These trees grow in savannas. They are found in Africa, in parts of Australia, and on the island of Madagascar.

Edible Parts: You can use the young leaves as a soup vegetable. The tender root of the young baobab tree is edible. The pulp and seeds of the fruit are also edible. Use one handful of pulp to about one cup of water for a refreshing drink. To obtain flour, roast the seeds, then grind them.

Other Uses: Drinking a mixture of pulp and water will help cure diarrhea. Often the hollow trunks are good sources of fresh water. The bark can be cut into strips and pounded to obtain a strong fiber for making rope.

Batoko plum
Flacourtia inermis

Description: This shrub or small tree has dark green, alternate, simple leaves. Its fruits are bright red and contain six or more seeds.

Habitat and Distribution: This plant is a native of the Philippines but is widely cultivated for its fruit in other areas. It can be found in clearings and at the edges of the tropical rain forests of Africa and Asia.

Edible Parts: Eat the fruit raw or cooked.

Bearberry or kinnikinnick
Arctostaphylos uvaursi

Description: This plant is a common evergreen shrub with reddish, scaly bark and thick, leathery leaves 4 centimeters long and 1 centimeter wide. It has white flowers and bright red fruits.

Habitat and Distribution: This plant is found in arctic, subarctic, and temperate regions, most often in sandy or rocky soil.

Edible Parts: Its berries are edible raw or cooked. You can make a refreshing tea from its young leaves.

Beech
Fagus species

Description: Beech trees are large (9 to 24 meters), symmetrical forest trees that have smooth, light-gray bark and dark green foliage. The character of its bark, plus its clusters of prickly seedpods, clearly distinguish the beech tree in the field.

Habitat and Distribution: This tree is found in the Temperate Zone. It grows wild in the eastern United States, Europe, Asia, and North Africa. It is found in moist areas, mainly in the forests. This tree is common throughout southeastern Europe and across temperate Asia. Beech relatives are also found in Chile, New Guinea, and New Zealand.

Edible Parts: The mature beechnuts readily fall out of the husklike seedpods. You can eat these dark brown triangular nuts by breaking the thin shell with your fingernail and removing the white, sweet kernel inside. Beechnuts are one of the most delicious of all wild nuts. They are a most useful survival food because of the kernel's high oil content. You can also use the beechnuts as a coffee substitue. Roast them so that the kernel becomes golden brown and quite hard. Then pulverize the kernel and, after boiling or steeping in hot water, you have a passable coffee substitute.

Bignay
Antidesma bunius

Description: Bignay is a shrub or small tree, 3 to 12 meters tall, with shiny, pointed leaves about 15 centimeters long. Its flowers are small, clustered, and green. It has fleshy, dark red or black fruit and a single seed. The fruit is about 1 centimeter in diameter.

Habitat and Distribution: This plant is found in rain forests and semievergreen seasonal forests in the tropics. It is found in open places and in secondary forests. It grows wild from the Himalayas to Ceylon and eastward through Indonesia to northern Australia. However, it may be found anywhere in the tropics in cultivated forms.

Edible Parts: The fruit is edible raw. Do not eat any other parts of the tree. In Africa, the roots are toxic. Other parts of the plant may be poisonous.

CAUTION

Eaten in large quantities, the fruit may have a laxative effect.

Blackberry, raspberry, and dewberry

Rubus species

Description: These plants have prickly stems (canes) that grow upward, arching back toward the ground. They have alternate, usually compound leaves. Their fruits may be red, black, yellow, or orange.

Habitat and Distribution: These plants grow in open, sunny areas at the margin of woods, lakes, streams, and roads throughout temperate regions. There is also an arctic raspberry.

Edible Parts: The fruits and peeled young shoots are edible. Flavor varies greatly.

Other Uses: Use the leaves to make tea. To treat diarrhea, drink a tea made by brewing the dried root bark of the blackberry bush.

Blueberry and huckleberry

Vaccinium and *Gaylussacia* species

Description: These shrubs vary in size from 30 centimeters to 3.7 meters tall. All have alternate, simple leaves. Their fruits may be dark blue, black, or red and have many small seeds.

Habitat and Distribution: These plants prefer open, sunny areas. They are found throughout much of the north temperate regions and at higher elevations in Central America.

Edible Parts: Their fruits are edible raw.

Breadfruit

Artocarpus incisa

Description: This tree may grow up to 9 meters tall. It has dark green, deeply divided leaves that are 75 centimeters long and 30 centimeters wide. Its fruits are large, green, ball-like structures up to 30 centimeters across when mature.

Habitat and Distribution: Look for this tree at the margins of forests and homesites in the humid tropics. It is native to the South Pacific region but has been widely planted in the West Indies and parts of Polynesia.

Edible Parts: The fruit pulp is edible raw. The fruit can be sliced, dried, and ground into flour for later use. The seeds are edible cooked.

Other Uses: The thick sap can serve as glue and caulking material. You can also use it as birdlime (to entrap small birds by smearing the sap on twigs where they usually perch).

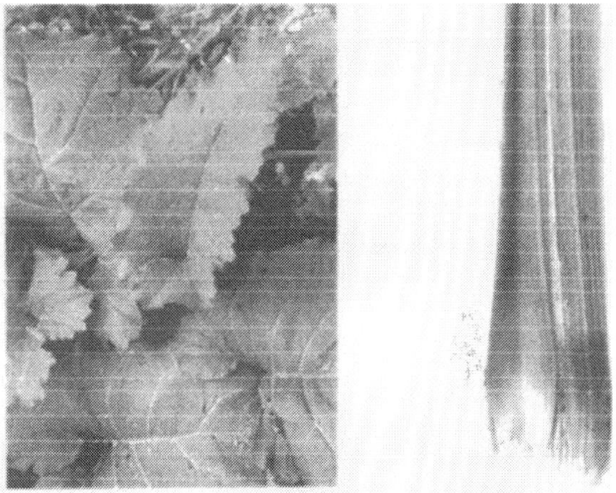

Burdock

Arctium lappa

Description: This plant has wavy-edged, arrow-shaped leaves and flower heads in burrlike clusters. It grows up to 2 meters tall, with purple or pink flowers and a large, fleshy root.

Habitat and Distribution: Burdock is found worldwide in the North Temperate Zone. Look for it in open waste areas during the spring and summer.

Edible Parts: Peel the tender leaf stalks and eat them raw or cook them like greens. The roots are also edible boiled or baked.

CAUTION

Do not confuse burdock with rhubarb that has poisonous leaves.

Other Uses: A liquid made from the roots will help to produce sweating and increase urination. Dry the root, simmer it in water, strain the liquid, and then drink the strained liquid. Use the fiber from the dried stalk to weave cordage.

Burl Palm

Corypha elata

Description: This tree may reach 18 meters in height. It has large, fan-shaped leaves up to 3 meters long and split into about 100 narrow segments. It bears flowers in huge dusters at the top of the tree. The tree dies after flowering.

Habitat and Distribution: This tree grows in coastal areas of the East Indies.

Edible Parts: The trunk contains starch that is edible raw. The very tip of the trunk is also edible raw or cooked. You can get large quantities of liquid by bruising the flowering stalk. The kernels of the nuts are edible.

> **CAUTION**
>
> The seed covering may cause dermatitis in some individuals.

Other Uses: You can use the leaves as weaving material.

Canna lily

Canna indica

Description: The canna lily is a coarse perennial herb, 90 centimeters to 3 meters tall. The plant grows from a large, thick, underground rootstock that is edible. Its large leaves resemble those of the banana plant but are not so large. The flowers of wild canna lily are usually small, relatively inconspicuous, and brightly colored reds, oranges, or yellows.

Habitat and Distribution: As a wild plant, the canna lily is found in all tropical areas, especially in moist places along streams, springs, ditches, and the margins of woods. It may also be found in wet temperate, mountainous regions. It is easy to recognize because it is commonly cultivated in flower gardens in the United States.

Edible Parts: The large and much branched rootstocks are full of edible starch. The younger parts may be finely chopped and then boiled or pulverized into a meal. Mix in the young shoots of palm cabbage for flavoring.

Carob tree

Ceratonia siliqua

Description: This large tree has a spreading crown. Its leaves are compound and alternate. Its seedpods, also known as Saint John's bread, are up to 45 centimeters long and are filled with round, hard seeds and a thick pulp.

Habitat and Distribution: This tree is found throughout the Mediterranean, the Middle East, and parts of North Africa.

Edible Parts: The young tender pods are edible raw or boiled. You can pulverize the seeds in mature pods and cook as porridge.

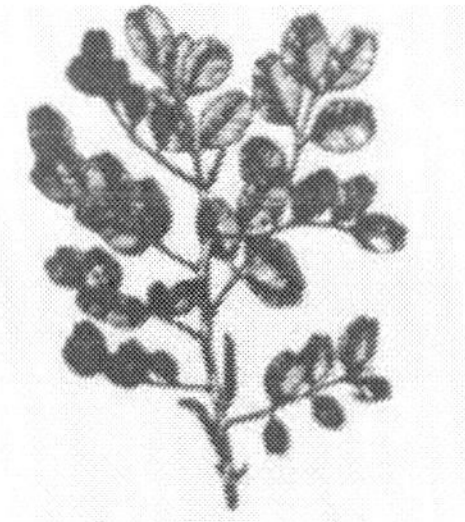

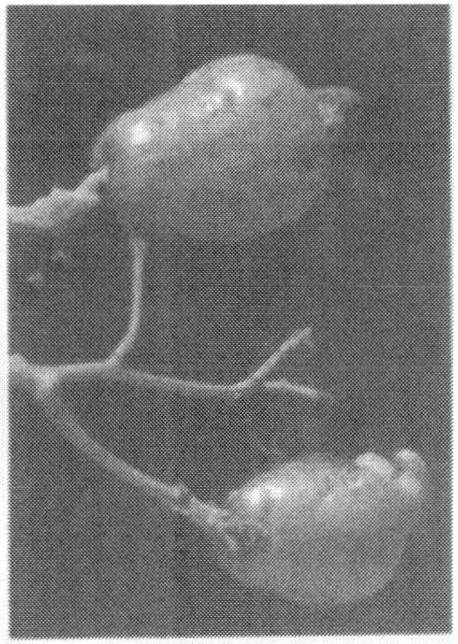

Cashew nut

Anacardium occidentale

Description: The cashew is a spreading evergreen tree growing to a height of 12 meters, with leaves up to 20 centimeters long and 10 centimeters wide. Its flowers are yellowish-pink. Its fruit is very easy to recognize because of its peculiar structure. The fruit is thick and pear-shaped, pulpy and red or yellow when ripe. This fruit bears a hard, green, kidney-shaped nut at its tip. This nut is smooth, shiny, and green or brown according to its maturity.

Habitat and Distribution: The cashew is native to the West Indies and northern South America, but transplantation has spread it to all tropical climates. In the Old World, it has escaped from cultivation and appears to be wild at least in parts of Africa and India.

Edible Parts: The nut encloses one seed. The seed is edible when roasted. The pear-shaped fruit is juicy, sweet-acid, and astringent. It is quite safe and considered delicious by most people who eat it.

> **CAUTION**
>
> The green hull surrounding the nut contains a resinous irritant poison that will blister the lips and tongue like poison ivy. Heat destroys this poison when roasting the nuts.

Cattail
Typha latifolia

Description: Cattails are grasslike plants with strap-shaped leaves 1 to 5 centimeters wide and growing up to 1.8 meters tall. The male flowers are borne in a dense mass above the female flowers. These last only a short time, leaving the female flowers that develop into the brown cattail. Pollen from the male flowers is often abundant and bright yellow.

Habitat and Distribution: Cattails are found throughout most of the world. Look for them in full sun areas at the margins of lakes, streams, canals, rivers, and brackish water.

Edible Parts: The young tender shoots are edible raw or cooked. The rhizome is often very tough but is a rich source of starch. Pound the rhizome to remove the starch and use as a flour. The pollen is also an exceptional source of starch. When the cattail is immature and still green, you can boil the female portion and eat it like corn on the cob.

Other Uses: The dried leaves are an excellent source of weaving material you can use to make floats and rafts. The cottony seeds make good pillow stuffing and insulation. The fluff makes excellent tinder. Dried cattails are effective insect repellents when burned.

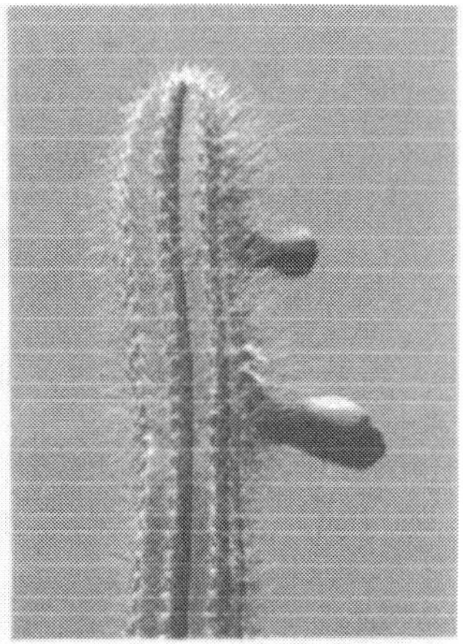

Cereus cactus
Cereus species

Description: These cacti are tall and narrow with angled stems and numerous spines.

Habitat and Distribution: They may be found in true deserts and other dry, open, sunny areas throughout the Caribbean region, Central America, and the western United States.

Edible Parts: The fruits are edible, but some may have a laxative effect.

Other Uses: The pulp of the cactus is a good source of water. Break open the stem and scoop out the pulp.

Chestnut

Castanea sativa

Description: The European chestnut is usually a large tree, up to 18 meters in height.

Habitat and Distribution: In temperate regions, the chestnut is found in both hardwood and coniferous forests. In the tropics, it is found in semievergreen seasonal forests. They are found over all of middle and south Europe and across middle Asia to China and Japan. They are relatively abundant along the edge of meadows and as a forest tree. The European chestnut is one of the most common varieties. Wild chestnuts in Asia belong to the related chestnut species.

Edible Parts: Chestnuts are highly useful as survival food. Ripe nuts are usually picked in autumn, although unripe nuts picked while green may also be used for food. Perhaps the easiest way to prepare them is to roast the ripe nuts in embers. Cooked this way, they are quite tasty, and you can eat large quantities. Another way is to boil the kernels after removing the outer shell. After being boiled until fairly soft, you can mash the nuts like potatoes.

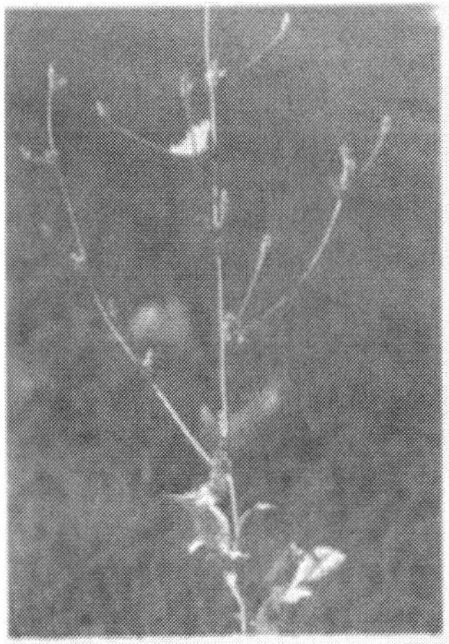

Chicory

Cichorium intybus

Description: This plant grows up to 1.8 meters tall. It has leaves clustered at the base of the stem and some leaves on the stem. The base leaves resemble those of the dandelion. The flowers are sky blue and stay open only on sunny days. Chicory has a milky juice.

Habitat and Distribution: Look for chicory in old fields, waste areas, weedy lots, and along roads. It is a native of Europe and Asia, but is also found in Africa and most of North America where it grows as a weed.

Edible Parts: All parts are edible. Eat the young leaves as a salad or boil to eat as a vegetable. Cook the roots as a vegetable. For use as a coffee substitute, roast the roots until they are dark brown and then pulverize them.

Chufa
Cyperus esculentus

Description: This very common plant has a triangular stem and grasslike leaves. It grows to a height of 20 to 60 centimeters. The mature plant has a soft furlike bloom that extends from a whorl of leaves. Tubers 1 to 2.5 centimeters in diameter grow at the ends of the roots.

Habitat and Distribution: Chufa grows in moist sandy areas throughout the world. It is often an abundant weed in cultivated fields.

Edible Parts: The tubers are edible raw, boiled, or baked. You can also grind them and use them as a coffee substitute.

Coconut
Cocos nucifera

Description: This tree has a single, narrow, tall trunk with a cluster of very large leaves at the top. Each leaf may be over 6 meters long with over 100 pairs of leaflets.

Habitat and Distribution: Coconut palms are found throughout the tropics. They are most abundant near coastal regions.

Edible Parts: The nut is a valuable source of food. The milk of the young coconut is rich in sugar and vitamins and is an excellent source of liquid. The nut meat is also nutritious but is rich in oil. To preserve the meat, spread it in the sun until it is completely dry.

Other Uses: Use coconut oil to cook and to protect metal objects from corrosion. Also use the oil to treat saltwater sores, sunburn, and dry skin. Use the oil in improvised torches. Use the tree trunk as building material and the leaves as thatch. Hollow out the large stump for use as a food container. The coconut husks are good flotation devices and the husk's fibers are used to weave ropes and other items. Use the gauzelike fibers at the leaf bases as strainers or use them to weave a bug net or to make a pad to use on wounds. The husk makes a good abrasive. Dried husk fiber is an excellent tinder. A smoldering husk helps to repel mosquitoes. Smoke caused by dripping coconut oil in a fire also repels mosquitoes. To render coconut oil, put the coconut meat in the sun, heat it over a slow fire, or boil it in a pot of water. Coconuts washed out to sea are a good source of fresh liquid for the sea survivor.

Common jujube

Ziziphus jujuba

Description: The common jujube is either a deciduous tree growing to a height of 12 meters or a large shrub, depending upon where it grows and how much water is available for growth. Its branches are usually spiny. Its reddish-brown to yellowish-green fruit is oblong to ovoid, 3 centimeters or less in diameter, smooth, and sweet in flavor, but has rather dry pulp around a comparatively large stone. Its flowers are green.

Habitat and Distribution: The jujube is found in forested areas of temperate regions and in desert scrub and waste areas worldwide. It is common in many of the tropical and subtropical areas of the Old World. In Africa, it is found mainly bordering the Mediterranean. In Asia, it is especially common in the drier parts of India and China. The jujube is also found throughout the East Indies. It can be found bordering some desert areas.

Edible Parts: The pulp, crushed in water, makes a refreshing beverage. If time permits, you can dry the ripe fruit in the sun like dates. Its fruits are high in vitamins A and C.

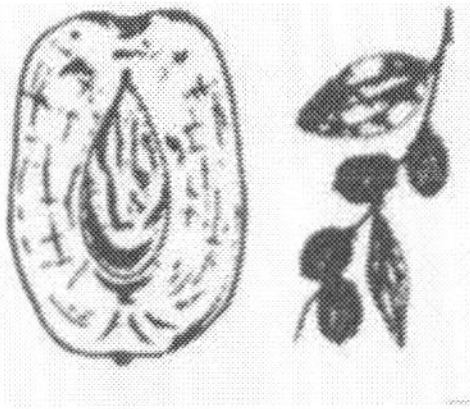

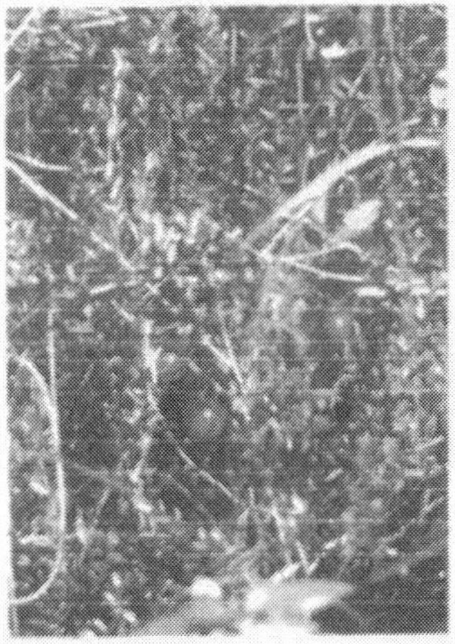

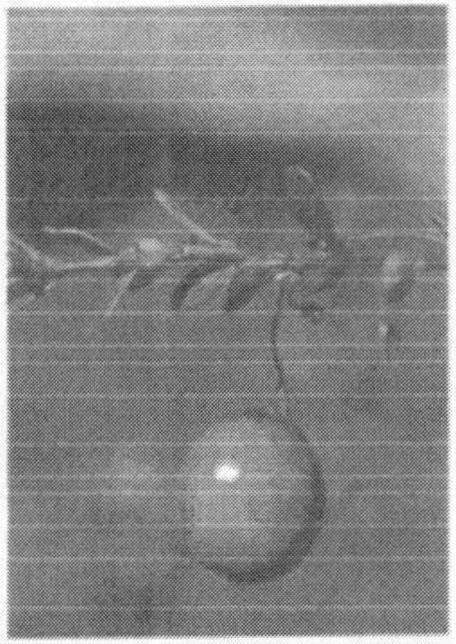

Cranberry

Vaccinium macrocarpon

Description: This plant has tiny leaves arranged alternately. Its stem creeps along the ground. Its fruits are red berries.

Habitat and Distribution: It only grows in open, sunny, wet areas in the colder regions of the Northern Hemisphere.

Edible Parts: The berries are very tart when eaten raw. Cook in a small amount of water and add sugar, if available, to make a jelly.

Other Uses: Cranberries may act as a diuretic. They are useful for treating urinary tract infections.

Crowberry

Empetrum nigrum

Description: This is a dwarf evergreen shrub with short needlelike leaves. It has small, shiny, black berries that remain on the bush throughout the winter.

Habitat and Distribution: Look for this plant in tundra throughout arctic regions of North America and Eurasia.

Edible Parts: The fruits are edible fresh or can be dried for later use.

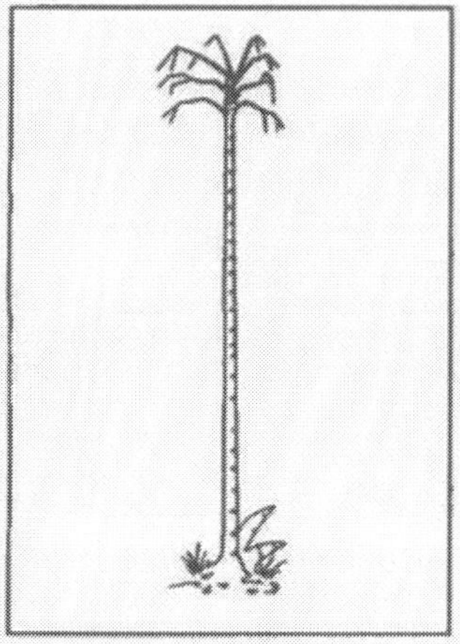

Cuipo tree

Cavanillesia platanifolia

Description: This is a very dominant and easily detected tree because it extends above the other trees. Its height ranges from 45 to 60 meters. It has leaves only at the top and is bare 11 months out of the year. It has rings on its bark that extend to the top to make is easily recognizable. Its bark is reddish or gray in color. Its roots are light reddish-brown or yellowish-brown.

Habitat and Distribution: The cuipo tree is located primarily in Central American tropical rain forests in mountainous areas.

Edible Parts: To get water from this tree, cut a piece of the root and clean the dirt and bark off one end, keeping the root horizontal. Put the clean end to your mouth or canteen and raise the other. The water from this tree tastes like potato water.

Other Uses: Use young saplings and the branches' inner bark to make rope.

Dandelion

Taraxacum officinale

Description: Dandelion leaves have a jagged edge, grow close to the ground, and are seldom more than 20 centimeters long. Its flowers are bright yellow. There are several dandelion species.

Habitat and Distribution: Dandelions grow in open, sunny locations throughout the Northern Hemisphere.

Edible Parts: All parts are edible. Eat the leaves raw or cooked. Boil the roots as a vegetable. Roots roasted and ground are a good coffee substitue. Dandelions are high in vitamins A and C and in calcium.

Other Uses: Use the white juice in the flower stems as glue.

Date palm

Phoenix dactylifera

Description: The date palm is a tall, unbranched tree with a crown of huge, compound leaves. Its fruit is yellow when ripe.

Habitat and Distribution: This tree grows in arid semitropical regions. It is native to North Africa and the Middle East but has been planted in the arid semitropics in other parts of the world.

Edible Parts: Its fruit is edible fresh but is very bitter if eaten before it is ripe. You can dry the fruits in the sun and preserve them for a long time.

Other Uses: The trunks provide valuable building material in desert regions where few other treelike plants are found. The leaves are durable and you can use them for thatching and as weaving material. The base of the leaves resembles coarse cloth that you can use for scrubbing and cleaning.

Daylily
Hemerocallis fulva

Description: This plant has unspotted, tawny blossoms that open for 1 day only. It has long, swordlike, green basal leaves. Its root is a mass of swollen and elongated tubers.

Habitat and Distribution: Daylilies are found worldwide in Tropic and Temperate Zones. They are grown as a vegetable in the Orient and as an ornamental plant elsewhere.

Edible Parts: The young green leaves are edible raw or cooked. Tubers are also edible raw or cooked. You can eat its flowers raw, but they taste better cooked. You can also fry the flowers for storage.

> **CAUTION**
> Eating excessive amounts of raw flowers may cause diarrhea.

Duchesnea or Indian strawberry
Duchesnea indica

Description: The duchesnea is a small plant that has runners and three-parted leaves. Its flowers are yellow and its fruit resembles a strawberry.

Habitat and Distribution: It is native to southern Asia but is a common weed in warmer temperate regions. Look for it in lawns, gardens, and along roads.

Edible Parts: Its fruit is edible. Eat it fresh.

Elderberry

Sambucus canadensis

Description: Elderberry is a many-stemmed shrub with opposite, compound leaves. It grows to a height of 6 meters. Its flowers are fragrant, white, and borne in large flat-topped clusters up to 30 centimeters across. Its berrylike fruits are dark blue or black when ripe.

Habitat and Distribution: This plant is found in open, usually wet areas at the margins of marshes, rivers, ditches, and lakes. It grows throughout much of eastern North America and Canada.

Edible Parts: The flowers and fruits are edible. You can make a drink by soaking the flower heads for 8 hours, discarding the flowers, and drinking the liquid.

> **CAUTION**
>
> All other parts of the plant are poisonous and dangerous if eaten.

Fireweed

Epilobium angustifolium

Description: This plant grows up to 1.8 meters tall. It has large, showy, pink flowers and lance-shaped leaves. Its relative, the dwarf fireweed (Epilobium latifolium), grows 30 to 60 centimeters tall.

Habitat and Ditstribution: Tall fireweed is found in open woods, on hillsides, on stream banks, and near seashores in arctic regions. It is especially abundant in burned-over areas. Dwarf fireweed is found along streams, sandbars, and lakeshores and on alpine and arctic slopes.

Edible Parts: The leaves, stems, and flowers are edible in the spring but become tough in summer. You can split open the stems of old plants and eat the pith raw.

Fishtail palm
Caryota urens

Description: Fishtail palms are large trees, at least 18 meters tall. Their leaves are unlike those of any other palm; the leaflets are irregular and toothed on the upper margins. All other palms have either fan-shaped or featherlike leaves. Its massive flowering shoot is borne at the top of the tree and hangs downward.

Habitat and Distribution: The fishtail palm is native to the tropics of India, Assam, and Burma. Several related species also exist in Southeast Asia and the Philippines. These palms are found in open hill country and jungle areas.

Edible Parts: The chief food in this palm is the starch stored in large quantities in its trunk. The juice from the fishtail palm is very nourishing and you have to drink it shortly after getting it from the palm flower shoot. Boil the juice down to get a rich sugar syrup. Use the same method as for the sugar palm to get the juice. The palm cabbage may be eaten raw or cooked.

Foxtail grass
Setaria species

Description: This weedy grass is readily recognized by the narrow, cylindrical head containing long hairs. Its grains are small, less than 6 millimeters long. The dense heads of grain often droop when ripe.

Habitat and Distribution: Look for foxtail grasses in open, sunny areas, along roads, and at the margins of fields. Some species occur in wet, marshy areas. Species of *Setaria* are found throughout the United States, Europe, western Asia, and tropical Africa. In some parts of the world, foxtail grasses are grown as a food crop.

Edible Parts: The grains are edible raw but are very hard and sometimes bitter. Boiling removes some of the bitterness and makes them easier to eat.

Goa bean

Psophocarpus tetragonolobus

Description: The goa bean is a climbing plant that may cover small shrubs and trees. Its bean pods are 22 centimeters long, its leaves 15 centimeters long, and its flowers are bright blue. The mature pods are 4-angled, with jagged wings on the pods.

Habitat and Distribution: This plant grows in tropical Africa, Asia, the East Indies, the Philippines, and Taiwan. This member of the bean (legume) family serves to illustrate a kind of edible bean oommon in the tropics of the Old World. Wild edible beans of this sort are most frequently found in clearings and around abandoned garden sites. They are more rare in forested areas.

Edible Parts: You can eat the young pods like string beans. The mature seeds are a valuable source of protein after parching or roasting them over hot coals. You can germinate the seeds (as you can many kinds of beans) in damp moss and eat the resultant sprouts. The thickened roots are edible raw. They are slightly sweet, with the firmness of an apple. You can also eat the young leaves as a vegetable, raw or steamed.

Hackberry

Celtis species

Description: Hackberry trees have smooth, gray bark that often has corky warts or ridges. The tree may reach 39 meters in height. Hackberry trees have long-pointed leaves that grow in two rows. This tree bears small, round berries that can be eaten when they are ripe and fall from the tree. The wood of the hackberry is yellowish.

Habitat and Distribution: This plant is widespread in the United States, especially in and near ponds.

Edible Parts: Its berries are edible when they are ripe and fall from the tree.

Hazelnut or wild filbert

Corylus species

Description: Hazelnuts grow on bushes 1.8 to 3.6 meters high. One species in Turkey and another in China are large trees. The nut itself grows in a very bristly husk that conspicuously contracts above the nut into a long neck. The different species vary in this respect as to size and shape.

Habitat and Distribution: Hazelnuts are found over wide areas in the United States, especially the eastern half of the country and along the Pacific coast. These nuts are also found in Europe where they are known as filberts. The hazelnut is common in Asia, especially in eastern Asia from the Himalayas to China and Japan. The hazelnut usually grows in the dense thickets along stream banks and open places. They are not plants of the dense forest.

Edible Parts: Hazelnuts ripen in the autumn when you can crack them open and eat the kernel. The dried nut is extremely delicious. The nut's high oil content makes it a good survival food. In the unripe stage, you can crack them open and eat the fresh kernel.

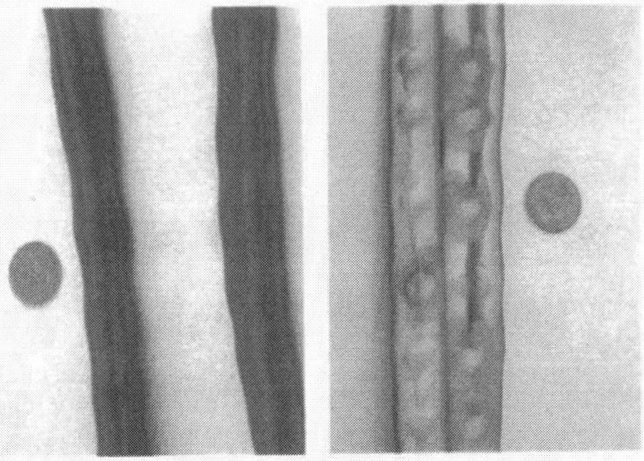

Horseradish tree

Moringa pterygosperma

Description: This tree grows from 4.5 to 14 meters tall. Its leaves have a fernlike appearance. Its flowers and long, pendulous fruits grow on the ends of the branches. Its fruit (pod) looks like a giant bean. Its 25- to 60-centimeter-long pods are triangular in cross section, with strong ribs. Its roots have a pungent odor.

Habitat and Distribution: This tree is found in the rain forests and semievergreen seasonal forests of the tropical regions. It is widespread in India, Southeast Asia, Africa, and Central America. Look for it in abandoned fields and gardens and at the edges of forests.

Edible Parts: The leaves are edible raw or cooked, depending on their hardness. Cut the young seedpods into short lengths and cook them like string beans or fry them. You can get oil for frying by boiling the young fruits of palms and skimming the oil off the surface of the water. You can eat the flowers as part of a salad. You can chew fresh, young seedpods to eat the pulpy and soft seeds. The roots may be ground as a substitute for seasoning similar to horseradish.

Iceland moss
Cetraria islandica

Description: This moss grows only a few inches high. Its color may be gray, white, or even reddish.

Habitat and Distribution: Look for it in open areas. It is found only in the arctic.

Edible Parts: All parts of the Iceland moss are edible. During the winter or dry season, it is dry and crunchy but softens when soaked. Boil the moss to remove the bitterness. After boiling, eat by itself or add to milk or grains as a thickening agent. Dried plants store well.

Indian potato or Eskimo potato
Claytonia species

Description: All *Claytonia* species are somewhat fleshy plants only a few centimeters tall, with showy flowers about 2.5 centimeters across.

Habitat and Distribution: Some species are found in rich forests where they are conspicuous before the leaves develop. Western species are found throughout most of the northern United States and in Canada.

Edible Parts: The tubers are edible but you should boil them before eating.

Juniper

Juniperus species

Description: Junipers, sometimes called cedars, are trees or shurbs with very small, scalelike leaves densely crowded around the branches. Each leaf is less than 1.2 centimeters long. All species have a distinct aroma resembling the well-known cedar. The berrylike cones are usually blue and covered with a whitish wax.

Habitat and Distribution: Look for junipers in open, dry, sunny areas throughout North America and northern Europe. Some species are found in southeastern Europe, across Asia to Japan, and in the mountains of North Africa.

Edible Parts: The berries and twigs are edible. Eat the berries raw or roast the seeds to use as a coffee substitute. Use dried and crushed berries as a seasoning for meat. Gather young twigs to make a tea.

CAUTION

Many plants may be called cedars but are not related to junipers and may be harmful. Always look for the berrylike structures, neddle leaves, and resinous, fragrant sap to be sure the plant you have is a juniper.

Lotus

Nelumbo species

Description: There are two species of lotus: one has yellow flowers and the other pink flowers. The flowers are large and showy. The leaves, which may float on or rise above the surface of the water, often reach 1.5 meters in radius. The fruit has a distinctive flattened shape and contains up to 20 hard seeds.

Habitat and Distribution: The yellow-flowered lotus is native to North America. The pink-flowered species, which is widespread in the Orient, is planted in many other areas of the world. Lotuses are found in quiet fresh water.

Edible Parts: All parts of the plant are edible raw or cooked. The underwater parts contain large quantities of starch. Dig the fleshy portions from the mud and bake or boil them. Boil the young leaves and eat them as a vegetable. The seeds have a pleasant flavor and are nutritious. Eat them raw, or parch and grind them into flour.

Malanga

Xanthosoma caracu

Description: This plant has soft, arrow-shaped leaves, up to 60 centimeters long. The leaves have no aboveground stems.

Habitat and Distribution: This plant grows widely in the Caribbean region. Look for it in open, sunny fields.

Edible Parts: The tubers are rich in starch. Cook them before eating to detroy a poi¬ son contained in all parts of the plant.

WARNING
Always cook before eating.

Mango

Mangifera indica

Description: This tree may reach 30 meters in height. It has alternate, simple, shiny, dark green leaves. Its flowers are small and inconspicuous. Its fruits have a large single seed. There are many cultivated varieties of mango. Some have red flesh, others yellow or orange, often with many fibers and a kerosene taste.

Habitat and Distribution: This tree grows in warm, moist regions. It is native to north¬ ern India, Burma, and western Malaysia. It is now grown throughout the tropics.

Edible Parts: The fruits area nutritious food source. The unripe fruit can be peeled and its flesh eaten by shredding it and eating it like a salad. The ripe fruit can be peeled and eaten raw. Roasted seed kernels are edible.

CAUTION
If you are sensitive to poison ivy, avoid eating mangoes, as they cause a severe reaction in sensitive individuals.

Manioc

Manihot utillissima

Description: Manioc is a perennial shrubby plant, 1 to 3 meters tall, with jointed stems and deep green, fingerlike leaves. It has large, fleshy rootstocks.

Habitat and Distribution: Manioc is widespread in all tropical climates, particularly in moist areas. Although cultivated extensively, it maybe found in abandoned gardens and growing wild in many areas.

Edible Parts: The rootstocks are full of starch and high in food value. Two kinds of manioc are known: bitter and sweet. Both are edible. The bitter type contains poisonous hydrocyanic acid. To prepare manioc, first grind the fresh manioc root into a pulp, then cook it for at least 1 hour to remove the bitter poison from the roots. Then flatten the pulp into cakes and bake as bread. Manioc cakes or flour will keep almost indefinitely if protected against insects and dampness. Wrap them in banana leaves for protection.

CAUTION

For safety, always cook the roots of either type.

Marsh marigold

Caltha palustris

Description: This plant has rounded, dark green leaves arising from a short stem. It has bright yellow flowers.

Habitat and Distribution: This plant is found in bogs, lakes, and slow-moving streams. It is abundant in arctic and subarctic regions and in much of the eastern region of the northern United States.

Edible Parts: All parts are edible if boiled.

CAUTION

As with all water plants, do not eat this plant raw. Raw water plants may carry dangerous organisms that are removed only by cooking.

Mulberry

Morus species

Description: This tree has alternate, simple, often lobed leaves with rough surfaces. Its fruits are blue or black and many seeded.

Habitat and Distribution: Mulberry trees are found in forests, along roadsides, and in abandoned fields in Temperate and Tropical Zones of North America, South America, Europe, Asia, and Africa.

Edible Parts: The fruit is edible raw or cooked. It can be dried for eating later.

CAUTION

When eaten in quantity, mulberry fruit acts as a laxative. Green, unripe fruit can be hallucinogenic and cause extreme nausea and cramps.

Other Uses: You can shred the inner bark of the tree and use it to make twine or cord.

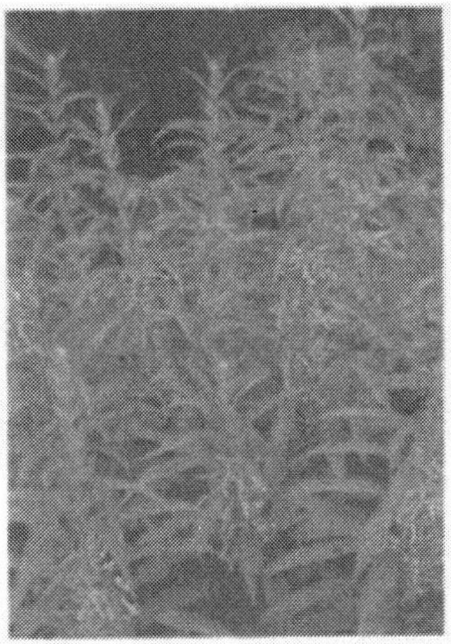

Nettle

Urtica and *Laportea* species

Description: These plants grow several feet high. They have small, inconspicuous flowers. Fine, hairlike bristles cover the stems, leafstalks, and undersides of leaves. The bristles cause a stinging sensation when they touch the skin.

Habitat and Distribution: Nettles prefer moist areas along streams or at the margins of forests. They are found throughout North America, Central America, the Caribbean, and northern Europe.

Edible Parts: Young shoots and leaves are edible. Boiling the plant for 10 to 15 minutes destroys the stinging element of the bristles. This plant is very nutritious.

Other Uses: Mature stems have a fibrous layer that you can divide into individual fibers and use to weave string or twine.

Nips palm

Nips fruticans

Description: This palm has a short, mainly underground trunk and very large, erect leaves up to 6 meters tall. The leaves are divided into leaflets. A flowering head forms on a short erect stem that rises among the palm leaves. The fruiting (seed) head is dark brown and may be 30 centimeters in diameter.

Habitat and Distribution: This palm is common on muddy shores in coastal regions throughout eastern Asia.

Edible Parts: The young flower stalk and the seeds provide a good source of water and food. Cut the flower stalk and collect the juice. The juice is rich in sugar. The seeds are hard but edible.

Other Uses: The leaves are excellent as thatch and coarse weaving material.

Oak

Quercus species

Description: Oak trees have alternate leaves and acorn fruits. There are two main groups of oaks: red and white. The red oak group has leaves with bristles and smooth bark in the upper part of the tree. Red oak acorns take 2 years to mature. The white oak group has leaves without bristles and a rough bark in the upper portion of the tree. White oak acorns mature in 1 year.

Habitat and Distribution: Oak trees are found in many habitats throughout North America, Central America, and parts of Europe and Asia.

Edible Parts: All parts are edible, but often contain large quantities of bitter sub¬stances. White oak acorns usually have a better flavor than red oak acorns. Gather and shell the acorns. Soak red oak acorns in water for 1 to 2 days to remove the bitter substance. You can speed up this process by putting wood ashes in the water in which you soak the acorns. Boil the acorns or grind them into flour and use the flour for baking. You can use acorns that you baked until very dark as a coffee substitute.

CAUTION

Tannic acid gives the acorns their bitter taste. Eating an excessive amount of acorns high in tannic acid can lead to kidney failure. Before eating acorns, leach out this chemical.

Other Uses: Oak wood is excellent for building or burning. Small oaks can be split and cut into long thin strips (3 to 6 millimeters thick and 1.2 centimeters wide) used to weave mats, baskets, or frameworks for packs, sleds, furniture, etc. Oak bark soaked in water produces a tanning solution used to preserve leather.

Orach

Atriplex species

Description: This plant is vinelike in growth and has arrowhead-shaped, alternate leaves up to 5 cenitmeters long. Young leaves maybe silver-colored. Its flowers and fruits are small and inconspicuous.

Habitat and Distribution: Orach species are entirety restricted to salty soils. They are found along North America's coasts and on the shores of alkaline lakes inland. They are also found along seashores from the Mediterranean countries to inland areas in North Africa and eastward to Turkey and central Siberia.

Edible Parts: The entire plant is edible raw or boiled.

Palmetto palm

Sabal palmetto

Description: The palmetto palm is a tall, unbranched tree with persistent leaf bases on most of the trunk. The leaves are large, simple, and palmately lobed. Its fruits are dark blue or black with a hard seed.

Habitat and Distribution: The palmetto palm is found throughout the coastal regions of the southeastern United States.

Edible Parts: The fruits are edible raw. The hard seeds may be ground into flour. The heart of the palm is a nutritious food source at any time. Cut off the top of the tree to obtain the palm heart.

Papaya or pawpaw

Carica papaya

Description: The papaya is a small tree 1.8 to 6 meters tall, with a soft, hollow trunk. When cut, the entire plant exudes a milky juice. The trunk is rough and the leaves are crowded at the trunk's apex. The fruit grows directly from the trunk, among and below the leaves. The fruit is green before ripening. When ripe, it turns yellow or remains greenish with a squashlike appearance.

Habitat and Distribution: Papaya is found in rain forests and semievergreen seasonal forests in tropical regions and in some temperate regions as well. Look for it in moist areas near clearings and former habitations. It is also found in open, sunny places in uninhabited jungle areas.

Edible Parts: The ripe fruit is high in vitamin C. Eat it raw or cock it like squash. Place green fruit in the sun to make it ripen quickly. Cook the young papaya leaves, flowers, and stems carefully, changing the water as for taro.

CAUTION

Be careful not to get the milky sap from the unripe fruit into your eyes. It will cause intense pain and temporary—sometimes even permanent—blindness.

Other Uses: Use the milky juice of the unripe fruit to tenderize tough meat. Rub the juice on the meat.

Persimmon

Diospyros virginiana and other species

Description: These trees have alternate, dark green, elliptic leaves with entire margins. The flowers are inconspicuous. The fruits are orange, have a sticky consistency, and have several seeds.

Habitat and Distribution: The persimmon is a common forest margin tree. It is wide spread in Africa, eastern North America, and the Far East.

Edible Parts: The leaves are a good source of vitamin C. The fruits are edible raw or baked. To make tea, dry the leaves and soak them in hot water. You can eat the roasted seeds.

CAUTION

Some persons are unable to digest persimmon pulp. Unripe persimmons are highly astringent and inedible.

Pincushion cactus

Mammilaria species

Description: Members of this cactus group are round, short, barrel-shaped, and without leaves. Sharp spines cover the entire plant.

Habitat and Distribution: These cacti are found throughout much of the desert regions of the western United States and parts of Central America.

Edible Parts: They are a good source of water in the desert.

Pine

Pinus species

Description: Pine trees are easily recognized by their needlelike leaves grouped in bundles. Each bundle may contain one to five needles, the number varying among species. The tree's odor and sticky sap provide a simple way to distinguish pines from similar looking trees with needlelike leaves.

Habitat and Distribution: Pines prefer open, sunny areas. They are found throughout North America, Central America, much of the Caribbean region, North Africa, the Middle East, Europe, and some places in Asia.

Edible Parts: The seeds of all species are edible. You can collect the young male cones, which grow only in the spring, as a survival food. Boil or bake the young cones. The bark of young twigs is edible. Peel off the bark of thin twigs. You can chew the juicy inner bark; it is rich in sugar and vitamins. Eat the seeds raw or cooked. Green pine needle tea is high in vitamin C.

Other Uses : Use the resin to waterproof articles. Also use it as glue. Collect the resin from the tree. If there is not enough resin on the tree, cut a notch in the bark so more sap will seep out. Put the resin in a container and heat it. The hot resin is your glue. Use it as is or add a small amount of ash dust to strengthen it. Use it immediately. You can use hardened pine resin as an emergency dental fillling.

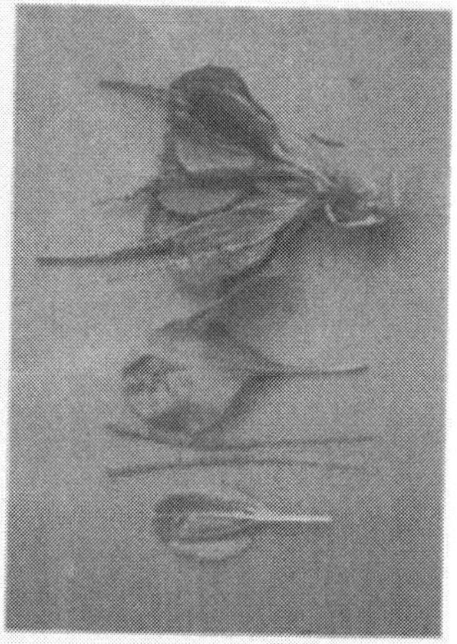

Plantain, broad and narrow leaf

Plantago species

Description: The broad leaf plantain has leaves over 2.5 centimeters across that grow close to the ground. The flowers are on a spike that rises from the middle of the cluster of leaves. The narrow leaf plantain has leaves up to 12 centimeters long and 2.5 centimeters wide, covered with hairs. The leaves form a rosette. The flowers are small and inconspicuous.

Habitat and Distribution: Look for these plants in lawns and along roads in the North Temperate Zone. This plant is a common weed throughout much of the world.

Edible Parts: The young tender leaves are edible raw. Older leaves should be cooked. Seeds are edible raw or roasted.

Other Uses: To relieve pain from wounds and sores, wash and soak the entire plant for a short time and apply it to the injured area. To treat diarrhea, drink tea made from 28 grams (1 ounce) of the plant leaves boiled in 0.5 liter of water. The seeds and seed husks act as laxatives.

Pokeweed

Phytolacca americana

Description: This plant may grow as high as 3 meters. Its leaves are elliptic and up to 1 meter in length. It produces many large clusters of purple fruits in late spring.

Habitat and Distribution: Look for this plant in open, sunny areas in forest clearings, in fields, and along roadsides in eastern North America, Central America, and the Caribbean.

Edible Parts: The young leaves and stems are edible cooked. Boil them twice, discarding the water from the first boiling. The fruits are edible if cooked.

CAUTION

All parts of this plant are poisonous if eaten raw. Never eat the underground portions of the plant as these contain the highest concentrations of the poisons. Do not eat any plant over 25 centimeters tall or when red is showing in the plant.

Other Uses: Use the juice of fresh berries as a dye.

Prickly pear cactus

Opuntia species

Description: This cactus has flat, padlike stems that are green. Many round, furry dots that contain sharp-pointed hairs cover these stems.

Habitat a nd Distribution: This cactus is found in arid and semiarid regions and in dry, sandy areas of wetter regions throughout most of the United States and Central and South America. Some species are planted in arid and semiarid regions of other parts of the world.

Edible Parts: All parts of the plant are edible. Peel the fruits and eat them fresh or crush them to prepare a refreshing drink. Avoid the tiny, pointed hairs. Roast the seeds and grind them to a flour.

CAUTION

Avoid any prickly pear cactus like plant with milky sap.

Other Uses: The pad is a good source of water. Peel it carefully to remove all sharp hairs before putting it in your mouth. You can also use the pads to promote healing. Split them and apply the pulp to wounds.

Purslane

Portulaca oleracea

Description: This plant grows close to the ground. It is seldom more than a few centimeters tall. Its stems and leaves are fleshy and often tinged with red. It has paddle-shaped leaves, 2.5 centimeter or less long, clustered at the tips of the stems. Its flowers are yellow or pink. Its seeds are tiny and black.

Habitat and Distribution: It grows in full sun in cultivated fields, field margins, and other weedy areas throughout the world.

Edible Parts: All parts are edible. Wash and boil the plants for a tasty vegetable or eat them raw. Use the seeds as a flour substitute or eat them raw.

Rattan palm

Calamus species

Description: The rattan palm is a stout, robust climber. It has hooks on the midrib of its leaves that it uses to remain attached to trees on which it grows. Sometimes, mature stems grow to 90 meters. It has alternate, compound leaves and a whitish flower.

Habitat and Distribution: The rattan palm is found from tropical Africa through Asia to the East Indies and Australia. It grows mainly in rain forests.

Edible Parts: Rattan palms hold a considerable amount of starch in their young stem tips. You can eat them roasted or raw. In other kinds, a gelatinous pulp, either sweet or sour, surrounds the seeds. You can suck out this pulp. The palm heart is also edible raw or cooked.

Other Uses: You can obtain large amounts of potable water by cutting the ends of the long stems (see Chapter 6). The stems can be used to make baskets and fish traps.

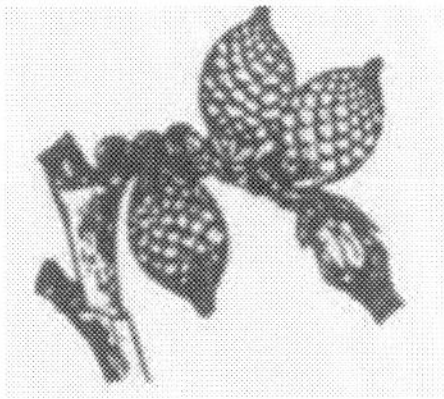

Reed

Phragmites australis

Description: This tall, coarse grass grows to 3.5 meters tall and has gray-green leaves about 4 centimeters wide. It has large masses of brown flower branches in early summer. These rarely produce grain and become fluffy, gray masses late in the season.

Habitat and Distribution: Look for reed in any open, wet area, especially one that has been disturbed through dredging. Reed is found throughout the temperate regions of both the Northern and Southern Hemispheres.

Edible Parts: All parts of the plant are edible raw or cooked in any season. Harvest the stems as they emerge from the soil and boil them. You can also harvest them just before they produce flowers, then dry and beat them into flour. You can also dig up and boil the underground stems, but they are often tough. Seeds are edible raw or boiled, but they are rarely found.

Reindeer moss

Cladonia rangiferina

Description: Reindeer moss is a low-growing plant only a few centimeters tall. It does not flower but does produce bright red reproductive structures.

Habitat and Distribution: Look for this lichen in open, dry areas. It is very common in much of North America.

Edible Parts: The entire plant is edible but has a crunchy, brittle texture. Soak the plant in water with some wood ashes to remove the bitterness, then dry, crush, and add it to milk or to other food.

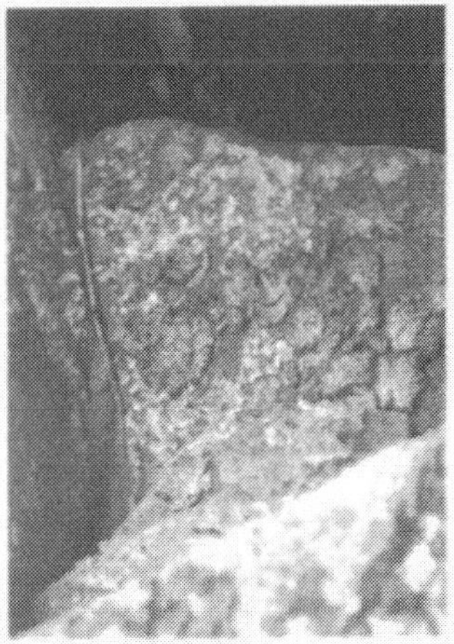

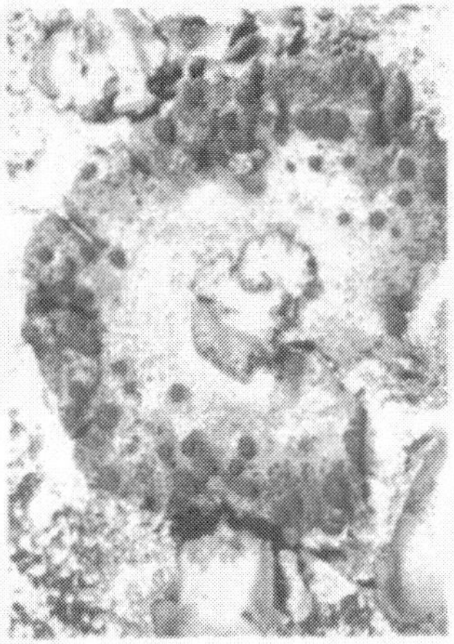

Rock tripe

Umbilicaria **species**

Description: This plant forms large patches with curling edges. The top of the plant is usually black. The underside is lighter in color.

Habitat and Distribution: Look on rocks and boulders for this plant. It is common throughout North America.

Edible Parts: The entire plant is edible. Scrape it off the rock and wash it to remove grit. The plant may be dry and crunchy; soak it in water until it becomes soft. Rock tripes may contain large quantities of bitter substances; soaking or boiling them in several changes of water will remove the bitterness.

> **CAUTION**
>
> There are some reports of poisoning from rock tripe, so apply the Universal Edibility Test.

Rose apple

Eugenia jambos

Description: This tree grows 3 to 9 meters high. It has opposite, simple, dark green, shiny leaves. When fresh, it has fluffy, yellowish-green flowers and red to purple egg-shaped fruit.

Habitat and Distribution: This tree is widely planted in all of the tropics. It can also be found in a semiwild state in thickets, waste places, and secondary forests.

Edible Parts: The entire fruit is edible raw or cooked.

Sago palm

Metroxylon sagu

Description: These palms are low trees, rarely over 9 meters tall, with a stout, spiny trunk. The outer rind is about 5 centimeters thick and hard as bamboo. The rind encloses a spongy inner pith containing a high proportion of starch. It has typical palmlike leaves clustered at the tip.

Habitat and Distribution: Sago palm is found in tropical rain forests. It flourishes in damp lowlands in the Malay Peninsula, New Guinea, Indonesia, the Philippines, and adjacent islands. It is found mainly in swamps and along streams, lakes, and rivers.

Edible Parts: These palms, when available, are of great use to the survivor. One trunk, cut just before it flowers, will yield enough sago to feed a person for 1 year. Obtain sago starch from nonflowering palms. To extract the edible sage, cut away the bark lengthwise from one half of the trunk, and pound the soft, whitish inner part (pith) as fine as possible. Knead the pith in water and strain it through a coarse cloth into a container. The fine, white sago will settle in the container. Once the sago settles, it is ready for use. Squeeze off the excess water and let it dry. Cook it as pancakes or oatmeal. Two kilograms of sago is the nutritional equivalent of 1.5 kilograms of rice. The upper part of the trunk's core does not yield sage, but you can roast it in lumps over a fire. You can also eat the young sago nuts and the growing shoots or palm cabbage.

Other Uses: Use the stems of tall sorghums as thatching materials.

Sassafras

Sassafras albidum

Description: This shrub or small tree bears different leaves on the same plant. Some leaves will have one lobe, some two lobes, and some no lobes. The flowers, which appear in early spring, are small and yellow. The fruits are dark blue. The plant parts have a characteristics root beer smell.

Habitat and Distribution: Sassafras grows at the margins of roads and forests, usually in open, sunny areas. It is a common tree throughout eastern North America.

Edible Parts: The young twigs and leaves are edible fresh or dried. You can add dried young twigs and leaves to soups. Dig the underground portion, peel off the bark, and let it dry. Then boil it in water to prepare sassafras tea.

Other Uses: Shred the tender twigs for use as a toothbrush.

Saxaul

Haloxylon ammondendron

Description: The saxaul is found either as a small tree or as a large shrub with heavy, coarse wood and spongy, water-soaked bark. The branches of the young trees are vivid green and pendulous. The flowers are small and yellow.

Habitat and Distribution: The saxaul is found in desert and arid areas. It is found on the arid salt deserts of Central Asia, particularly in the Turkestan region and east of the Caspian Sea.

Edible Parts: The thick bark acts as a water storage organ. You can get drinking water by pressing quantities of the bark. This plant is an important some of water in the arid regions in which it grows.

Screw pine

Pandanus species

Description: The screw pine is a strange plant on stilts, or prop roots, that support the plant above-ground so that it appears more or less suspended in midair. These plants are either shrubby or treelike, 3 to 9 meters tall, with stiff leaves having sawlike edges. The fruits are large, roughened balls resembling pineapples, but without the tuft of leaves at the end.

Habitat and Distribution: The screw pine is a tropical plant that grows in rain forests and semievergreen seasonal forests. It is found mainly along seashores, although certain kinds occur inland for some distance, from Madagascar to southern Asia and the islands of the southwestern Pacific. There are about 180 types.

Edible Parts: Knock the ripe fruit to the ground to separate the fruit segments from the hard outer covering. Chew the inner fleshy part. Cook fruit that is not fully ripe in an earth oven. Before cooking, wrap the whole fruit in banana leaves, breadfruit leaves, or any other suitable thick, leathery leaves. After cooking for about 2 hours, you can chew fruit segments like ripe fruit. Green fruit is inedible.

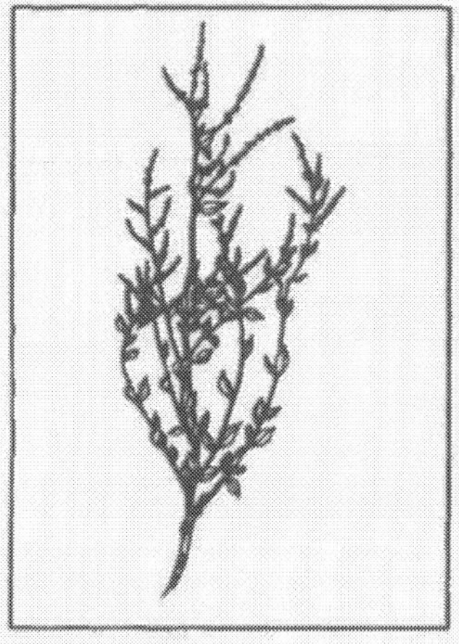

Sea orach

Atriplex halimus

Description: The sea orach is a sparingly branched herbaceous plant with small, gray-colored leaves up to 2.5 centimeters long. Sea orach resembles lamb's quarter, a common weed in most gardens in the United States. It produces its flowers in narrow, densely compacted spikes at the tips of its branches.

Habitat and Distribution: The sea orach is found in highly alkaline and salty areas along seashores from the Mediterranean countries to inland areas in North Africa and eastward to Turkey and central Siberia. Generally, it can be found in tropical scrub and thorn forests, steppes in temperate regions, and most desert scrub and waste areas.

Edible Parts: Its leaves are edible. In the areas where it grows, it has the healthy reputation of being one of the few native plants that can sustain man in times of want.

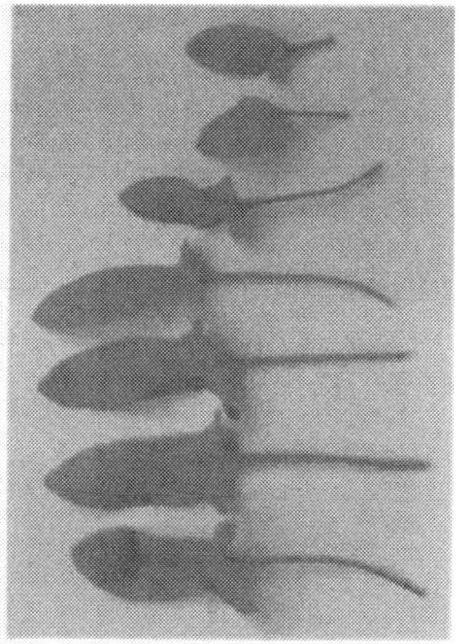

Sheep sorrel

Rumex acerosella

Description: These plants are seldom more than 30 centimeters tall. They have alternate leaves, often with arrowlike bases, very small flowers, and frequently reddish stems.

Habitat and Distribution: Look for these plants in old fields and other disturbed areas in North America and Europe.

Edible Parts: The plants are edible raw or cooked.

> **CAUTION**
>
> These plants contain oxalic acid that can be damaging if too many plants are eaten raw. Cooking seems to destroy the chemical.

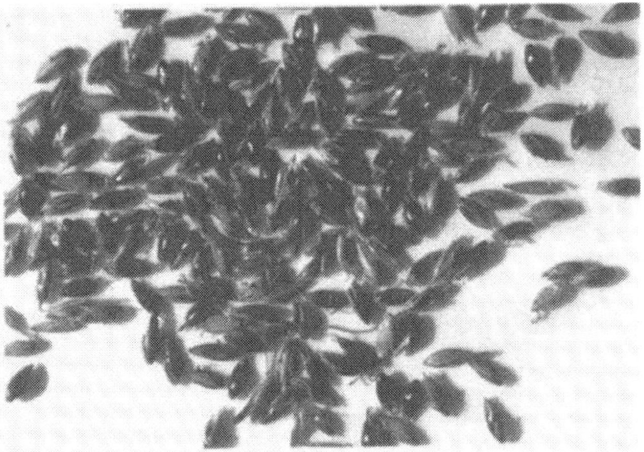

Sorghum

Sorghum species

Description: There are many different kinds of sorghum, all of which bear grains in heads at the top of the plants. The grains are brown, white, red, or black. Sorghum is the main food crop in many parts of the world.

Habitat and Distribution: Sorghum is found worldwide, usually in warmer climates. All species are found in open, sunny areas.

Edible Parts: The grains are edible at any stage of development. When young, the grains are milky and edible raw. Boil the older grains. Sorghum is a nutritious food.

Other Uses: Use the stems of tall sorghum as building materials.

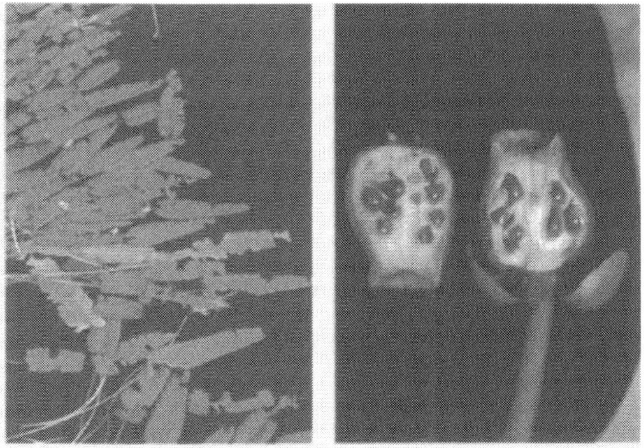

Spatterdock or yellow water lily

Nuphar species

Description: This plant has leaves up to 60 centimeters long with a triangular notch at the base. The shape of the leaves is somewhat variable. The plant's yellow flowers are 2.5 centimeter across and develop into bottle-shaped fruits. The fruits are green when ripe.

Habitat and Distribution: These plants grow throughout most of North America. They are found in quiet, fresh, shallow water (never deeper than 1.8 meters).

Edible Parts: All parts of the plant are edible. The fruits contain several dark brown seeds you can parch or roast and then grind into flour. The large rootstock contains starch. Dig it out of the mud, peel off the outside, and boil the flesh. Sometimes the rootstock contains large quantities of a very bitter compound. Boiling in several changes of water may remove the bitterness.

Sterculia

Sterculia foetida

Description: Sterculias are tall trees, rising in some instances to 30 meters. Their leaves are either undivided or palmately lobed. Their flowers are red or purple. The fruit of all sterculias is similar in aspect, with a red, segmented seedpod containing many edible black seeds.

Habitat and Distribution: There are over 100 species of sterculias distributed through all warm or tropical climates. They are mainly forest trees.

Edible Parts: The large, red pods produce a number of edible seeds. The seeds of all sterculias are edible and have a pleasant taste similar to cocoa. You can eat them like nuts, either raw or roasted.

CAUTION

Avoid eating large quantities. The seeds may have a laxative effect.

Strawberry

Fregaria species

Description: Strawberry is a small plant with a three-leaved growth pattern. It has small, white flowers usually produced during the spring. Its fruit is red and fleshy.

Habitat and Distribution: Strawberries are found in the North Temperate Zone and also in the high mountains of the southern Western Hemisphere. Strawberries prefer open, sunny areas. They are commonly planted.

Edible Parts: The fruit is edible fresh, cooked, or dried. Strawberries are a good source of vitamin C. You can also eat the plant's leaves or dry them and make a tea with them.

WARNING

Eat only white-flowering true strawberries. Other similar plants without white flowers can be poisonous.

Sugarcane

Saccharum officinarum

Description: This plant grows up to 4.5 meters tall. It is a grass and has grasslike leaves. Its green or reddish stems are swollen where the leaves grow. Cultivated sugarcane seldom flowers.

Habitat and Distribution: Look for sugarcane in fields. It grows only in the tropics (throughout the world). Because it is a crop, it is often found in large numbers.

Edible Parts: The stem is an excellent source of sugar and is very nutritious. Peel the outer portion off with your teeth and eat the sugarcane raw. You can also squeeze juice out of the sugarcane.

Sugar palm

Arenga pinnata

Description: This tree grows about 15 meters high and has huge leaves up to 6 meters long. Needlelike structures stick out of the bases of the leaves. Flowers grow below the leaves and form large conspicuous dusters from which the fruits grow.

Habitat and Distribution: This palm is native to the East Indies but has been planted in many parts off the tropics. It can be found at the margins of forests.

Edible Parts: The chief use of this palm is for sugar. However, its seeds and the tip of its stems are a survival food. Bruise a young flower stalk with a stone or similar object and collect the juice as it comes out. It is an excellent source of sugar. Boil the seeds. Use the tip of the stems as a vegetable.

> **CAUTION**
> The flesh covering the seeds may cause dermatitis.

Other Uses: The shaggy material at the base of the leaves makes an excellent rope as it is strong and resists decay.

Sweetsop

Annona squamosa

Description: This tree is small, seldom more than 6 meters tall, and multi-branched. It has alternate, simple, elongate, dark green leaves. Its fruit is green when ripe, round in shape, and covered with protruding bumps on its surface. The fruit's flesh is white and creamy.

Habitat and Distribution: Look for sweetsop at margins of fields, near villages, and around homesites in tropical regions.

Edible Parts: The fruit flesh is edible raw.

Other Uses: You can use the finely ground seeds as an insecticide.

> **CAUTION**
> The ground seeds are extremely dangerous to the eyes.

Tamarind

Tamarindus indica

Description: The tamarind is a large, densely branched tree, up to 25 meters tall. Its has pinnate leaves (divided like a feather) with 10 to 15 pairs of leaflets.

Habitat and Distribution: The tamarind grows in the drier parts of Africa, Asia, and the Philippines. Although it is thought to be a native of Africa, it has been cultivated in India for so long that it looks like a native tree. It it also found in the American tropics, the West Indies, Central America, and tropical South America.

Edible Parts: The pulp surrounding the seeds is rich in vitamin C and is an important survival food. You can make a pleasantly acid drink by mixing the pulp with water and sugar or honey and letting the mixture mature for several days. Suck the pulp to relieve thirst. Cook the young, unripe fruits or seedpods with meat. Use the young leaves in soup. You must cook the seeds. Roast them above a fire or in ashes. Another way is to remove the seed coat and soak the seeds in salted water and grated coconut for 24 hours, then cook them. You can peel the tamarind bark and chew it.

Taro, cocoyam, elephant ears, eddo, dasheen

Colocasia and *Alocasia* species

Description: All plants in these groups have large leaves, sometimes up to 1.8 meters tall, that grow from a very short stem. The rootstock is thick and fleshy and filled with starch.

Habitat and Distribution: These plants grow in the humid tropics. Look for them in fields and near homesites and villages.

Edible Parts: All parts of the plant are edible when boiled or roasted. When boiling, change the water once to get rid of any poison.

> **CAUTION**
>
> If eaten raw, these plants will cause a serious inflammation of the mouth and throat.

Thistle

Cirsium species

Description: This plant may grow as high as 1.5 meters. Its leaves are long-pointed, deeply lobed, and prickly.

Habitat and Distribution: Thistles grow worldwide in dry woods and fields.

Edible Parts: Peel the stalks, cut them into short sections, and boil them before eating. The roots are edible raw or cooked.

> **CAUTION**
>
> Some thistle species are poisonous.

Other Uses: Twist the tough fibers of the stems to make a strong twine.

Ti

Cordyline terminalis

Description: The ti has unbranched stems with straplike leaves often clustered at the tip of the stem. The leaves vary in color and may be green or reddish. The flowers grow at the plant's top in large, plumelike clusters. The ti may grow up to 4.5 meters tall.

Habitat and Distribution: Look for this plant at the margins of forests or near home-sites in tropical areas. It is native to the Far East but is now widely planted in tropical areas worldwide.

Edible Parts: The roots and very tender young leaves are good survival food. Boil or bake the short, stout roots found at the base of the plant. They are a valuable source of starch. Boil the very young leaves to eat. You can use the leaves to wrap other food to cook over coals or to steam.

Other Uses: Use the leaves to cover shelters or to make a rain cloak. Cut the leaves into liners for shoes; this works especially well if you have a blister. Fashion temporary sandals from the ti leaves. The terminal leaf, if not completely unfurled, can be used as a sterile bandage. Cut the leaves into strips, then braid the strips into rope.

Tree fern

Various genera

Description: Tree ferns are tall trees with long, slender trunks that often have a very rough, barklike covering. Large, lacy leaves uncoil from the top of the trunk.

Habitat and Distribution: Tree ferns are found in wet, tropical forests.

Edible Parts: The young leaves and the soft inner portion of the trunk are edible. Boil the young leaves and eat as greens. Eat the inner portion of the trunk raw or bake it.

Tropical almond

Terminalia catappa

Description: This tree grows up to 9 meters tall. Its leaves are evergreen, leathery, 45 centimeters long, 15 centimeters wide, and very shiny. It has small, yellowish-green flowers. Its fruit is flat, 10 centimeters long, and not quite as wide. The fruit is green when ripe.

Habitat and Distribution: This tree is usually found growing near the ocean. It is a common and often abundant tree in the Caribbean and Central and South America. It is also found in the tropical rain forests of southeastern Asia, northern Australia, and Polynesia.

Edible Parts: The seed is a good source of food. Remove the fleshy, green covering and eat the seed raw or cooked.

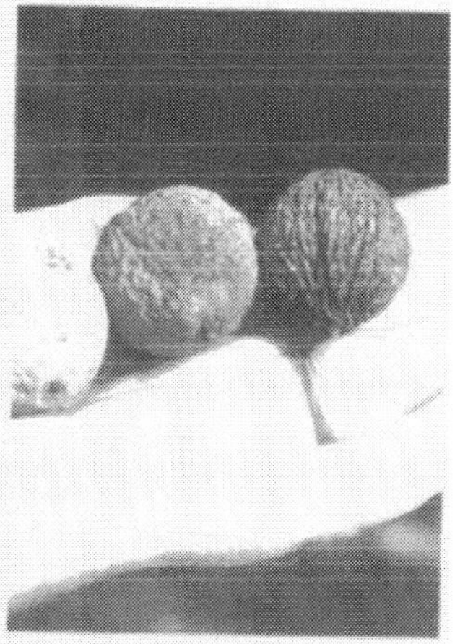

Walnut

Juglans species

Description: Walnuts grow on very large trees, often reaching 18 meters tall. The divided leaves characterize all walnut spades. The walnut itself has a thick outer husk that must be removed to reach the hard inner shell of the nut.

Habitat and Distribution: The English walnut, in the wild state, is found from southeastern Europe across Asia to China and is abundant in the Himalayas. Several other species of walnut are found in China and Japan. The black walnut is common in the eastern United States.

Edible Parts: The nut kernel ripens in the autumn. You get the walnut meat by cracking the shell. Walnut meats are highly nutritious because of their protein and oil content.

Other Uses: You can boil walnuts and use the juice as an antifungal agent. The husks of "green" walnuts produce a dark brown dye for clothing or camouflage. Crush the husks of "green" black walnuts and sprinkle them into sluggish water or ponds for use as fish poison.

Water chestnut

Trapa natans

Description: The water chestnut is an aquatic plant that roots in the mud and has finely diviided leaves that grow underwater. Its floating leaves are much larger and coarsely toothed. The fruits, borne underwater, have four sharp spines on them.

Habitat and Distribution: The water chestnut is a freshwater plant only. It is a native of Asia but has spread to many parts of the world in both temperate and tropical areas.

Edible Parts: The fruits are edible raw and cooked. The seeds are also a source of food.

Water lettuce

Ceratopteris species

Description: The leaves of water lettuce are much like lettuce and are very tender and succulent. One of the easiest ways of distinguishing water lettuce is by the little plantlets that grow from the margins of the leaves. These little plantlets grow in the shape of a rosette. Water lettuce plants often cover large areas in the regions where they are found.

Habitat and Distribution: Found in the tropics throughout the Old World in both Africa and Asia. Another kind is found in the New World tropics from Florida to South America. Water lettuce grows only in very wet places and often as a floating water plant. Look for water lettuce in still lakes, ponds, and the backwaters of rivers.

Edible Parts: Eat the fresh leaves like lettuce. Be careful not to dip the leaves in the contaminated water in which they are growing. Eat only the leaves that are well out of the water.

> **CAUTION**
>
> This plant has carcinogenic properties and should only be used as a last resort.

Water lily

Nymphaea odorata

Description: These plants have large, triangular leaves that float on the water's surface, large, fragrant flowers that are usually white, or red, and thick, fleshy rhizomes that grow in the mud.

Habitat and Distribution: Water lilies are found throughout much of the temperate and subtropical regions.

Edible Parts: The flowers, seeds, and rhizomes are edible raw or cooked. To prepare rhizomes for eating, peel off the corky rind. Eat raw, or slice thinly, allow to dry, and then grind into flour. Dry, parch, and grind the seeds into flour.

Other Uses: Use the liquid resulting from boiling the thickened root in water as a medicine for diarrhea and as a gargle for sore throats.

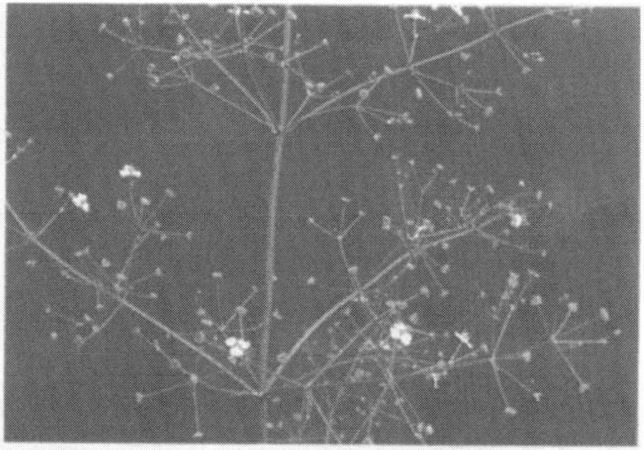

Water plantain

Alisma plantago-aquatica

Description: This plant has small, white flowers and heart-shaped leaves with pointed tips. The leaves are clustered at the base of the plant.

Habitat and Distribution: Look for this plant in fresh water and in wet, full sun areas in Temperate and Tropical Zones.

Edible Parts: The rootstocks are a good source of starch. Boil or soak them in water to remove the bitter taste.

> **CAUTION**
>
> To avoid parasites, always cook aquatic plants.

Wild caper

Capparis aphylla

Description: This is a thorny shrub that loses its leaves during the dry seson. Its stems are gray-green and its flowers pink.

Habitat and Distribution: These shrubs form large stands in scrub and thorn forests and in desert scrub and waste. They are common throughout North Africa and the Middle East.

Edible Parts: The fruit and the buds of young shoots are edible raw.

Wild crab apple or wild apple

Malus species

Description: Most wild apples look enough like domestic apples that the survivor can easily recognize them. Wild apple varieties are much smaller than cultivated kinds; the largest kinds usually do not exceed 5 to 7.5 centimeters in diameter, and most often less. They have small, alternate, simple leaves and often have thorns. Their flowers are white or pink and their fruits reddish or yellowish.

Habitat and Distribution: They are found in the savanna regions of the tropics. In temperate areas, wild apple varieties are found mainly in forested areas. Most frequently, they are found on the edge of woods or in fields. They are found throughout the Northern Hemisphere.

Edible Parts: Prepare wild apples for eating in the same manner as cultivated kinds. Eat them fresh, when ripe, or cooked. Should you need to store food, cut the apples into thin slices and dry them. They are a good source of vitamins.

CAUTION

Apple seeds contain cyanide compounds. Do not eat.

Wild desert gourd or colocynth

Citrullus colocynthis

Description: The wild desert gourd, a member of the watermelon family, produces an 2.4- to 3-meter-long ground-trailing vine. The perfectly round gourds are as large as an orange. They are yellow when ripe.

Habitat and Distribution: This creeping plant can be found in any climatic zone, generally in desert scrub and waste areas. It grows abundantly in the Sahara, in many Arab countries, on the southeastern coast of India, and on some of the islands of the Aegean Sea. The wild desert gourd will grow in the hottest localities.

Edible Parts: The seeds inside the ripe gourd are edible after they are completely separated from the very bitter pulp. Roast or boil the seeds - their kernels are rich in oil. The flowers are edible. The succulent stem tips can be chewed to obtain water.

Wild dock and wild sorrel

Rumex crispus and *Rumex acetosella*

Description: Wild dock is a stout plant with most of its leaves at the base of its stem that is commonly 15 to 30 centimeters brig. The plants usually develop from a strong, fleshy, carrotlike taproot. Its flowers are usually very small, growing in green to purplish plumelike clusters. Wild sorrel similar to the wild dock but smaller. Many of the basal leaves are arrow-shaped but smaller than those of the dock and contain a sour juice.

Habitat and Distribution: These plants can be found in almost all climatic zones of the world, in areas of high as well as low rainfall. Many kinds are found as weeds in fields, along roadsides, and in waste places.

Edible Parts: Because of tender nature of the foliage, the sorrel and the dock are useful plants, especially in desert areas. You can eat their succulent leaves fresh or slightly cooked. To take away the strong taste, change the water once or twice during cooking. This latter tip is a useful hint in preparing many kinds of wild greens.

Wild fig

Ficus species

Description: These trees have alternate, simple leaves with entire margins. Often, the leaves are dark green and shiny. All figs have a milky, sticky juice. The fruits vary in size depending on the species, but are usually yellow-brown when ripe.

Habitat and Distribution: Figs are plants of the tropics and semitropics. They grow in several different habitats, including dense forests, margins of forests, and around human settlements.

Edible Parts: The fruits are edible raw or cooked. Some figs have little flavor.

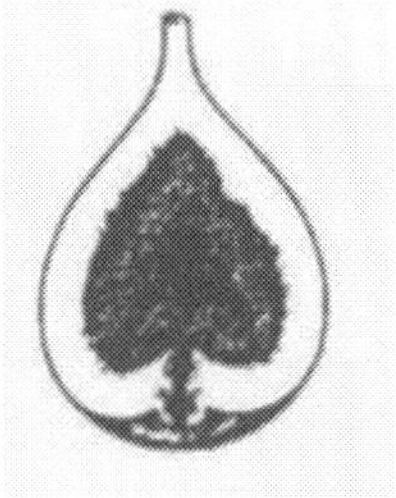

Wild gourd or luffa sponge

Luffa cylindrica

Description: The luffs sponge is widely distributed and fairly typical of a wild squash. There are several dozen kinds of wild squashes in tropical regions. Like most squashes, the luffa is a vine with leaves 7.5 to 20 centimeters across having 3 lobes. Some squashes have leaves twice this size. Luffs fruits are oblong or cylindrical, smooth, and many-seeded. Luffs flowers are bright yellow. The luffa fruit, when mature, is brown and resembles the cucumber.

Habitat and Distribution: A member of the squash family, which also includes the watermelon, cantaloupe, and cucumber, the luffa sponge is widely cultivated throughout the Tropical Zone. It may be found in a semiwild state in old clearings and abandoned gardens in rain forests and semievergreen seasonal forests.

Edible Parts: You can boil the young green (half-ripe) fruit and eat them as a vegetable. Adding coconut milk will improve the flavor. After ripening, the luffa sponge develops an inedible spongelike texture in the interior of the fruit. You can also eat the tender shoots, flowers, and young leaves after cooking them. Roast the mature seeds a little and eat them like peanuts.

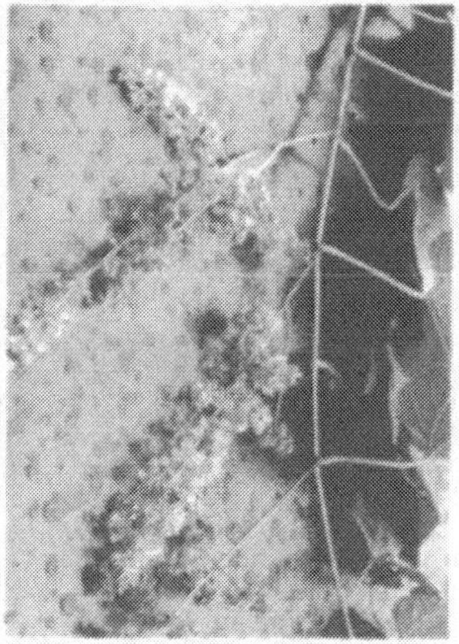

Wild grape vine

Vitis species

Description: The wild grape vine climbs with the aid of tendrils. Most grape vines produce deeply lobed leaves similar to the cultivated grape. Wild grapes grow in pyramidal, hanging bunches and are black-blue to amber, or white when ripe.

Habitat and Distribution: Wild grapes are distributed worldwide. Some kinds are found in deserts, others in temperate forests, and others in tropical areas. Wild grapes are commonly found throughout the eastern United States as well as in the southwestern desert areas. Most kinds are rampant climbers over other vegetation. The best place to look for wild grapes is on the edges of forested areas. Wild grapes are also found in Mexico. In the Old World, wild grapes are found from the Mediterranean region eastward through Asia, the East Indies, and to Australia. Africa also has several kinds of wild grapes.

Edible Parts: The ripe grape is the portion eaten. Grapes are rich in natural sugars and, for this reason, are much sought after as a source of energy-giving wild food. None are poisonous.

Other Uses: You can obtain water from severed grape vine stems. Cut off the vine at the bottom and place the cut end in a container. Make a slant-wise cut into the vine about 1.8 meters upon the hanging part. This cut will allow water to flow from the bottom end. As water diminishes in volume, make additional cuts further down the vine.

CAUTION

To avoid poisoning, do not eat grapelike fruits with only a *single* seed (moonseed).

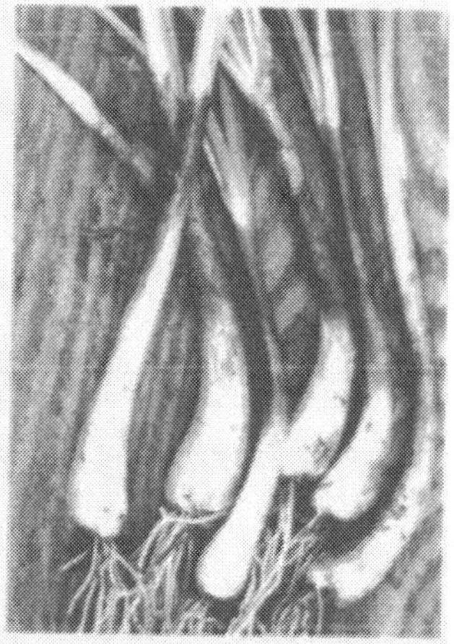

Wild onion and garlic

Allium species

Description: *Allium cernuum* is an example of the many species of wild onions and garlics, all easily recognized by their distinctive odor.

Habitat and Distribution: Wild onions and garlics are found in open, sunny areas throughout the temperate regions. Cultivated varieties are found anywhere in the world.

Edible Parts: The bulbs and young leaves are edible raw or cooked. Use in soup or to flavor meat.

> **CAUTION**
>
> There are several plants with onionlike bulbs that are extremely poisonous. Be certain that the plant you are using is a true onion or garlic. Do not eat bulbs with no onion smell.

Wild pistachio

Pistacia species

Description: Some kinds of pistachio trees are evergreen, while others lose their leaves during the dry season. The leaves alternate on the stem and have either three large leaves or a number of leaflets. The fruits or nuts are usually hard and dry at maturity.

Habitat and Distribution: About seven kinds of wild pistachio nuts are found in desert, or semidesert areas surrounding the Mediterranean Sea to Turkey and Afghanistan. It is generally found in evergreen scrub forests or scrub and thorn forests.

Edible Parts: You can eat the oil nut kernels after parching them over coals.

Wild rice

Zizania aquatica

Description: Wild rice is a tall grass that averages 1 to 1.5 meters in height, but may reach 4.5 meters. Its grain grows in very loose heads at the top of the plant and is dark brown or blackish when ripe.

Habitat and Distribution: Wild rice grows only in very wet areas in tropical and temperate regions.

Edible Parts: During the spring and summer, the central portion of the lower stems and root shoots are edible. Remove the tough covering before eating. During the late summer and fall, collect the straw-covered husks. Dry and parch the husks, break them, and remove the rice. Boil or roast the rice and then beat it into flour.

Wild rose

Rosa species

Description: This shrub grows 60 centimeters to 2.5 meters high. It has alternate leaves and sharp prickles. Its flowers may be red, pink, or yellow. Its fruit, called rose hip, stays on the shrub year-round.

Habitat and Distribution: Look for wild roses in dry fields and open woods throughout the Northern Hemisphere.

Edible Parts: The flowers and buds are edible raw or boiled. In an emergency, you can peel and eat the young shoots. You can boil fresh, young leaves in water to make a tea. After the flower petals fall, eat the rose hips; the pulp is highly nutritious and an excellent source of vitamin C. Crush or grind dried rose hips to make flour.

CAUTION

Eat only the outer portion of the fruit as the seeds of some species are quite prickly and can cause internal distress.

Wood sorrel

Oxalis species

Description: Wood sorrel resembles shamrock or four-leaf clover, with a bell-shaped pink, yellow, or white flower.

Habitat and Distribution: Wood sorrel is found in Temperate Zones worldwide, in lawns, open areas, and sunny woods.

Edible Parts: Cook the entire plant.

> **CAUTION**
>
> Eat only small amounts of this plant as it contains a fairly high concentration of oxalic acid that can be harmful.

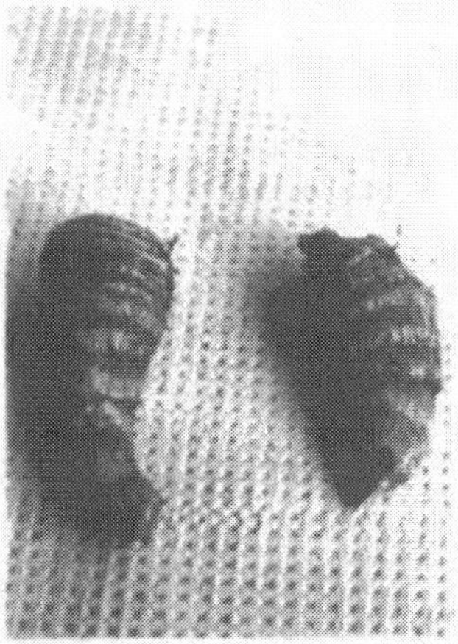

Yam

Dioscorea species

Description: These plants are vines that creep along the ground. They have alternate, heart- or arrow-shaped leaves. Their rootstock may be very large and weigh many kilograms.

Habitat and Distribution: True yams are restricted to tropical regions where they are an important food crop. Look for yams in fields, clearings, and abandoned gardens. They are found in rain forests, semievergreen seasonal forests, and scrub and thorn forests in the tropics. In warm temperate areas, they are found in seasonal hardwood or mixed hardwood-coniferous forests, as well as some mountainous areas.

Edible Parts: Boil the rootstock and eat it as a vegetable.

Yam bean
Pachyrhizus erosus

Description: The yam bean is a climbing plant of the bean family, with alternate, three-parted leaves and a turniplike root. The bluish or purplish flowers are pealike in shape. The plants are often so rampant that they cover the vegetation upon which they are growing.

Habitat and Distribution: The yam bean is native to the American tropics, but it was carried by man years ago to Asia and the Pacific islands. Now it is commonly cultivated in these places, and is also found growing wild in forested areas. This plant grows in wet areas of tropical regions.

Edible Parts: The tubers are about the size of a turnip and they are crisp, sweet, and juicy and have a nutty flavor. They are nourishing and at the same time quench the thirst. Eat them raw or boiled. To make flour, slice the raw tubers, let them dry in the sun, and grind into a flour that is high in starch and may be used to thicken soup.

CAUTION
The raw seeds are poisonous.

APPENDIX C

POISONOUS PLANTS

Plants basically poison on contact, ingestion, or by absorption or inhalation. They cause painful skin irritations upon contact, they cause internal poisoning when eaten, and they poison through skin absorption or inhalation in respiratory system. Many edible plants have deadly relatives and look-alikes. Preparation for military missions includes learning to identify those harmful plants in the target area. Positive identification of edible plants will eliminate the danger of accidental poisoning. There is no room for experimentation where plants are concerned, especially in unfamiliar territory.

Castor bean, castor-oil plant, palma Christi

Ricinus communis

Spurge *(Euphorbiaceae)* Family

Description: The castor bean is a semiwoody plant with large, alternate, starlike leaves that grows as a tree in tropical regions and as an annual in temperate regions. Its flowers are very small and inconspicuous. Its fruits grow in clusters at the tops of the plants.

> **CAUTION**
>
> All parts of the plant are very poisonous to eat. The seeds are large and may be mistaken for a beanlike food.

Habitat and Distribution: This plant is found in all tropical regions and has been introduced to temperate regions.

Chinaberry

Melia azedarach

Mahogany *(Meliaceae)* Family

Description: This tree has a spreading crown and grows up to 14 meters tall. It has alternate, compound leaves with toothed leaflets. Its flowers are light purple with a dark center and grow in ball-like masses. It has marble-sized fruits that are light orange when first formed but turn lighter as they become older.

> **CAUTION**
>
> All parts of the tree should be considered dangerous if eaten. Its leaves are a natural insecticide and will repel insects from stored fruits and grains. Take care not to eat leaves mixed with the stored food.

Habitat and Distribution: Chinaberry is native to the Himalayas and eastern Asia but is now planted as an ornamental tree throughout the tropical and subtropical regions. It has been introduced to the southern United States and has escaped to thickets, old fields, and disturbed areas.

Cowhage, cowage, cowitch
Mucuna pruritum
Leguminosae *(Fabaceae)* Family

Description: A vinelike plant that has oval leaflets in groups of three and hairy spikes with dull purplish flowers. The seeds are brown, hairy pods.

> **CAUTION**
>
> Contact with the pods and flowers causes irritation and blindness if in the eyes.

Death camas, death lily
Zigadenus species
Lily *(Liliaceae)* Family

Description: This plant arises from a bulb and may be mistaken for an onionlike plant. Its leaves are grasslike. Its flowers are six-parted and the petals have a green, heart-shaped structure on them. The flowers grow on showy stalks above the leaves.

> **CAUTION**
>
> All parts of this plant are very poisonous. Death camas does not have the onion smell.

Habitat and Distribution: Death camas is found in wet, open, sunny habitats, although some species favor dry, rocky slopes. They are common in parts of the western United States. Some species are found in the eastern United States and in parts of the North American western subarctic and eastern Siberia.

Lantana

Lantana camara

Vervain *(Verbenaceae)* Family

Description: Lantana is a shrublike plant that may grow up to 45 centimeters high. It has opposite, round leaves and flowers borne in flat-topped clusters. The flower color (which varies in different areas) maybe white, yellow, orange, pink, or red. It has a dark blue or black berrylike fruit. A distinctive feature of all parts of this plant is its strong scent.

CAUTION

All parts of this plant are poisonous if eaten and can be fatal. This plant causes dermatitis in some individuals.

Habitat and Distribution: Lantana is grown as an ornamental in tropical and temperate areas and has escaped cultivation as a weed along roads and old fields.

Manchineel

Hippomane mancinella

Spurge *(Euphorbiaceae)* Family

Description: Manchineel is a tree reaching up to 15 meters high with alternate, shiny green leaves and spikes of small greenish flowers. Its fruits are green or greenish-yellow when ripe.

CAUTION

This tree is extremly toxic. It causes severe dermatitis in most individuals after only .5 hour. Even water dripping from the leaves may cause dermatitis. The smoke from burning it irritates the eyes. No part of this plant should be considered a food.

Habitat and Distribution: The tree prefers coastal regions. Found in south Florida, the Caribbean, Central America, and northern South America.

Oleander

Nerium oleander

Dogbane *(Apocynaceae)* Family

Description: This shrub or small tree grows to about 9 meters, with alternate, very straight, dark green leaves. Its flowers may be white, yellow, red, pink, or intermediate colors. Its fruit is a brown, podlike structure with many small seeds.

> **CAUTION**
> All parts of the plant are very poisonous. Do not use the wood for cooking; it gives off poisonous fumes that can poison food.

Habitat and Distribution: This native of the Mediterranean area is now grown as an ornamental in tropical and temperate regions.

Pangi

Pangium edule

Pangi Family

Description: This tree, with heart-shaped leaves in spirals, reaches a height of 18 meters. Its flowers grow in spikes and are green in color. Its large, brownish, pear-shaped fruits grow in clusters.

> **CAUTION**
> All parts are poisonous, especially the fruit.

Habitat and Distribution: Pangi trees grow in southeast Asia

Physic nut

Jatropha curcas

Spurge *(Euphoriaceae)* Family

Description: This shrub or small tree has large, 3- to 5-parted alternate leaves. It has small, greenish-yell low flowers and its yellow, apple-sized fruits contain three large seeds.

> **CAUTION**
>
> The seeds taste sweet but their oil is violently purgative. All parts of the physic nut are poisonous.

Habitat and Distribution: Throughout the tropics and southern United States.

Poison hemlock, fool's parsley

Conium maculatum

Parsley *(Apiaceae)* Family

Description: This biennial herb may grow to 2.5 meters high. The smooth, hollow stem may or may not be purple or red striped or mottled. Its white flowers are small and grow in small groups that tend to form flat umbels. Its long, turniplike taproot is solid.

> **CAUTION**
>
> This plant is very poisonous and even a very small amount may cause death. This plant is easy to confuse with wild carrot or Queen Anne's lace, especially in its first stage of growth. Wild carrot or Queen Anne's lace has hairy leaves and stems and smells like carrot. Poison hemlock does not.

Habitat and Distribution: Poison hemlock grows in wet or moist ground like swamps, wet meadows, stream banks, and ditches. Native to Eurasia, it has been introduced to the United States and Canada.

Poison ivy and poison oak

Toxicodendron radicans and Toxicodendron diversibba

Cashew *(Anacardiacese)* Family

Description: These two plants are quite similar in appearance and will often crossbreed to make a hybrid. Both have alternate, compound leaves with three leaflets. The leaves of poison ivy are smooth or serrated. Poison oak's leaves are lobed and resemble oak leaves. Poison ivy grows as a vine along the ground or climbs by red feeder roots. Poison oak grows like a bush. The greenish-white flowers are small and inconspicuous and are followed by waxy green berries that turn waxy white or yellow, then gray.

> **CAUTION**
>
> All parts, at all times of the year, can cause serious contact dermatitis.

Habitat and Distribution: Poison ivy and oak can be found in almost any habitat in North America.

Poison sumac

Toxicodendron vernix

Cashew *(Anacardiacese)* Family

Description: Poison sumac is a shrub that grows to 8.5 meters tall. It has alternate, pinnately compound leafstalks with 7 to 13 leaflets. Flowers are greenish-yellow and inconspicuous and are followed by white or pale yellow berries.

> **CAUTION**
>
> All parts can cause serious contact dermatitis at all times of the year.

Habitat and Distribution: Poison sumac grows only in wet, acid swamps in North America.

Renghas tree, rengas tree, marking nut, black-varnish tree

Gluta

Cashew *(Anacardiacese)* Family

Description: This family comprises about 48 species of trees or shrubs with alternating leaves in terminal or axillary panicles. Flowers are similar to those of poison ivy and oak.

> **CAUTION**
> Can cause contact dermatitis similar to poison ivy and oak.

Habitat and Distribution: India, east to Southeast Asia.

Rosary pea or crab's eyes

Abrus precatorius

Leguminosae *(Fabaceae)* Family

Description: This plant is a vine with alternate compound leaves, light purple flowers, and beautiful seeds that are red and black.

> **CAUTION**
> This plant is one of the most danderous plants. One seed may contain enough poison to kill an adult.

Habitat and Distribution: This is a common weed in parts of Africa, southern Florida, Hawaii, Guam, the Caribbean, and Central and South America.

Strychnine tree
Nux vomica
Logania *(Loganiaceae)* Family

Description: The strychnine tree is a medium-sized evergreen, reaching a height of about 12 meters, with a thick, frequently crooked trunk. Its deeply veined oval leaves grow in alternate pairs. Small, loose clusters of greenish flowers appear at the ends of branches and are followed by fleshy, orange-red berries about 4 centimeters in diameter.

CAUTION

The berries contain the dislike seeds that yield the poisonous substance strychnine. All parts of the plant are poisonous.

Habitat and Distribution: A native of the tropics and subtropics of southeastern Asia and Australia.

Trumpet vine or trumpet creeper
Campsis radicans
Trumpet creeper *(Bignoniaceae)* Family

Description: This woody vine may climb to 15 meters high. It has pealike fruit capsules. The leaves are pinnately compound, 7 to 11 toothed leaves per leaf stock. The trumpet-shaped flowers are orange to scarlet in color.

CAUTION

This plant causes contact dermatitis.

Habitat and Distribution: This vine is found in wet woods and thickets throughout eastern and central North America.

Water hemlock or spotted cowbane

Cicuta maculata

Parsley *(Apiaceae)* Family

Description: This perennial herb may grow to 1.8 meters high. The stem is hollow and sectioned off like bamboo. It may or may not be purple or red striped or mottled. Its flowers are small, white, and grow in groups that tend to form flat umbels. Its roots may have hollow air chambers and, when cut, may produce drops of yellow oil.

> **CAUTION**
>
> This plant is very poisonous and even a very small amount of this plant may cause death. Its roots have been mistaken for parsnips.

Habitat and Distribution: Water hemlock grows in wet or moist ground like swamps, wet meadows, stream banks, and ditches throughout the Unites States and Canada.

APPENDIX D

DANGEROUS INSECTS AND ARACHNIDS

Insects are often overlooked as a danger to the survivor. More people in the United States die each year from bee stings, and resulting anaphylactic shock, than from snake bites. A few other insects are venomous enough to kill, but often the greatest danger is the transmission of disease.

Scorpion

Scorpionidae order

Description: Dull brown, yellow, or black. Have 7.5- to 20-centimeter-long lobsterlike pincers and jointed tail usually held over the back. There are 800 species of scorpions.

Habitat: Decaying matter, under debris, logs, and rocks. Feeds at night. Sometimes hides in boots.

CAUTION

Scorpions sting with their tails, causing local pain, swelling, possible incapacitation, and death.

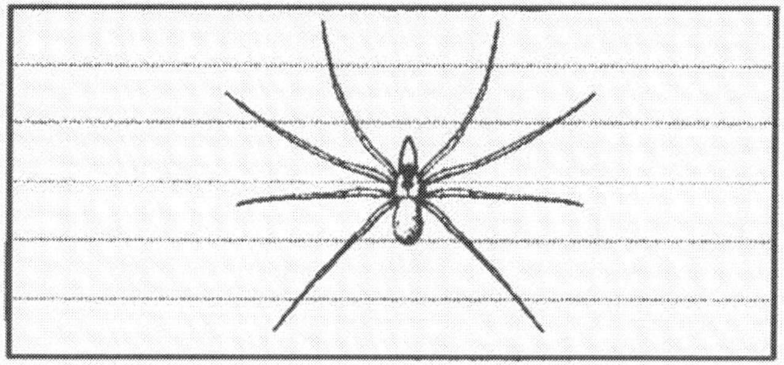

Brown house spider or brown recluse spider

Laxosceles reclusa

Description: Brown to black with obvious "fiddle" on back of head and thorax. Chunky body with long, slim legs 2.5 to 4 centimeters long.

Habitat Under debris, rocks, and logs. In caves and dark places.

Distribution: North America.

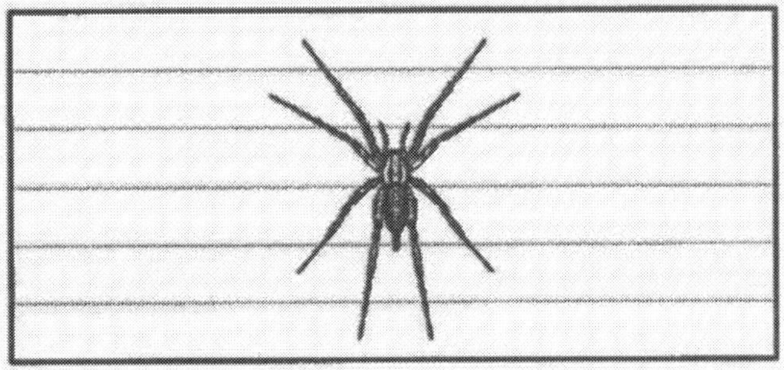

Funnelweb spider

Atrax species (*A. robustus, A. formidablis*)

Description: Large, brown, bulky spiders. Aggressive when disturbed.

Habitat: Woods, jungles, and brushy areas. Web has funnellike opening.

Distribution: Australia. (Other nonvenomous species worldwide.)

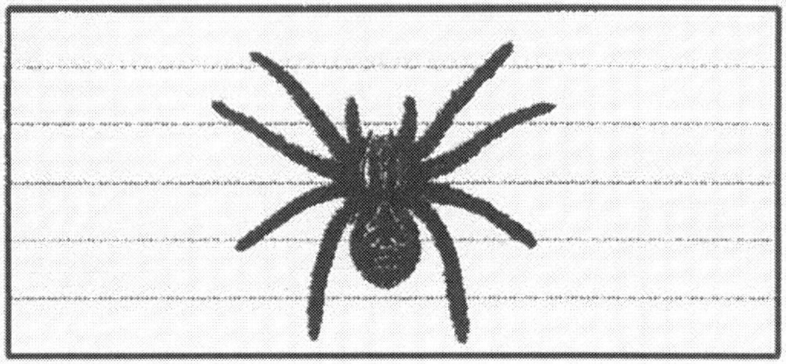

Tarantula

Theraphosidae and *Lycosa* species

Description: Very large, brown, black, reddish hairy spiders. Large fangs inflict painful bite.

Habitat Desert areas, tropics.

Distribution: Americas, southern Europe.

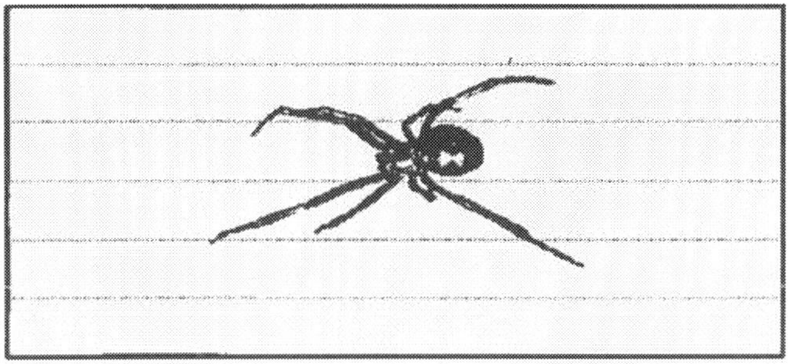

Widow spider

Latrodectus species

Description: Dark spiders with light red or orange markings on female's abdomen.

Habitat: Under logs, rocks, and debris. In shaded places.

Distribution: Varied species worldwide. Black widow in United States, red widow in Middle East, and brown widow in Australia.

Note: Females are the poisonous gender. Red widow in the Middle East is the only spider known to be deadly to man.

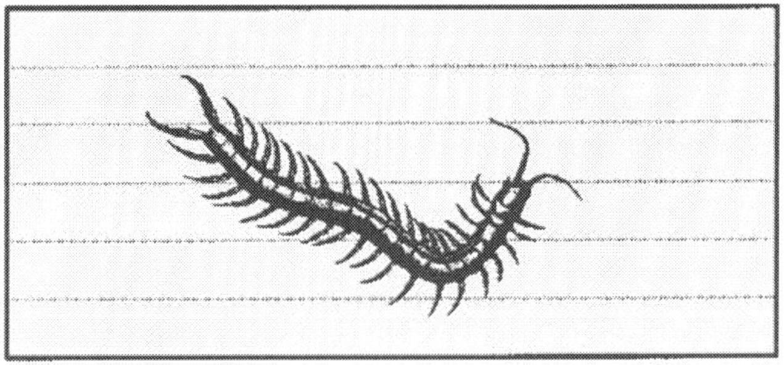

Centipede

Description: Multijointed body to 30 centimeters long. Dull orange to brown, with black point eyes at base of antennae. There are 2,800 species worldwide.

Habitat: Under bark and stones by day. Active at night.

Distribution: Worldwide.

Bee

Description: Insect with brown or black, thick, hairy bodies. Generally found in colonies. Many build wax combs.

Habitat Hollow trees, caves, dwellings. Near water in desert areas.

Distribution: Worldwide.

Note: Bees have barbed stingers and die after stinging because their venom sac and internal organs are pulled out during the attack.

Tick

Description: Round body from size of pinhead to 2.5 centimeters. Has 8 legs and sucking mouth parts. There are 850 species worldwide.

Habitat: Mainly in forests and grasslands. Also in urban areas and farmlands.

Distribution: Worldwide.

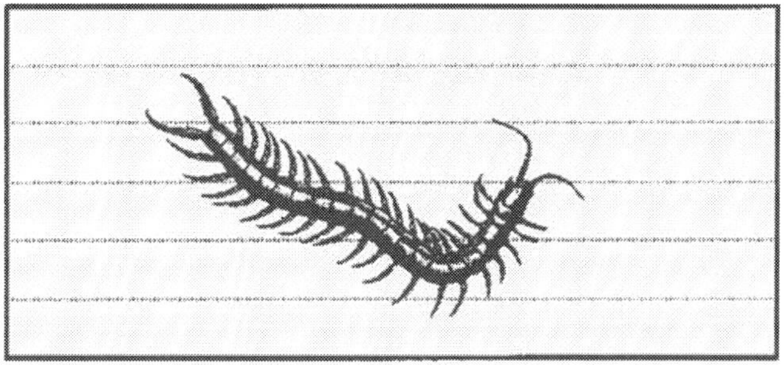

Centipede

Description: Multijointed body to 30 centimeters long. Dull orange to brown, with black point eyes at base of antennae. There are 2,800 species worldwide.

Habitat Under bark and stones by day. Active at night.

Distribution: Worldwide.

WAYS TO AVOID SNAKEBITE

Snakes are widely distributed. They are found in all tropical, subtropical, and most temperate regions. Some species of snakes have specialized glands that contain a toxic venom and long hollow fangs to inject their venom.

Although venomous snakes use their venom to secure food, they also use it for self-defense. Human accidents occur when you don't see or hear the snake, when you step on them, or when you walk too close to them.

Follow these simple rules to reduce the chance of accidental snakebite:

- Don't sleep next to brush, tall grass, large boulders, or trees. They provide hiding places for snakes. Place your sleeping bag in a clearing. Use mosquito netting tucked well under the bag. This netting should provide a good barrier.

  Don't put your hands into dark places, such as rock crevices, heavy brush, or hollow logs, without first investigating.
- Don't step over a fallen tree. Step on the log and look to see if there is a snake resting on the other side.
- Don't walk through heavy brush or tall grass without looking down. Look where you are walking.
- Don't pick up any snake unless you are absolutely positive it is not venomous.

  Don't pick up freshly killed snakes without first severing the head. The nervous system may still be active and a dead snake can deliver a bite.

SNAKE GROUPS

Snakes dangerous to man usually fall into two groups: proteroglypha and solenoglypha. Their fangs and their venom best describe these two groups (Figure E-l).

Fangs

The proteroglypha have, in front of the upper jaw and preceding the ordinary teeth, permanently erect fangs. These fangs are called fixed fangs.

The solenoglypha have erectile fangs; that is, fangs they can raise to an erect position. These fangs are called folded fangs.

Venom

The fixed-fang snakes (proteroglypha) usually have neurotoxic venoms. These venoms affect the nervous system, making the victim unable to breathe.

The folded-fang snakes (solenoglypha) usually have hemotoxic venoms. These venoms affect the circulatory system, destroying blood cells, damaging skin tissues, and causing internal hemorrhaging.

Remember, however, that most poisonous snakes have both neurotoxic and hemotoxic venom. Usually one type of venom in the snake is dominant and the other is weak.

Poisonous Versus Nonpoisonous Snakes

No single characteristic distinguishes a poisonous snake from a harmless one except the presence of poison fangs and glands. Only in dead specimens can you determine the presence of these fangs and glands without danger.

DESCRIPTIONS OF POISONOUS SNAKES

There are many different poisonous snakes throughout the world. It is unlikely you will see many except in a zoo. This manual describes only a few poisonous snakes. You should, however, be able to spot a poisonous snake if you-

Learn about the two groups of snakes and the families in which they fall (Figure E-2).

Examine the pictures and read the descriptions of snakes in this appendix.

Group	Fang Type	Venom Type
Proteroglypha	Fixed	Usually dominant neurotoxic
Solenoglypha	Folded	Usually dominant hemotoxic

Figure E-1. Snake group characteristics.

Tick

Description: Round body from size of pinhead to 2.5 centimeters. Has 8 legs and sucking mouth parts. There are 850 species worldwide.

Habitat: Mainly in forests and grasslands. Also in urban areas and farmlands.

Distribution: Worldwide.

The fangs of this group of snakes are movable. These snakes fold their fangs into the roof of their mouths. When they strike, their fangs come forward, stabbing the victim. The snake controls the movement of its fangs; fang movement is not automatic. The venom is usually hemotoxic. There are, however, several species that have large quantities of neurotoxic elements, thus making them even more dangerous. The vipers are responsible for many human fatalities around the world.

Crotalidae

The crotalids, or pit vipers (Figure E-4), may be either slender or thick-bodied. Their heads are usually much wider than their necks. These snakes take their name from the deep pit located between the eye and the nostril. They are commonly brown with dark blotches, though some kinds are green.

Rattlesnakes, copperheads, cottonmouths, and several species of dangerous snakes from Central and South America, Asia, China, and India fall into the pit viper group. The pit is a highly sensitive organ capable of picking up the slightest temperature variance. Most pit vipers are nocturnal. They hunt for food at night with the aid of these specialized pits

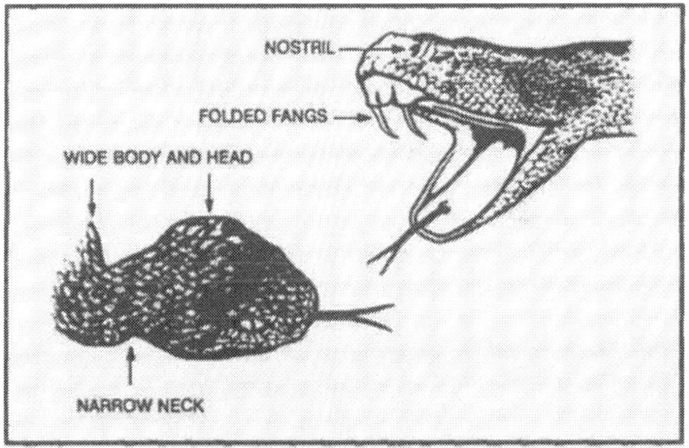

Figure E-3. Positive identification of vipers.

that let them locate prey in total darkness. Rattlesnakes are the only pit vipers that possess a rattle at the tip of the tail.

India has about 12 species of these snakes. You find them in trees or on the ground in all types of terrain. The tree snakes are slender; the ground snakes are heavy-bodied. All are dangerous.

China has a pit viper similar to the cottonmouth found in North America. You find it in the rocky areas of the remote mountains of South China. It reaches a length of 1.4 meters but is not vicious unless irritated. You can also find a small pit viper, about 45 centimeters long, on the plains of eastern China. It is too small to be dangerous to a man wearing shoes.

There are about 27 species of rattlesnakes in the United States and Mexico. They vary in color and may or may not have spots or blotches. Some are small while others, such as the diamondbacks, may grow to 2.5 meters long.

There are five kinds of rattlesnakes in Central and South America, but only the tropical rattlesnake is widely distributed. The rattle on the tip of the tail is sufficient identification for a rattlesnake.

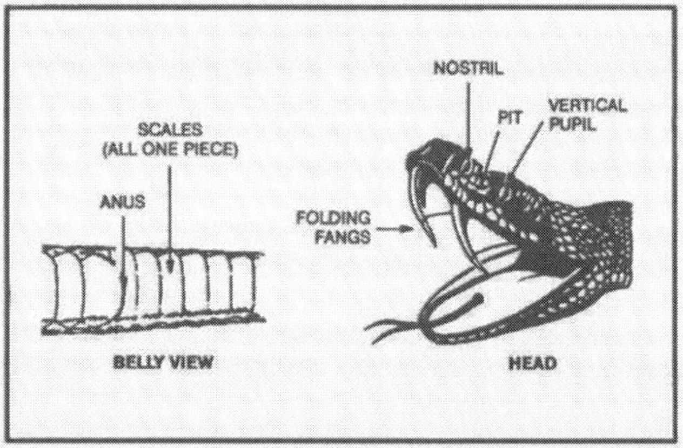

Figure E-4. Positive identification of pit vipers.

Most will try to escape without a fight when approached, but there is always a chance one will strike at a passerby. They do not always give a warning; they may strike first and rattle afterwards or not at all.

The genus Trimeresurus is a subgroup of the crotalidae. These are Asian pit vipers. These pit vipers are normally tree-loving snakes with a few species living on the ground. They basically have the same characteristics of the crotalidae-slender build and very dangerous. Their bites usually are on the upper extremities-head, neck, and shoulders. Their venom is largely hemotoxic.

Elapidae

A group of highly dangerous snakes with powerful neurotoxic venom that affects the nervou system, causing respiratory paralysis. Included in this family are coral snakes, cobras, mambas, and all the Australian venomous snakes. The coral snake is small and has caused human fatalities. The Australian death adder, tiger, taipan, and king brown snakes are among the most venomous in the world, causing many human fatalities.

Only by examining a dead snake can you positively determine if it is a cobra or a near relative (Figure E-5). On cobras, kraits, and coral

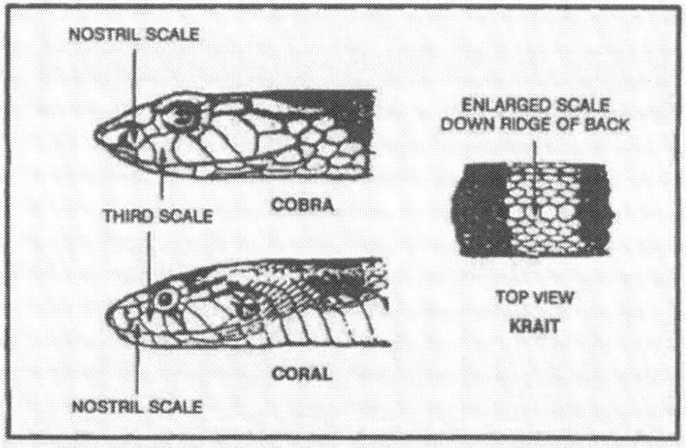

Figure E-5. Positive identification of cobras, kraits, and coral snakes.

snakes, the third scale on the upper lip touches both the nostril scale and the eye. The krait also has a row of enlarged scales down its ridged back.

You can find the cobras of Africa and the Near East in almost any habitat. One kind may live in or near water, another in trees. Some are aggressive and savage. The distance a cobra can strike in a forward direction is equal to the distance its head is raised above the ground. Some cobras, however, can spit venom a distance of 3 to 3.5 meters. This venom is harmless unless it gets into your eyes; then it may cause blindness if not washed out immediately. Poking around in holes and rock piles is dangerous because of the chance of encountering a spitting cobra.

Laticaudinae and Hydrophidae

A subfamily of elapidae, these snakes are specialized in that they found a better environment in the oceans. Why they are in the oceans is not clear to science.

Sea snakes differ in appearance from other snakes in that they have an oarlike tail to aid in swimming. Some species of sea snakes have venom several times more toxic than the cobra's. Because of their marine environment, sea snakes seldom come in contact with humans. The exceptions are fisherman who capture these dangerous snakes in fish nets and scuba divers who swim in waters where sea snakes are found.

There are many species of sea snakes. They vary greatly in color and shape. Their scales distinguish them from eels that have no scales.

Sea snakes occur in salt water along the coasts throughout the Pacific. There are also sea snakes on the east coast of Africa and in the Persian Gulf. There are no sea snakes in the Atlantic Ocean.

There is no need to fear sea snakes. They have not been known to attack a man swimming. Fishermen occasionally get bit by a sea snake caught in a net. The bite is dangerous.

Colubridae

The largest group of snakes worldwide. In this family there are species that are rear-fanged; however, most are completely harmless to man. They have a venom-producing gland and enlarged, grooved rear fangs that allow venom to flow into the wound. The inefficient venom apparatus and the specialized venom is effective on cold-blooded animals (such as frogs and lizards) but not considered a threat to human life. The boomslang and the twig snake of Africa have, however, caused human deaths.

Viperidae	Common adder
	Long-nosed adder
	Gaboon viper
	Levant viper
	Horned desert viper
	McMahon's viper
	Mole viper
	Palestinian viper
	Puff adder
	Rhinoceros viper
	Russell's viper
	Sand viper
	Saw-scaled viper
	Ursini's viper
Elapidae	Australian copperhead
	Common cobra
	Coral snake
	Death adder
	Egyptian cobra
	Green mamba
	King cobra
	Krait
	Taipan
	Tiger snake
Crotalidae	American copperhead
	Boomslang
	Bush viper
	Bushmaster
	Cottonmouth
	Eastern diamondback rattlesnake
	Eyelash pit viper
	Fer-de-lance
	Green tree pit viper
	Habu pit viper
	Jumping viper
	Malayan pit viper
	Mojave rattlesnake
	Pallas' viper
	Tropical rattlesnake
	Wagler's pit viper
	Western diamondback rattlesnake
Hydrophidae	Banded sea snake
	Yellow-bellied sea snake

LIZARDS

There is little to fear from lizards as long as you follow the same precautions as for avoiding snakebite. Usually, there are only two poisonous lizards: the Gila monster and the Mexican beaded lizard. The venom of both these lizards is neurotoxic. The two lizards are in the same family, and both are slow moving with a docile nature.

The komodo dragon *(Varanus komodoensis),* although not poisonous, can be dangerous due to its large size. These lizards can reach lengths of 3 meters and weigh over 115 kilograms. Do not try to capture this lizard.

POISONOUS SNAKES OF THE AMERICAS

American copperhead
Agkistrodon contortrix

Description: Chestnut color dominates overall, with darker crossbands of rich browns that become narrower on top and widen at the bottom. The top of the head is a coppery color.

Characteristic: Very common over much of its range, with a natural camouflage ability to blend in the environment. Copperheads are rather quiet and inoffensive in disposition but will defend themselves vigorously. Bites occur when the snakes are stepped on or when a victim is lying next to one. A copperhead lying on a bed of dead leaves becomes invisible. Its venom is hemotoxic.

Habitat: Found in wooded and rocky areas and mountainous regions.

Length: Average 60 centimeters, maximum 120 centimeters.

Distribution: Eastern Gulf States, Texas, Arkansas, Maryland, North Florida, Illinois, Oklahoma, Kansas, Ohio, New York, Alabama, Tennessee, and Massachusetts.

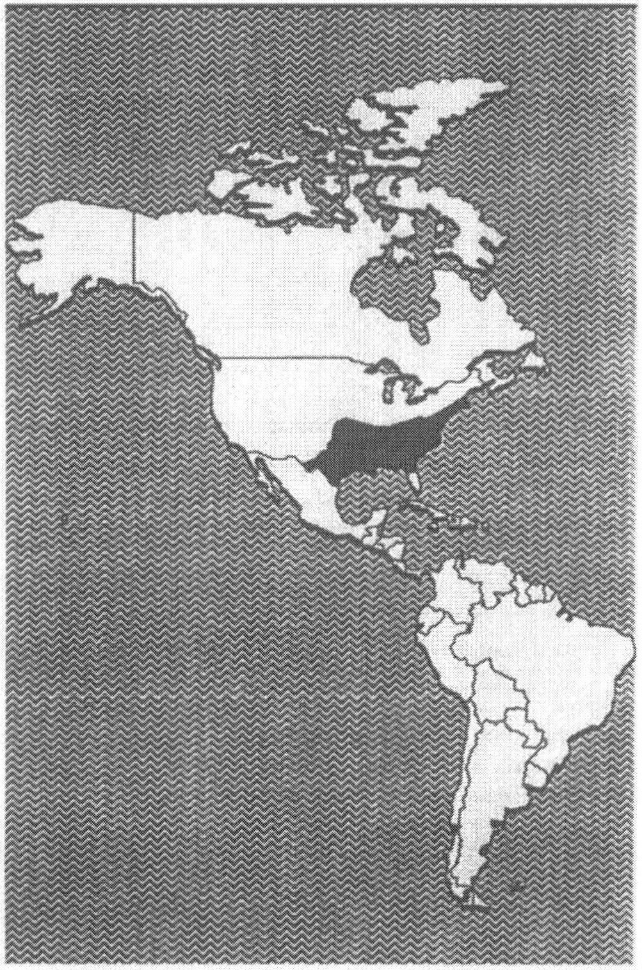

Bushmaster

Lachesis mutus

Description: The body hue is rather pale brown or pinkish, with a series of large bold dark brown or black blotches extending along the body. Its scales are extremely rough.

Characteristics: The world's largest pit viper has a bad reputation. This huge venomous snake is not common anywhere in its range. It lives in remote and isolated habitats and is largely nocturnal in its feeding habits; it seldom bites anyone, so few bites are recorded. A bite from one would indeed be very serious and fatal if medical aid was not immediately available. Usually, the bites occur in remote, dense jungles, many kilometers and several hours or even days away from medical help. Bushmaster fangs are long. In large bushmasters, they can measure 3.8 centimeters. Its venom is a powerful hemotoxin.

Habitat: Found chiefly in tropical forests in their range.

Length: Average 2.1 meters, maximum 3.7 meters.

Distribution: Nicaragua, Costa Rica, Panama, Trinidad, and Brazil.

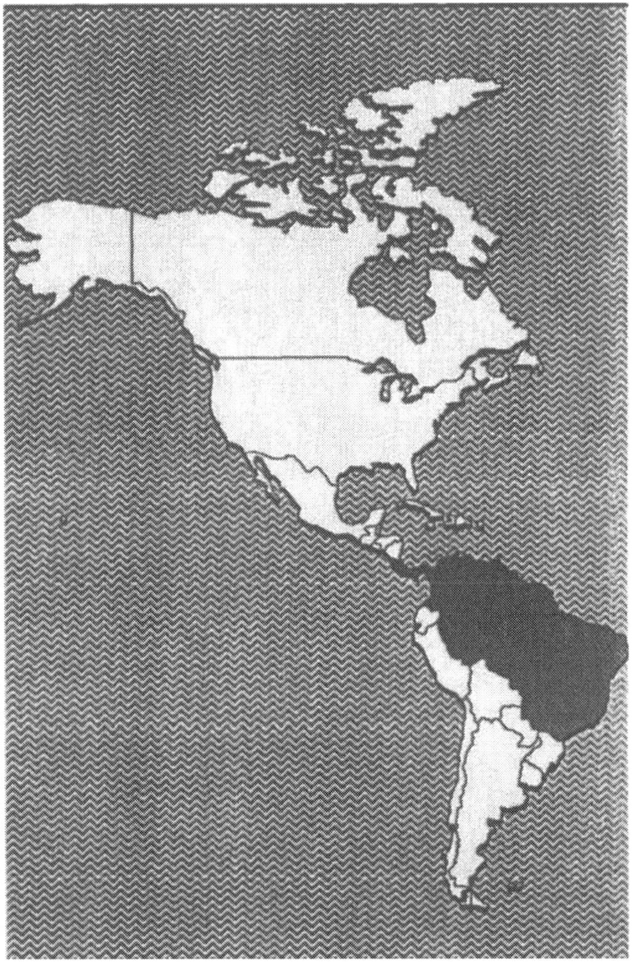

Coral snake

Micrurus fulvius

Description: Beautifully marked with bright blacks, reds, and yellows. To identify the species, remember that when red touches yellow it is a coral snake.

Characteristics: Common over range, but secretive in its habits, therefore seldom seen. It has short fanges that are fixed in an erect position. It often chews to release its venom into a wound. Its venom is very powerful. The venom is neurotoxic, causing respiratory paralysis in the victim, who succumbs to suffocation.

Habitat: Found in a variety of habitats including wooded areas, swamps, palmetto and scrub areas. Coral snakes often venture into residential locations.

Length: Average 60 centimeters, maximum 115 centimeters.

Distribution: Southeast North Carolina, Gulf States, west central Mississippi, Florida, Florida Keys, and west to Texas. Another genus of coral snake is found in Arizona. Coral snakes are also found throughout Central and most South America.

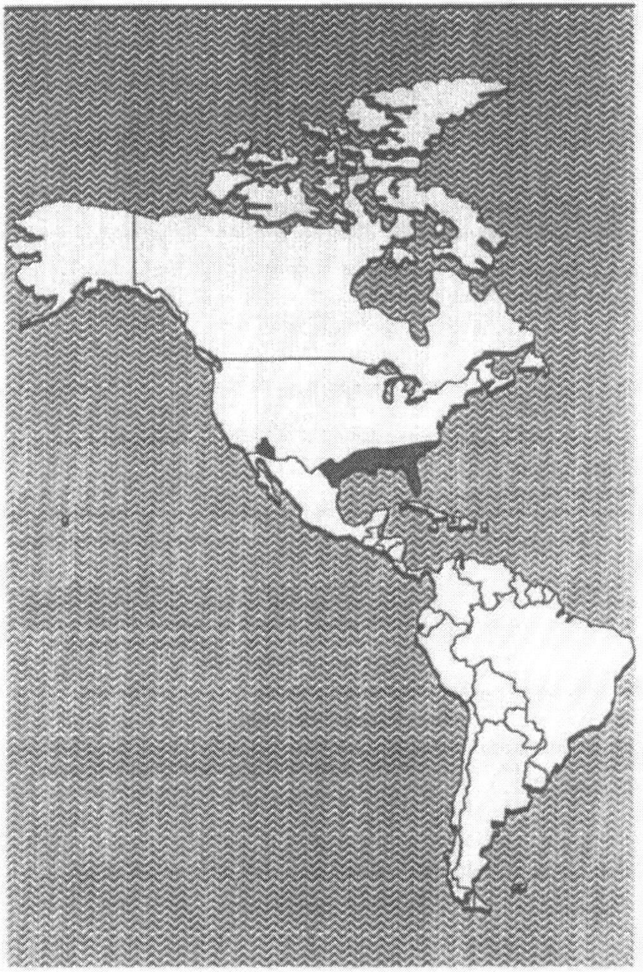

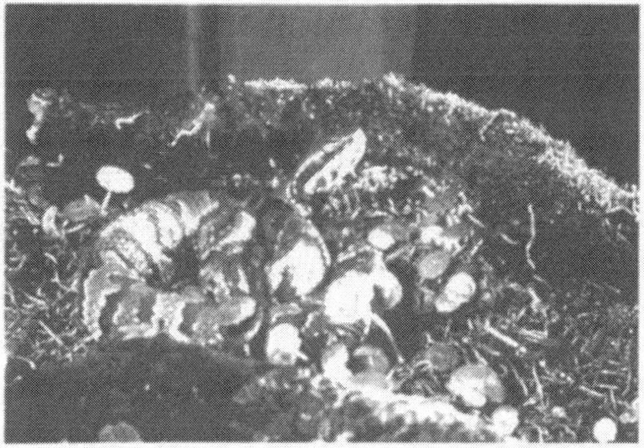

Cottonmouth

Agkistrodon piscivorus

Description: Colors are variable. Adults are uniformly olive brown or black. The young and subadults are strongly crossbanded with dark brown.

Characteristics: These dangerous semiaquatic snakes closely resemble harmless water snakes that have the same habitat. Therefore, it is best to leave all water snakes alone. Cottonmouths often stand their ground. An aroused cottonmouth will draw its head close to its body and open its mouth showing its white interior. Cottonmouth venom is hemotoxic and potent. Bites are prone to gangrene.

Habitat: Found in swamps, lakes, rivers, and ditches.

Length: Average 90 centimeters, maximum 1.8 meters.

Distribution: Southeast Virginia, west central Alabama, south Georgia, Illinois, east central Kentucky, south central Oklahoma, Texas, North and South Carolina, Florida, and the Florida Keys.

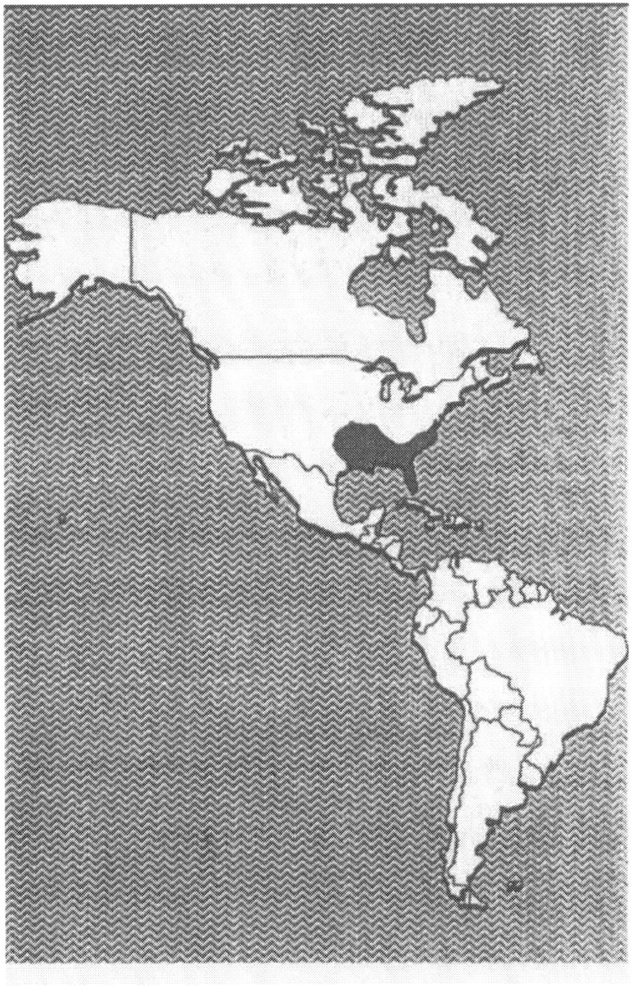

Eastern diamondback rattlesnake

Crotalus adamanteus

Description: Diamonds are dark brown or black, outlined by a row of cream or yellowish scales. Ground color is olive to brown.

Characteristics: The largest venomous snake in the United States. Large individual snakes can have fangs that measure 2.5 centimeters in a straight line. This species has a sullen disposition, ready to defend itself when threatened. Its venom is potent and hemotoxic, causing great pain and damage to tissue.

Habitat: Found in palmettos and scrubs, swamps, pine woods, and flatwoods. It has been observed swimming many miles out in the Gulf of Mexico, reaching some of the islands off the Florida coast.

Length: Average 1.4 meters, maximum 2.4 meters.

Distribution: Coastal areas of North Carolina, South Carolina, Louisiana, Florida, and the Florida Keys.

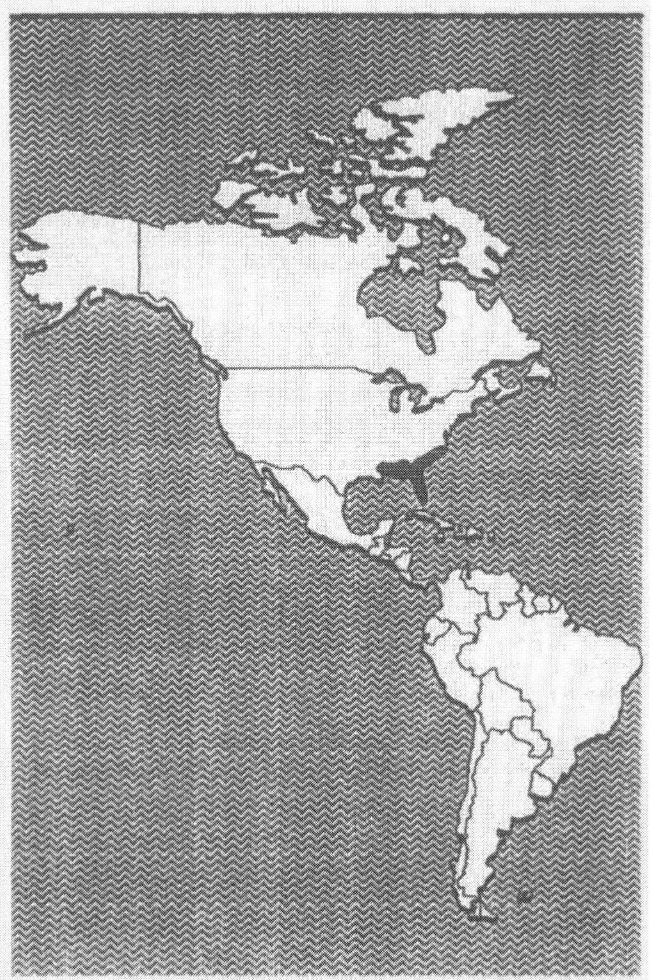

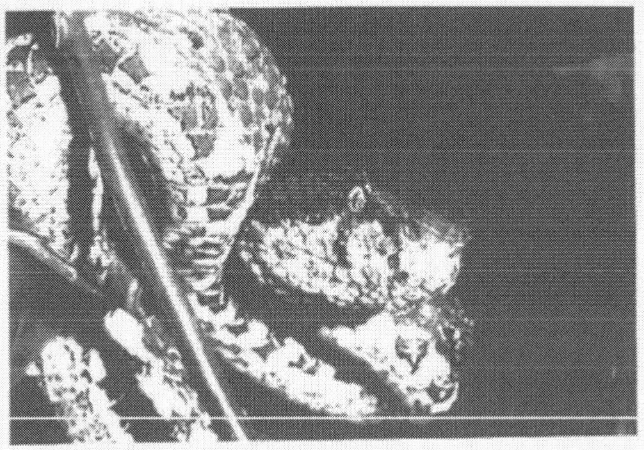

Eyelash pit viper

Bothrops schlegeli

Description: Identified by several spiny scales over each eye. Color is highly variable, from bright yellow over its entire body to reddish-yellow spots throughout the body.

Characteristics: Arboreal snake that seldom comes to the ground. It feels more secure in low-hanging trees where it looks for tree frogs and birds. It is a dangerous species because most of its bites occur on the upper extremities. It has an irritable disposition. It will strike with little provocation. Its venom is hemotoxic, causing severe tissue damage. Deaths have occurred from the bites of these snakes.

Habitat: Tree-loving species found in rain forests; common on plantations and in palm trees.

Length: Average 45 centimeters, maximum 75 centimeters.

Distribution: Southern Mexico, throughout Central America, Columbia, Ecuador, and Venezuela.

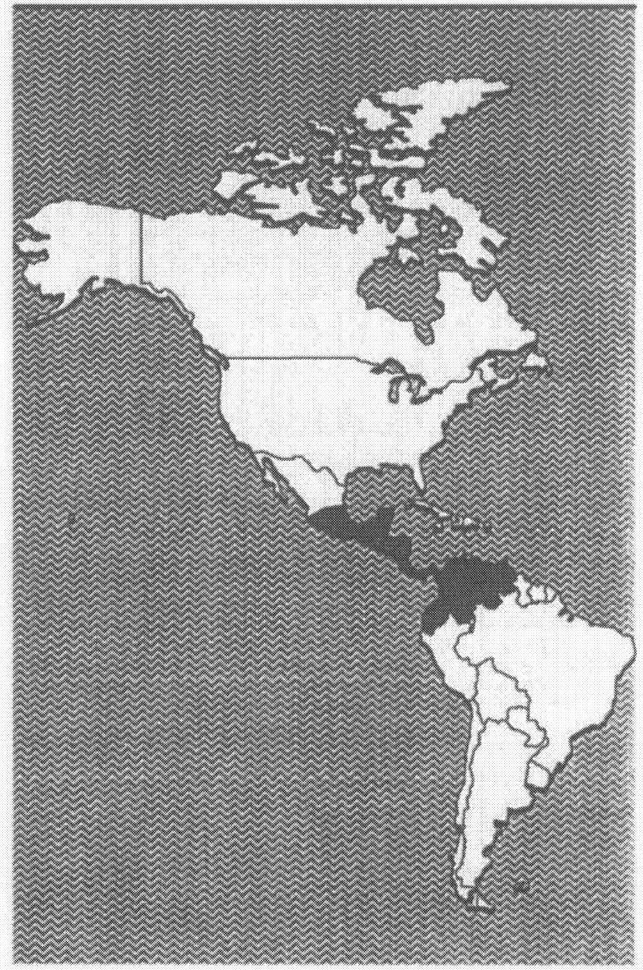

Fer-de-lance
Bothrops atrox

There are several closely related species in this group. All are very dangerous to man.

Description: Variable coloration, from gray to olive, brown, or reddish, with dark triangles edged with light scales. Triangles are narrow at the top and wide at the bottom.

Characteristics: This highly dangerous snake is responsible for a high mortality rate. It has an irritable disposition, ready to strike with little provocation. The female fer-de-lance is highly prolific, producing up to 60 young born with a dangerous bite. The venom of this species is hemotoxic, painful, and hemorrhagic (causing profuse internal bleeding). The venom causes massive tissue destruction.

Habitat: Found on cultivated land and farms, often entering houses in search of rodents.

Length: Average 1.4 meters, maximum 2.4 meters.

Distribution: Southern Mexico, throughout Central and South America.

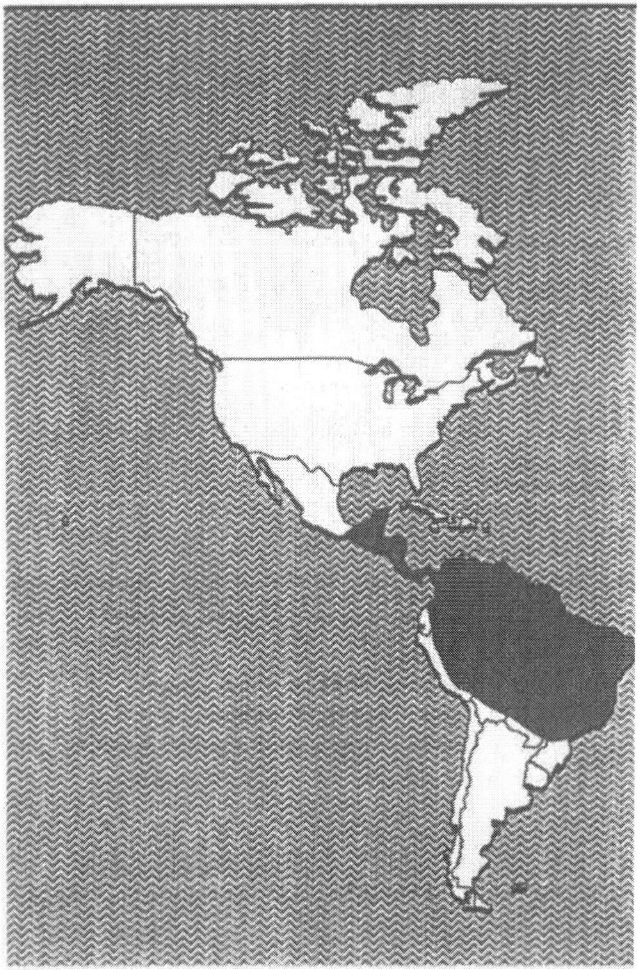

Jumping viper
Bothrops nummifer

Jumping viper
Bothrops nummifer

Description: It has a stocky body. Its ground color varies from brown to gray and it has dark brown or black dorsal blotches. It has no pattern on its head.

Characteristics: It is chiefly a nocturnal snake. It comes out in the early evening hours to feed on lizards, rodents, and frogs. As the name implies, this species can strike with force as it actually leaves the ground. Its venom is hemotoxic. Humans have died from the bites inflicted by large jumping vipers. They often hide under fallen logs and piles of leaves and are difficult to see.

Habitat: Found in rain forests, on plantations, and on wooded hillsides.

Length: Average 60 centimeters, maximum 120 centimeter.

Distribution: Southern Mexico, Honduras, Guatemala, Costa Rica, Panama, and El Salvador.

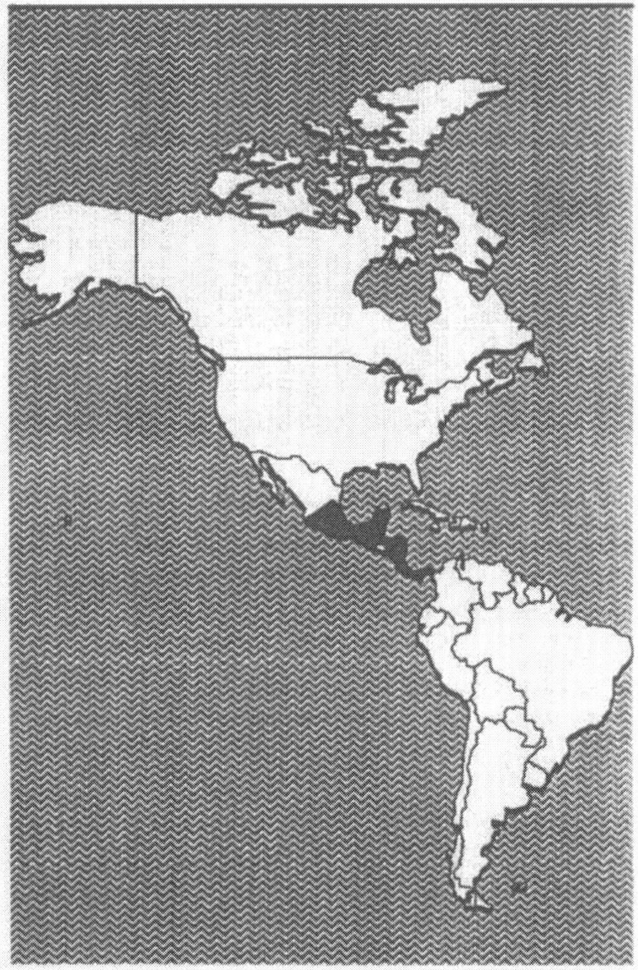

Mojave rattlesnake
Crotalus scutulatus

Description: This snake's entire body is a pallid or sandy odor with darker diamond-shaped markings bordered by lighter-colored scales and black bands around the tail.

Characteristics: Although this rattlesnake is of moderate size, its bite is very serious. Its venom has quantities of neurotoxic elements that affect the central nervous system. Deaths have resulted from this snake's bite.

Habitat: Found in arid regions, deserts, and rocky hillsides from sea level to 2400-meter elevations.

Length: Average 75 centimeters, maximum 1.2 meters.

Distribution: Mojave Desert in California, Nevada, southwest Arizona, and Texas into Mexico.

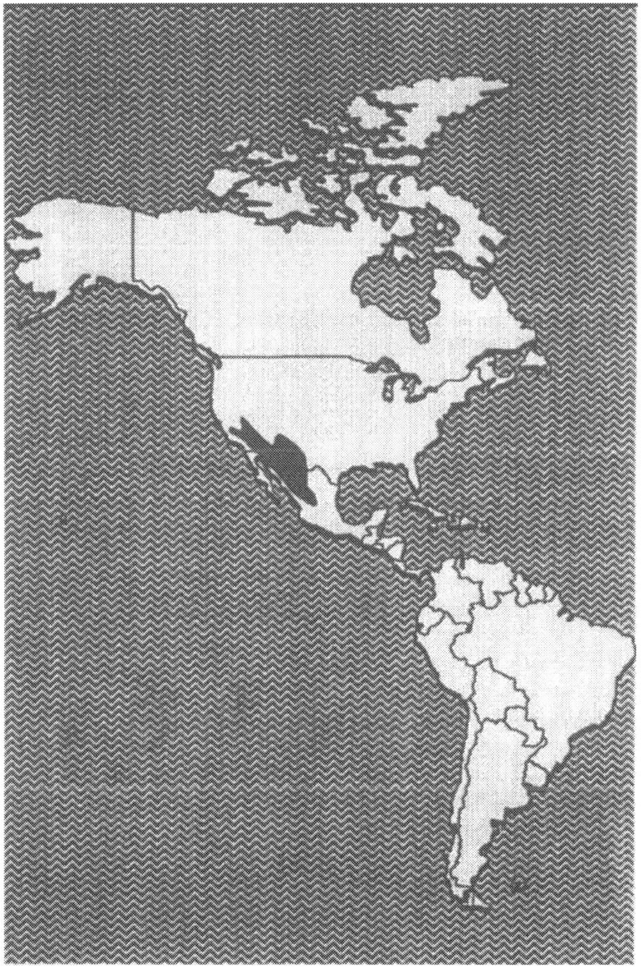

Tropical rattlesnake
Crotalus terrificus

Description: Coloration is light to dark brown with a series of darker rhombs or diamonds bordered by a buff color.

Characteristics: Extremely dangerous with an irritable disposition, ready to strike with little or no warning (use of its rattle). This species has a highly toxic venom containing neurotoxic and hemotoxic components that paralyze the central nervous system and cause great damage to tissue.

Habitat: Found in sandy places, plantations, and dry hillsides.

Length: Average 1.4 meters, maximum 2.1 meters.

Distribution: Southern Mexico, Central America, and Brazil to Argentina.

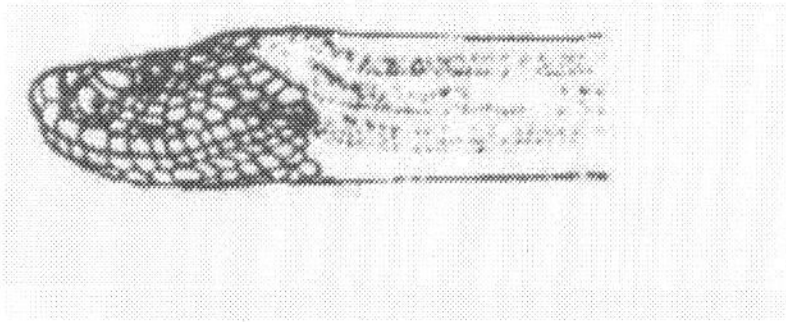

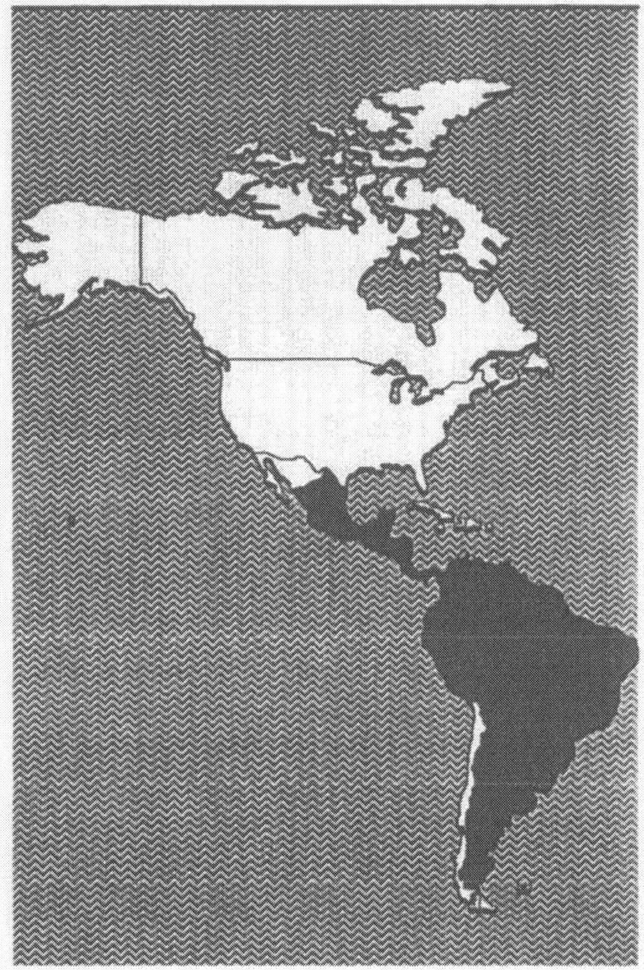

Western diamondback rattlesnake
Crotalus atrox

Description: The body is a light buff color with darker brown diamond-shaped markings. The tail has heavy black and white bands.

Characteristics: This bold rattlesnake holds its ground. When coiled and rattling, it is ready to defend itself. It injects a large amount of venom when it bites, making it one of the most dangerous snakes. Its venom is hemotoxic, causing considerable pain and tissue damage.

Habitat: It is a very common snake over its range. It is found in grasslands, deserts, woodlands, and canyons.

Length: Average 1.5 meters, maximum 2 meters.

Distribution: Southeast California, Oklahoma, Texas, New Mexico, and Arizona.

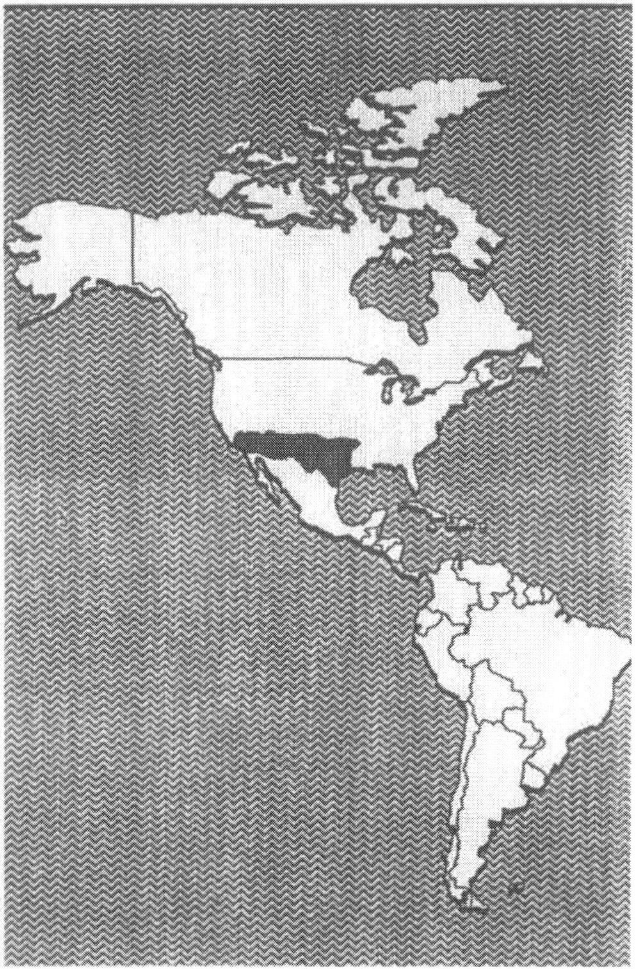

POISONOUS SNAKES OF EUROPE

Common adder
Vipera berus

Description: Its color is variable. Some adult specimens are completely black while others have a dark zigzag pattern running along the back.

Characteristics: The common adder is a small true viper that has a short temper and often strikes without hesitation. Its venom is hemotoxic, destroying blood cells and causing tissue damage. Most injuries occur to campers, hikers, and field workers.

Habitat: Common adders are found in a variety of habitats, from grassy fields to rocky slopes, and on farms and cultivated lands.

Length: Average 45 centimeters, maximum 60 centimeters.

Distribution: Very common throughout most of Europe.

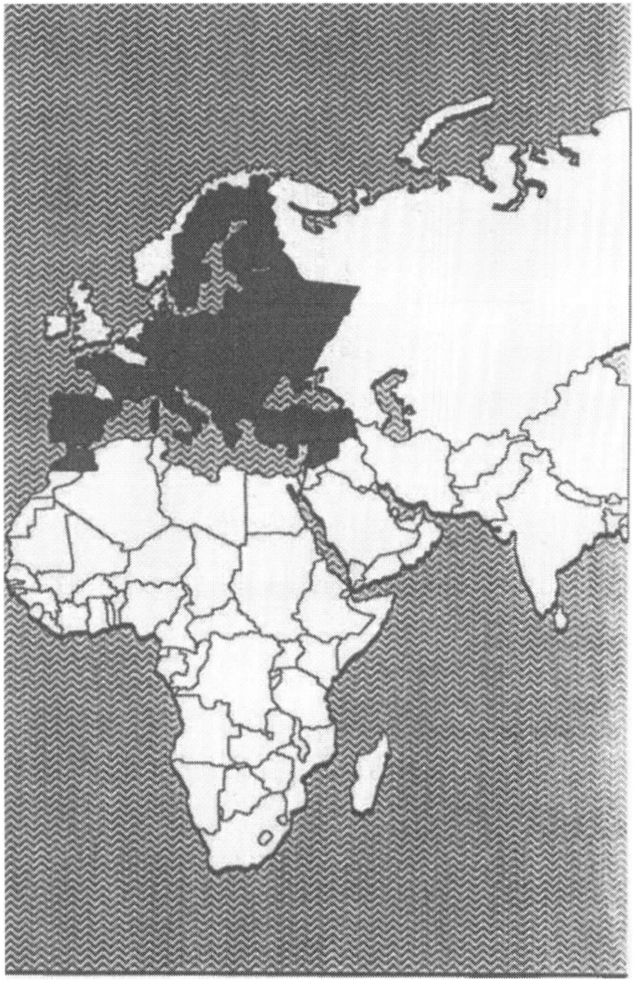

Long-nosed adder
Vipera ammodytes

Description: Coloration is gray, brown, or reddish with a dark brown or black zigzag pattern running the length of its back. A dark stripe is usually found behind each eye.

Characteristics: A small snake commonly found in much of its range. The term "long-nosed" comes from the projection of tiny scales located on the tip of its nose. This viper is responsible for many bites. Deaths have been recorded. Its venom is hemotoxic, causing severe pain and massive tissue damage. The rate of survival is good with medical aid.

Habitat: Open fields, cultivated lands, farms, and rocky slopes.

Length: Average 45 centimeters, maximum 90 centimeters.

Distribution: Italy, Yugoslavia, northern Albania, and Romania.

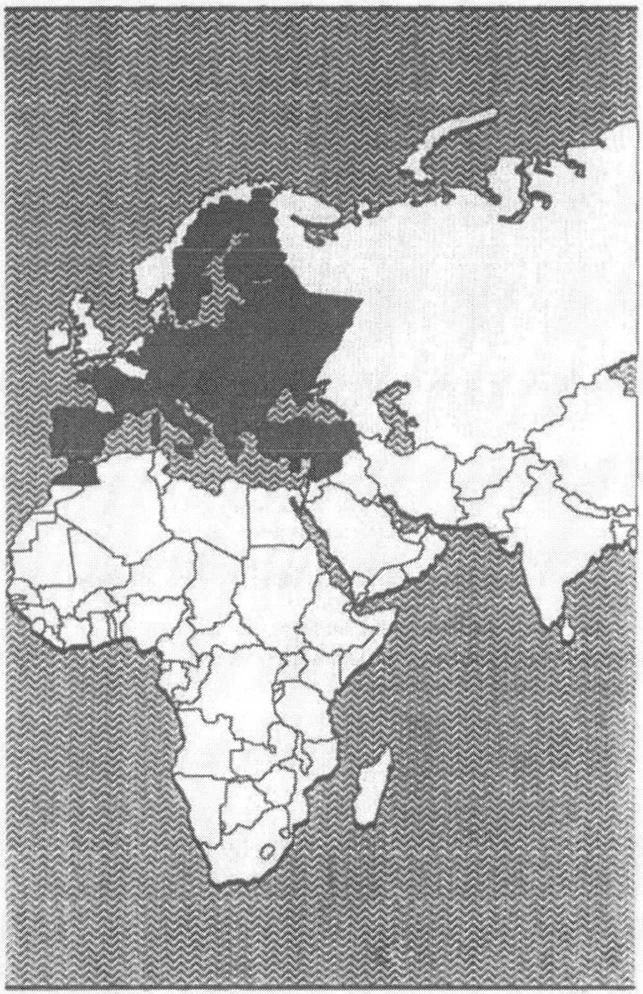

JOHN H. TASHJIAN/BERND VON SCHROEDER

Pallas' viper

Agkistrodon halys

Description: Coloration is gray, tan, or yellow, with markings similar to those of the American copperhead.

Characteristics: This snake is timid and rarely strikes. Its venom is hemotoxic but rarely fatal.

Habitat: Found in open fields, hillsides, and farming regions.

Length: Average 45 centimeters, maximum 90 centimeters.

Distribution: Throughout southeastern Europe.

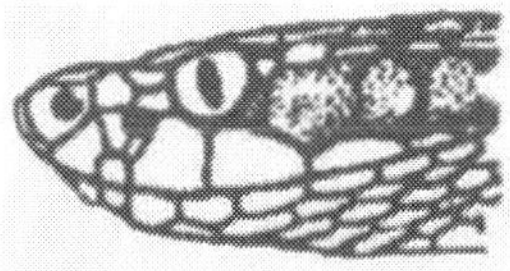

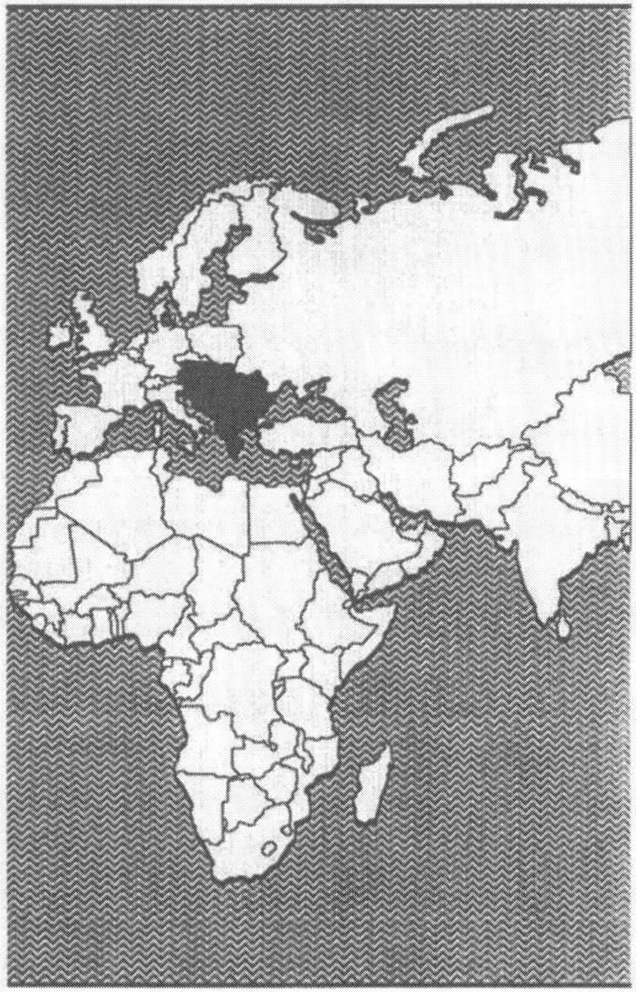

Ursini's viper
Vipera ursinii

JOHN H. TASHJIAN/BORJE FLARDH

Description: The common adder, long-nosed adder, and Ursini's viper basically have the same coloration and dorsal zigzag pattern. The exception among these adders is that the common adder and Ursini's viper lack the projection of tiny scales on the tip of the nose.

Characteristics: These little vipers have an irritable disposition. They will readily strike when approached. Their venom is hemotoxic. Although rare, deaths from the bites of these vipers have been recorded.

Habitat: Meadows, farmlands, rocky hillsides, and open, grassy fields.

Length: Average 45 centimeters, maximum 90 centimeters.

Distribution: Most of Europe, Greece, Germany, Yugoslavia, France, Italy, Hungary, Romania, Bulgaria, and Albania.

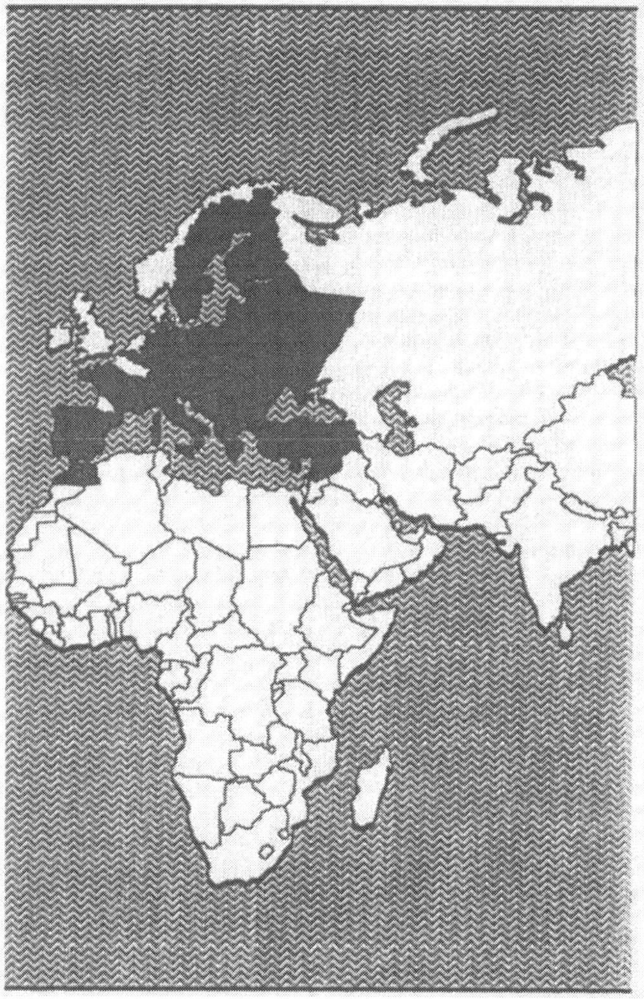

JOHN JASHJIAN/CALIFORNIA ACADEMY OF SCIENCE

Boomslang
Dispholidus typus

Description: Coloration varies but is generally green or brown, which makes it very hard to see in its habitat.

Characteristics: Will strike if molested. Its venom is hemotoxic; even small amounts cause severe hemorrhaging, making it dangerous to man.

Habitat: Found in forested areas. It will spend most of its time in trees or looking for chameleons and other prey in bushes.

Length: Generally less than 60 centimeters.

Distribution: Found throughout sub-Saharan Africa.

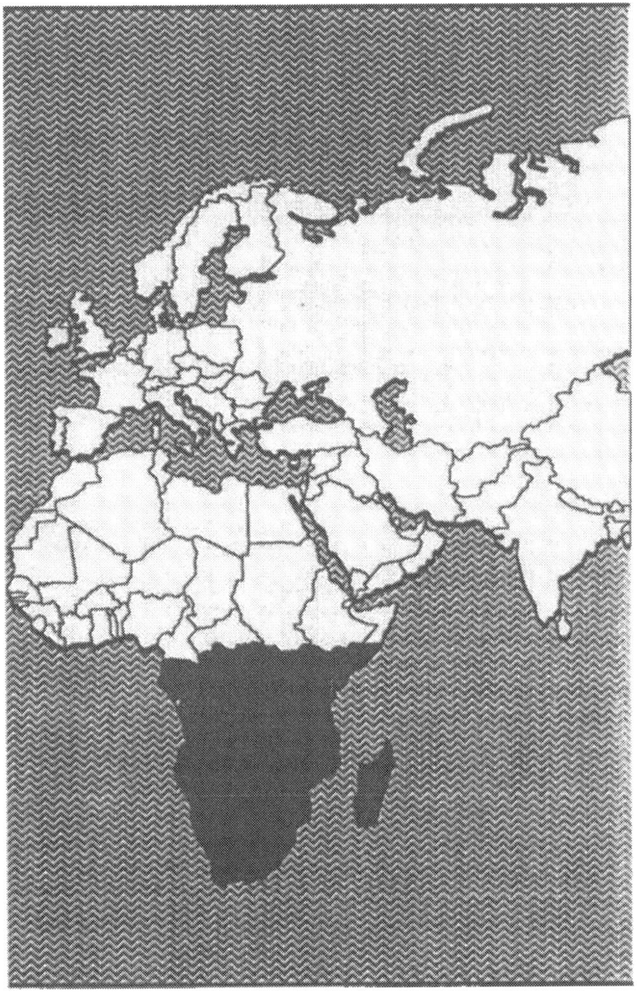

Bush viper

Atheris squamiger

Description: Often called leaf viper, its color varies from ground colors of pale green to olive, brown, or rusty brown. It uses it prehensile tail to secure itself to branches.

Characteristics: An aboreal species that often comes down to the ground to feed on small rodents. It is not aggressive, but it will defend itself when molested or touched. Its venom is hemotoxic; healthy adults rarely die from its bite.

Habitat: Found in rain forests and woodlands bordering swamps and forests. Often found in trees, low-hanging branches, or brush.

Length: Average 45 centimeters, maximum 75 centimeters.

Distrubition: Most of Africa, Angola, Cameroon, Uganda, Kenya, and Zaire.

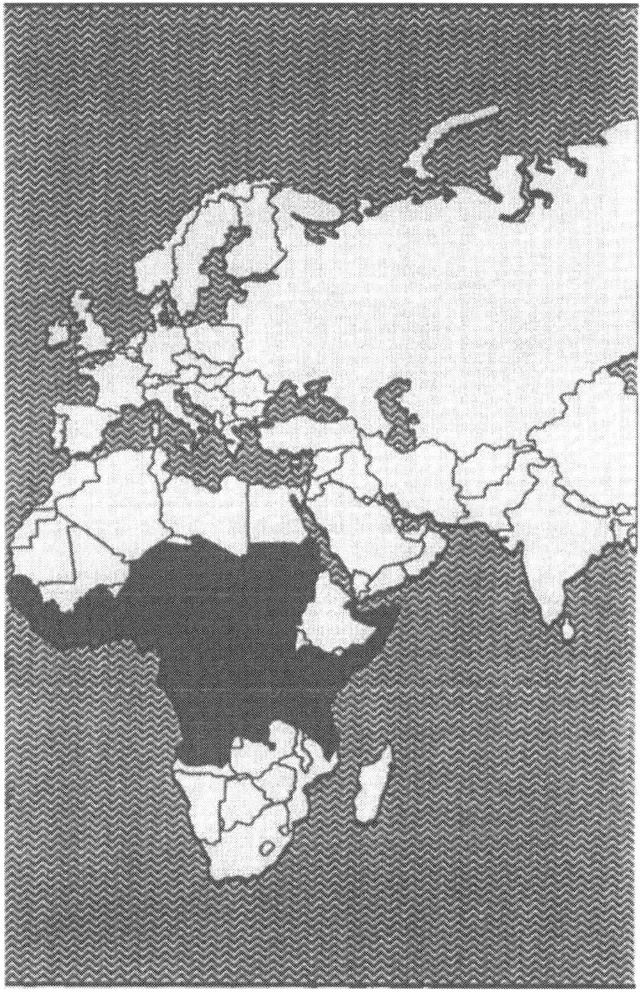

Common cobra

Wa/a na/a

Description: Also known as the Asiatic cobra. Usually slate gray to brown overall. The back of the hood may or may not have a pattern.

Characteristics: A very common species responsibie for many deaths each year. When aroused or threatened, the cobra will lift its head off the ground and spread its hood, making it more menacing. Its venom is highly neurotoxic, causing respiratory paralysis with some tissue damage. The cobra would rather retreat if possible, but if escape is shut off, it will be a dangerous creature to deal with.

Habitat: Found in any habitat cultivated farms, swamps, open fields, and human dwelling where it searches for rodents.

Length: Average 1.2 meters, maximum 2.1 meters.

Distribution: All of Asia.

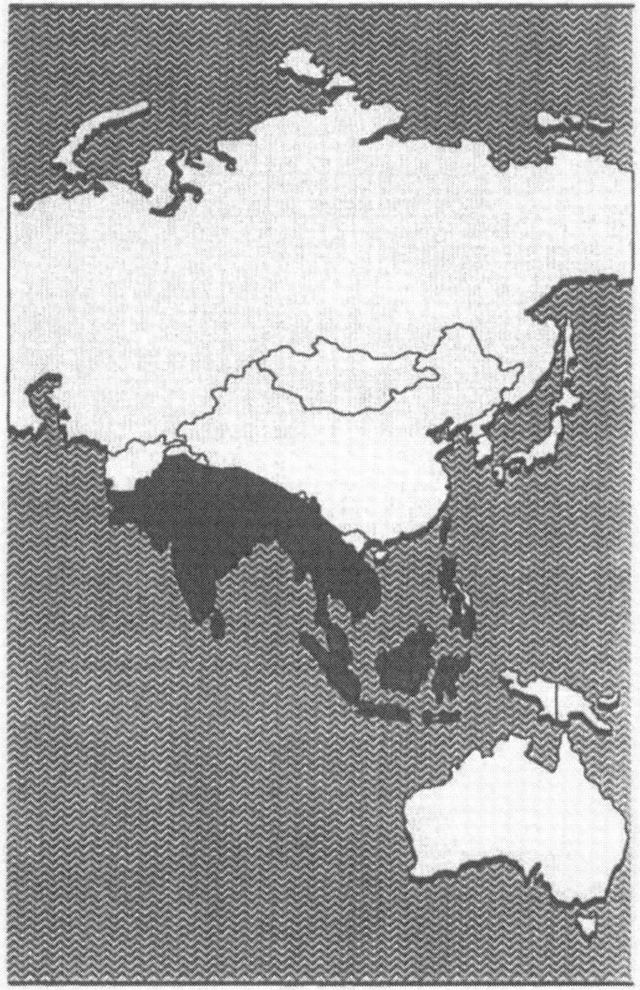

Egyptian cobra

Naja haje

Description: Yellowish, dark brown, or black uniform top with brown crossbands. Its head is sometimes black.

Characteristics: It is extremely dangerous. It is responsible for many human deaths. Once aroused or threatened, it will attack and continue the attack until it feels an escape is possible. Its venom is neurotoxic and much stronger than the common cobra. Its venom causes paralysis and death due to respiratory failure.

Habitat: Cultivated farmlands, open fields, and arid countrysides. It is often seen around homes searching for rodents.

Length: Average 1.5 meters, maximum 2.5 meters.

Distribution: Africa, Iraq, Syria, and Saudi Arabia.

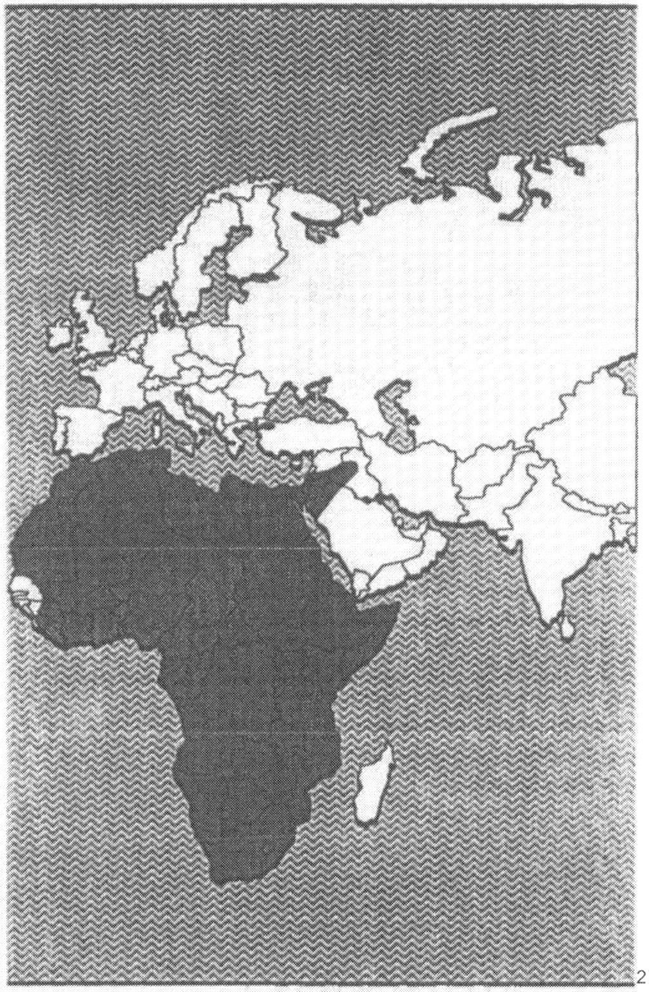

Gaboon viper
Bitis gabonica

Description: Pink to brown with a vertebral series of elongated yellowish or light brown spots connected by hourglass-shaped markings on each side. It has a dark brown stripe behind each eye. This dangerous viper is almost invisible on the forest floor. A 1.8-meter-long Gaboon viper could weigh 16 kilograms.

Characteristics: The largest and heaviest of all true vipers, having a very large triangular head. It comes out in the evening to feed. Fortunately, it is not aggressive, but it will stand its ground if approached. It bites when molested or stepped on. Its fangs are enormous, often measuring 5 centimeters long. It injects a large amount of venom when it strikes. Its venom is neurotoxic and hemotoxic.

Habitat: Dense rain forests. Occasionally found in open country.

Length: Average 1.2 meters, maximum 1.8 meters.

Distribution: Most of Africa.

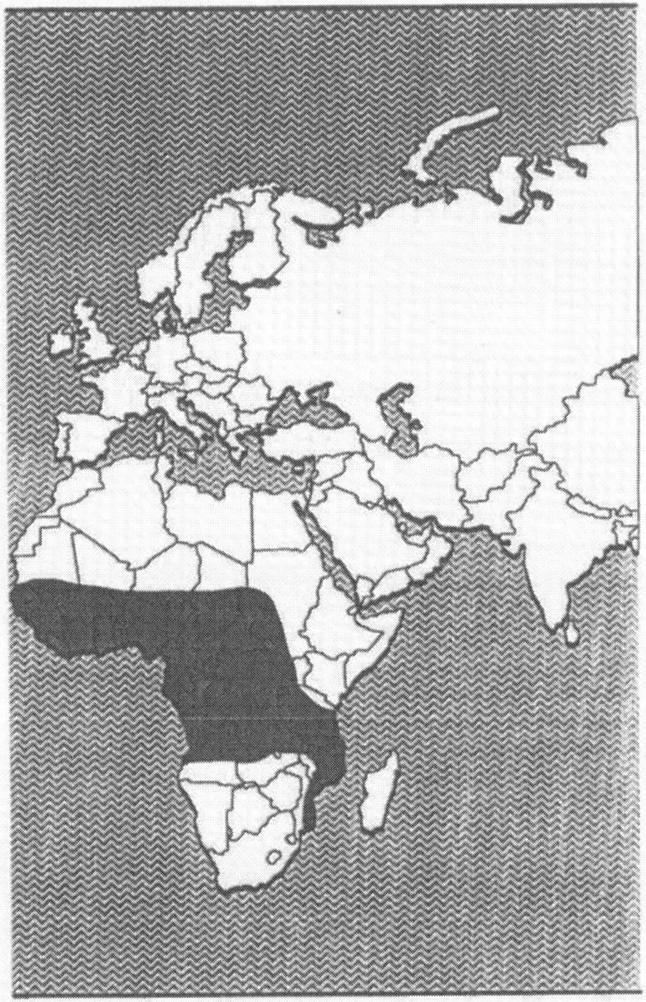

Green mamba

Dendraspis angusticeps

Green mamba

Dendraspis angusticeps

Description: Most mambas are uniformly bright green over their entire body. The black mamba, the largest of the species, is uniformly olive to black.

Characteristics: The mamba is the dreaded snake species of Africa. Treat it with great respect. It is considered one of the most dangerous snakes known. Not only is it highly venomous but it is aggressive and its victim has little chance to escape from a bite. Its venom is highly neurotoxic.

Habitat: Mambas are at home in brush, trees, and low-hanging branches looking for birds, a usual diet for this species.

Length: Average 1.8 meters, maximum 3.7 meters.

Distribution: Most of Africa.

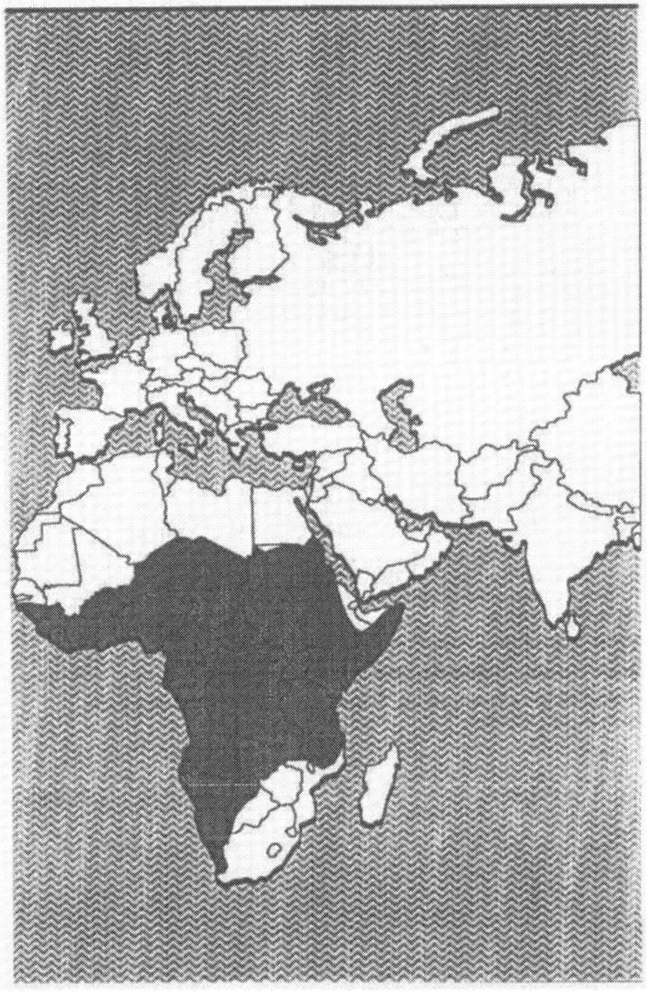

Green tree pit viper
Trimeresurus gramineus

Description: Uniform bright or dull green with light yellow on the facial lips.

Characteristics: A small arboreal snake of some importance, though not considered a deadly species. It is a dangerous species because most of its bites occur in the head, shoulder, and neck areas. It seldom comes to the ground. It feeds on young birds, lizards, and tree frogs.

Habitat: Found in dense rain forests and plantations.

Length: Average 45 centimeters, maximum 75 centimeters.

Distribution: India, Burma, Malaya, Thailand, Laos, Cambodia, Vietnam, China, Indonesia, and Formosa.

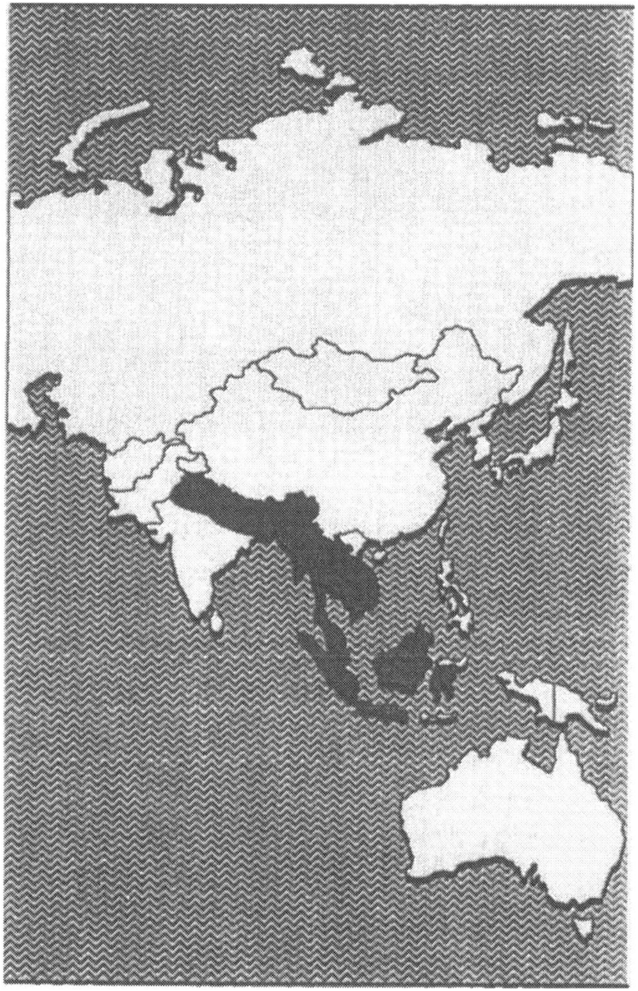

Habu pit viper

Trimeresurus flavoviridis

Description: Light brown or olive-yellow with black markings and a yellow or greenish-white belly.

Characteristics: This snake is responsible for biting many humans and its bite could be fatal. It is an irritable species ready to defend itself. Its venom is hemotoxic, causing pain and considerable tissue damage.

Habitat: Found in a variety of habitats, ranging from lowlands to mountainous regions. Often encountered in old houses and rock walls surroundings buildings.

Length: Average 1 meter, maximum 1.5 meters.

Distribution: Okinawa and neighboring islands and Kyushu.

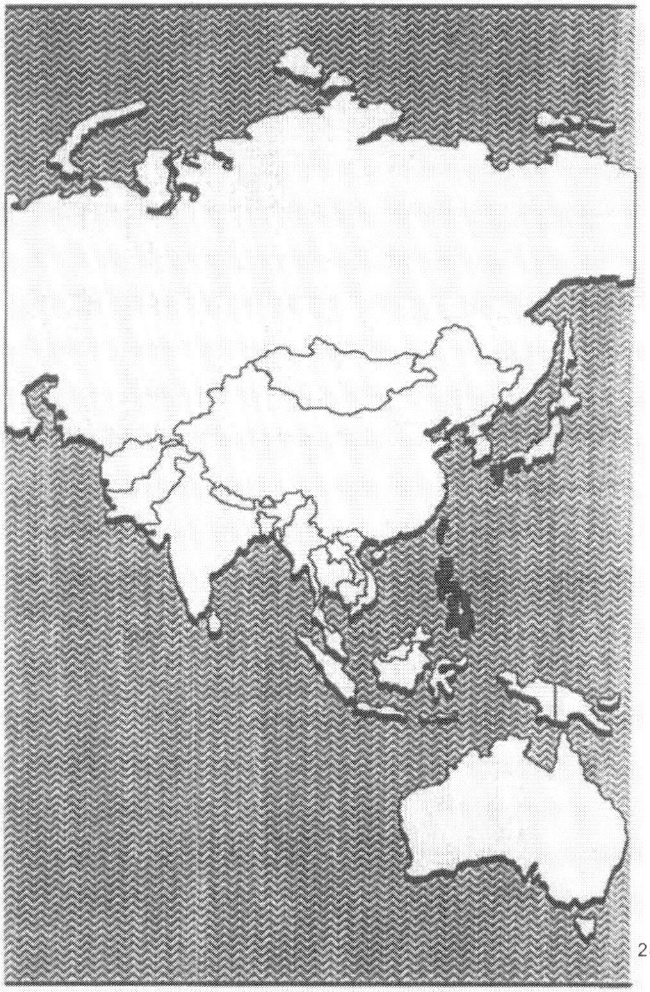

Horned desert viper
Cerastes cerastes

Horned desert viper
Cerastes cerastes

Description: Pale buff color with obscure markings and a sharp spine (scale) over each eye.

Characteristics: As with all true vipers that live in the desert, it finds refuge by burrowing in the heat of the day, coming out at night to feed. It is difficult to detect when buried; therefore, many bites result from the snake being accidentally stepped on. Its venom is hemotoxic, causing severe damage to blood cells and tissue.

Habitat: Only found in very arid places within its range.

Length: Average 45 centimeters, maximum 75 centimeters.

Distribution: Arabian Peninsula, Africa, Iran, and Iraq.

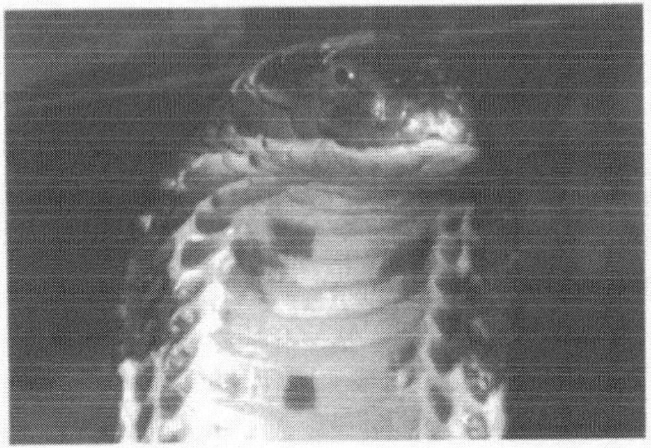

King cobra
Ophiophagus hannah

Description: Uniformly olive, brown, or green with ringlike crossbands of black.

Characteristics: Although it is the largest venomous snake in the world and it has a disposition to go with this honor, it causes relatively few bites on humans. It appears to have a degree of intelligence. It avoids attacking another venomous snake for fear of being bitten. It feeds exclusively on harmless species. The female builds a nest then deposits her eggs. Lying close by, she guards the nest and is highly aggressive toward anything that closely approaches the nest. Its venom is a powerful neurotoxin. Without medical aid, death is certain for its victims.

Habitat: Dense jungle and cultivated fields.

Length: Average 3.5 meters, maximum 5.5 meters.

Distibution: Thailand, southern China, Malaysia Peninsula, and Phillipines.

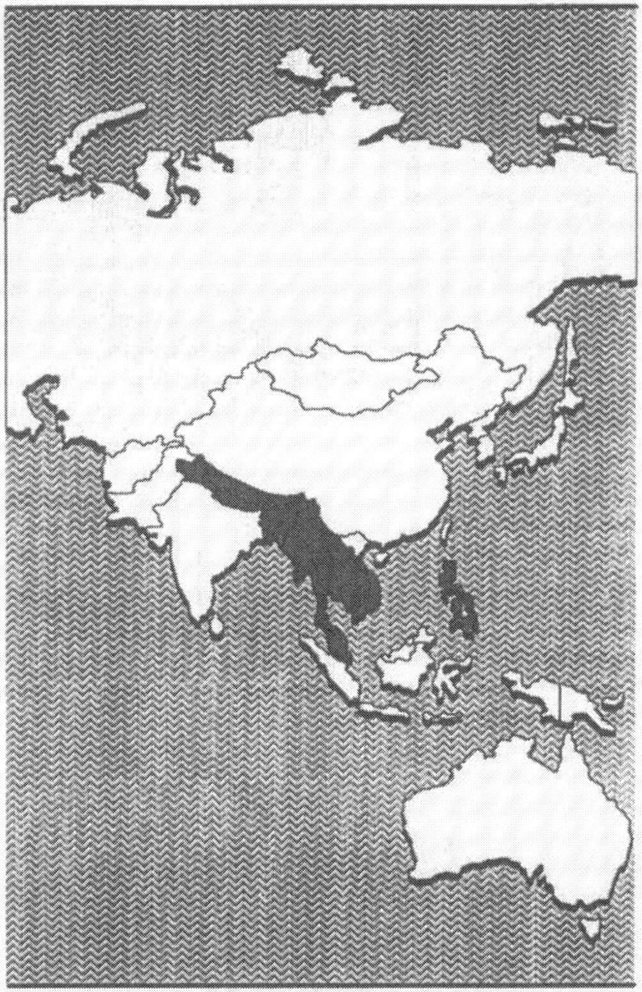

Krait
Bungarus caeruleus

Description: Black or bluish-black with white narrow crossbands and a narrow head.

Characteristics: Kraits are found only in Asia. This snake is of special concern to man. It is deadly - about 15 times more deadly than the common cobra It is active at night and relatively passive during the day. The native people often step on kraits while walking through their habitats. The krait has a tendency to seek shelter in sleeping bags, boots, and tents. Its venom is a powerful neurotoxin that causes respiratory failure.

Habitat: Open fields, human settlements, and dense jungle.

Length: Average 90 centimeters, maximum 1.5 meters.

Distribution: India, Sri Lanka, and Pakistan.

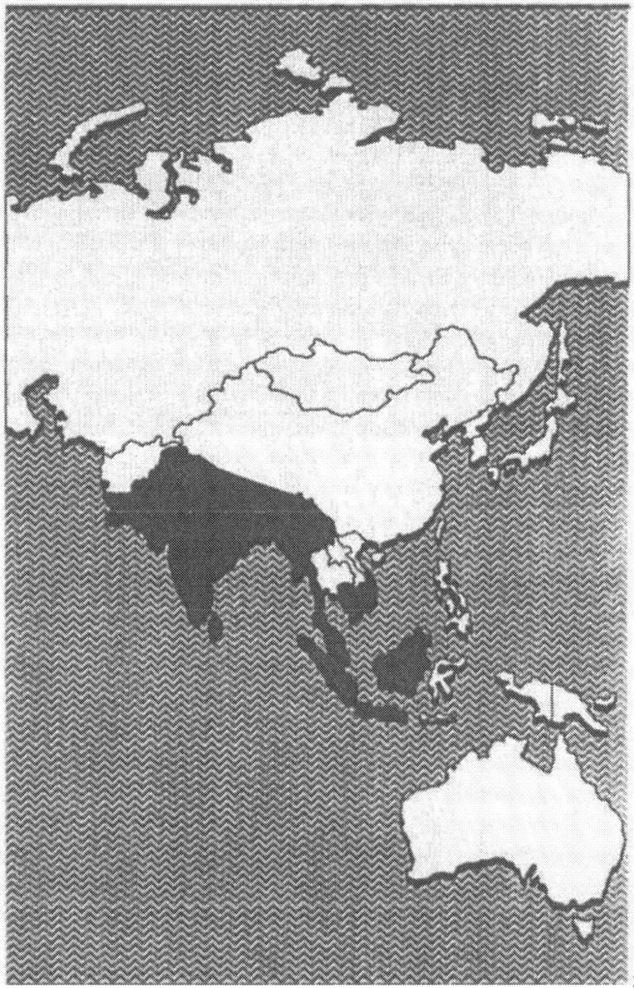

Levant viper

Vipera lebetina

Description: Gray to pale brown with large dark brown spots on the top of the black and a " Λ " mark on top of the head.

Characteristics: This viper belongs to a large group of true vipers. Like its cousins, it is large and dangerous. Its venom is hemotoxic. Many deaths have been reported from bites of this species. It is a strong snake with an irritable disposition; it hisses loudly when ready to strike.

Habitat: Varies greatly, from farmlands to mountainous areas.

Length: Average 1 meter, maximum 1.5 meters.

Distribution: Greece, Iraq, Syria, Lebanon, Turkey, Afganistan, lower portion of the former USSR, and Saudi Arabia.

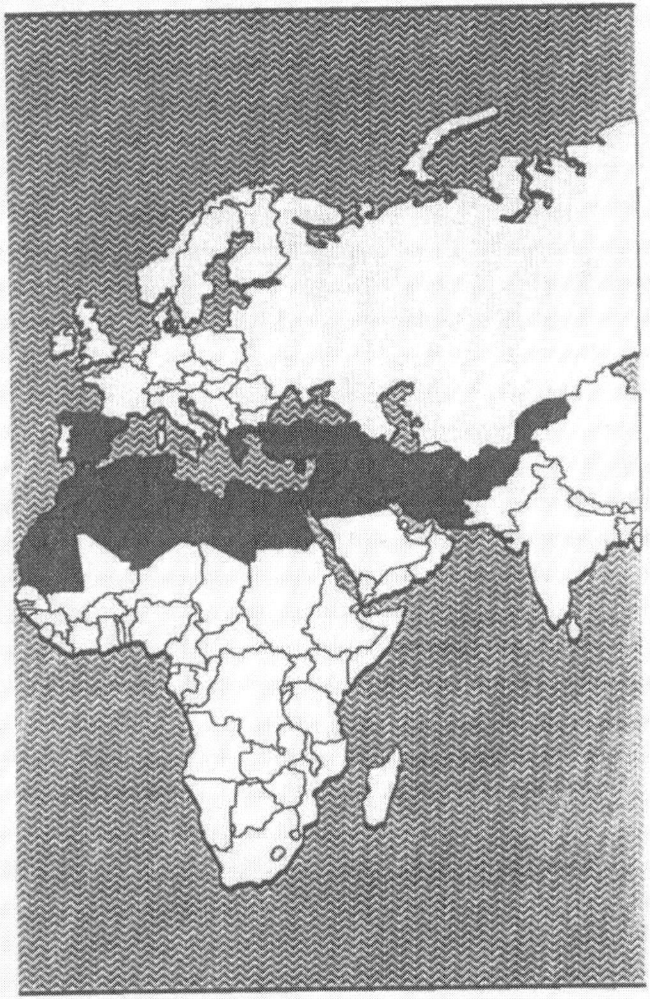

Malayan pit viper
Callaselasma rhostoma

Description: Reddish running into pink tinge toward the belly with triangular-shaped, brown markings bordered with light-colored scales. The base of the triangular-shaped markings end at the midline. It has dark brown, arrow-shaped markings on the top and each side of its head.

Characteristics: This snake has long fangs, is ill-tempered, and is responsible for many bites. Its venom is hemotoxic, destroying blood cells and tissue, but a victim's chances of survival are good with medical aid. This viper is a ground dweller that moves into many areas in search of food. The greatest danger is in stepping on the snake with bare feet.

Habitat: Rubber plantations, farms, rural villages, and rain forests.

Length: Average 60 centimeters, maximum 1 meter.

Distribution: Thailand, Laos, Cambodia, Java, Sumatra, Malaysia, Vietnam, Burma, and China.

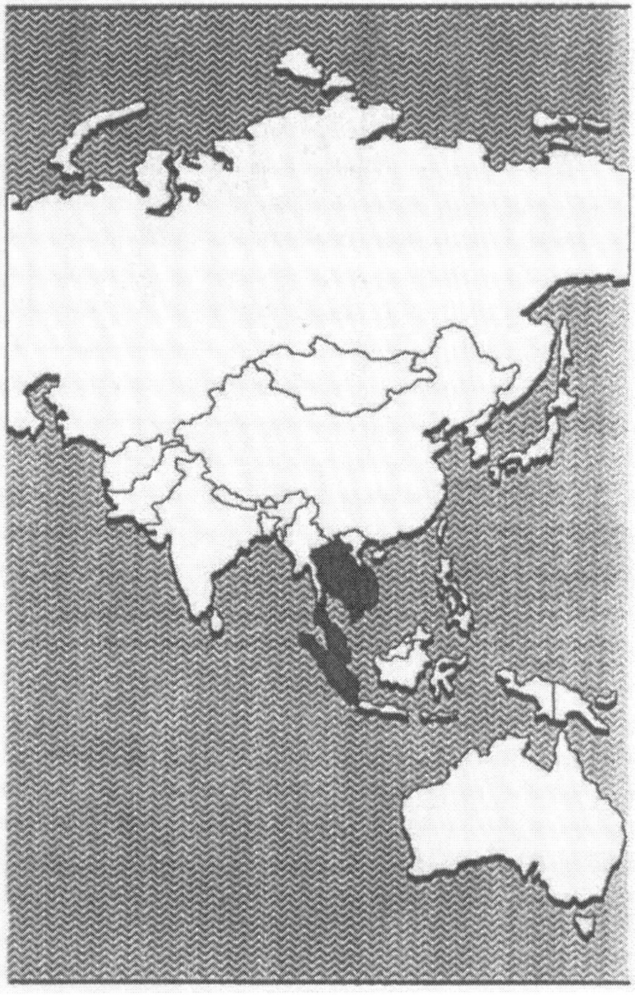

McMahon's viper

Eristicophis macmahonii

Description: Sandy buff color dominates the body with darker brown spots on the side of the body. Its nose shield is broad, aiding in burrowing.

Characteristics: Very little is known about this species. It apparently is rare or seldom seen. This viper is very irritable; it hisses, coils, and strikes at any intruder that ventures too close. Its venom is highly hemotoxic, causing great pain and tissue damage.

Habitat: Arid or semidesert. It hides during the day's sun, coming out only at night to feed on rodents.

Length: Average 45 centimeters, maximum 1 meter.

Distribution: West Pakistan and Afghanistan.

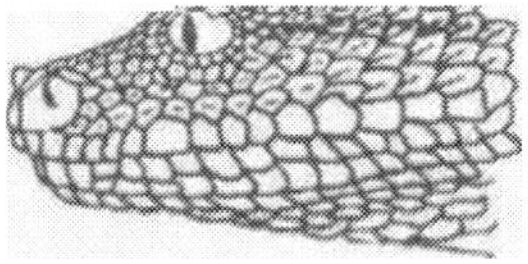

Mole viper or burrowing viper
Atracaspis microlepidota

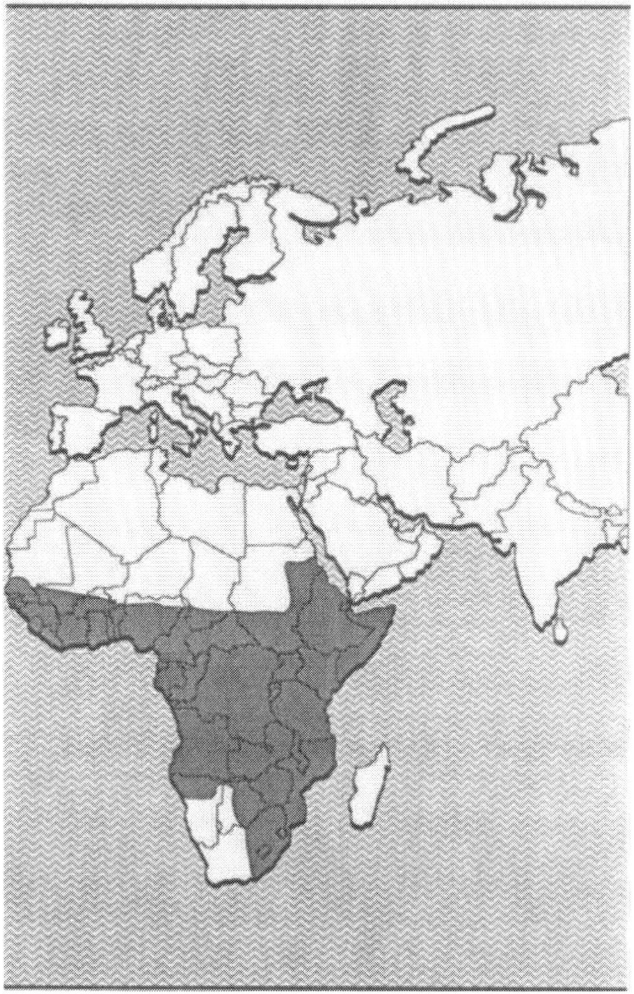

Palestinian viper
Vipera palaestinae

Description: Olive to rusty brown with a dark V-shaped mark on the head and a brown, zigzag band along the back.

Characteristics: The Palestinian viper is closely related to the Russell's viper of Asia. Like its cousin, it is extremely dangerous. It is active and aggressive at night but fairly placid during the day. When threatened or molested, it will tighten its coils, hiss loudly, and strike quickly.

Habitat: Arid regions, but may be found around barns and stables. It has been seen entering houses in search of rodents.

Length: Average 0.8 meter, maximum 1.3 meters.

Distribution: Turkey, Syria, Palestine, Israel, Lebanon, and Jordan.

Puff adder
Bitis arietans

Puff adder

Bitis arietans

Description: Yellowish, light brown, or orange with chevron-shaped dark brown or black bars.

Characteristics: The puff adder is the second largest of the dangerous vipers. It is one of the most common snakes in Africa. It is largely nocturnal, hunting at night and seeking shelter during the day's heat. It is not shy when approached. It draws its head close to its coils, makes a loud hissing sound, and is quick to strike any intruder. Its venom is strongly hemotoxic, destroying bloods cells and causing extensive tissue damage.

Habitat: And regions to swamps and dense forests. Common around human settlements.

Length: Average 12 meters, maximum 1.8 meters.

Distribution: Most of Africa, Saudi Arabia, Iraq, Lebanon, Israel, and Jordan.

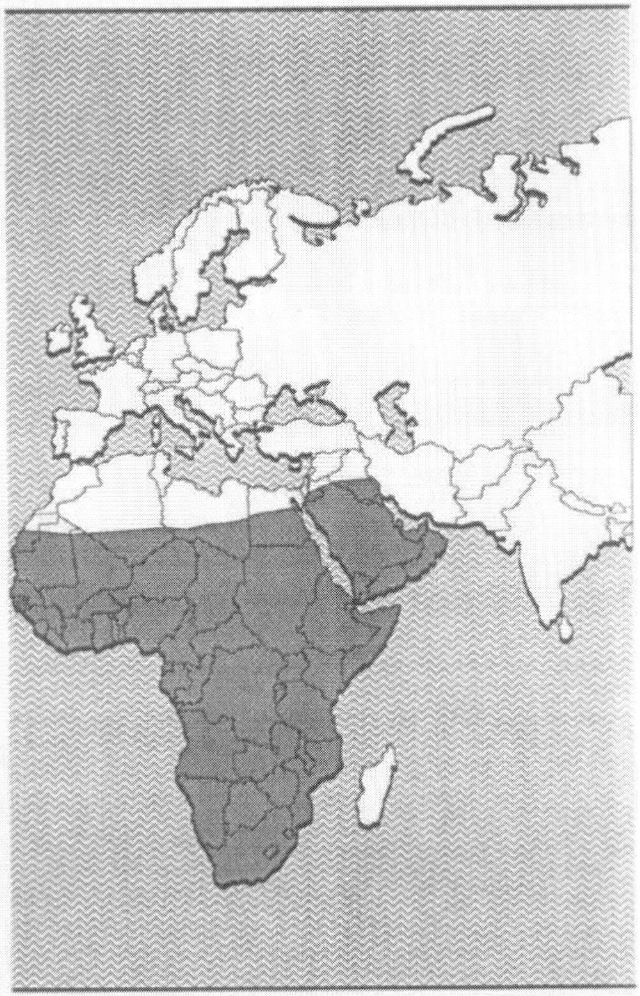

Rhinoceros viper or river jack

Bitis nasicornis

Description: Brightly colored with purplish to reddish-brown markings and black and light olive markings along the back. On its head it has a triangular marking that starts at the tip of the nose. It has a pair of long horns (scales) on the tip of its nose.

Characteristics: Its appearance is awesome; its horns and very rough scales give it a sinister look. It has an irritable disposition. It is not aggressive but will stand its ground ready to strike if disturbed. Its venom is neurotoxic and hemotoxic.

Habitat: Rain forests, along waterways, and in swamps.

Length: Average 75 centimeters, maximum 1 meter.

Distribution: Equatorial Africa.

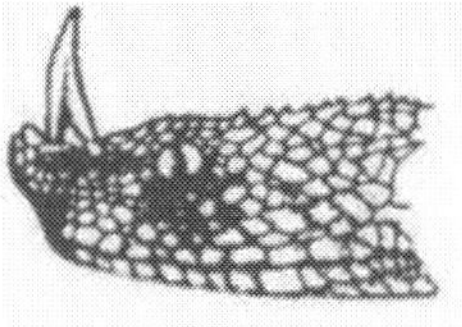

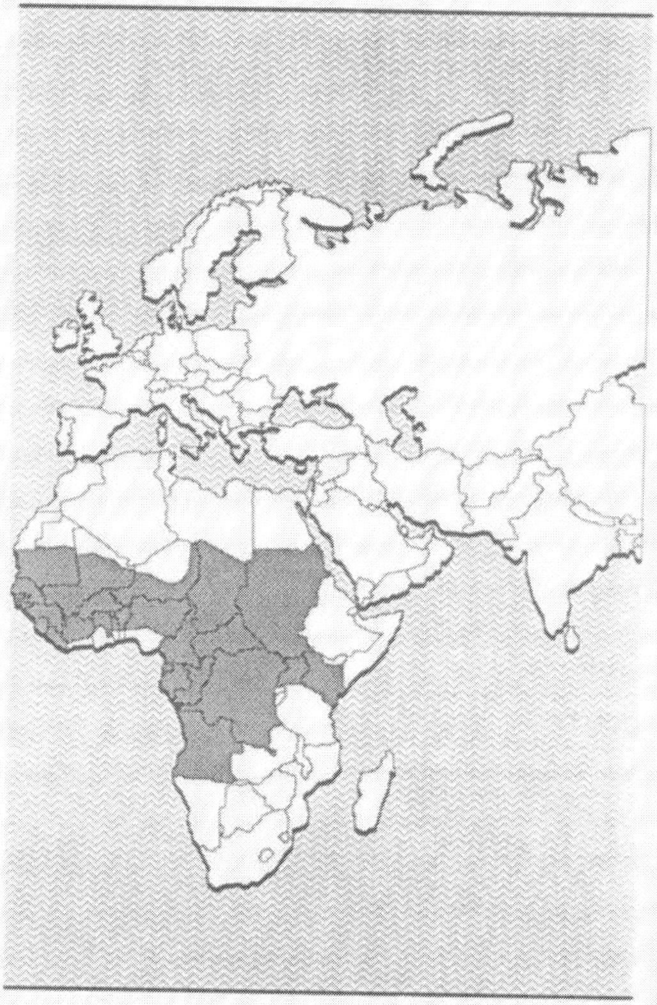

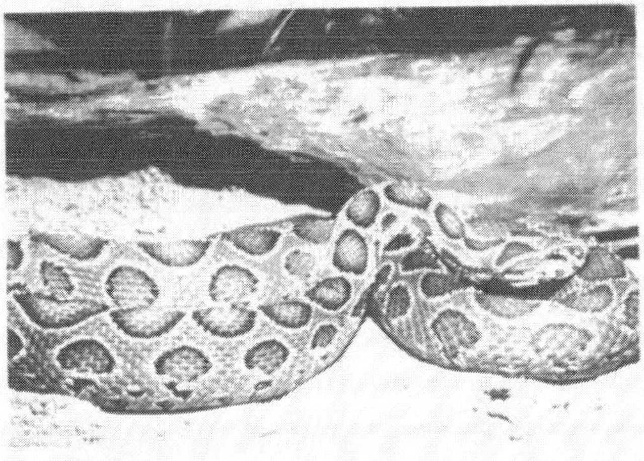

Russell's viper
Vipera russellii

Russell's viper

Vipera russellii

Description: Light brown body with three rows of dark brown or black splotches bordered with white or yellow extending its entire length.

Characteristics: This dangerous species is abundant over its entire range. It is responsible for more human fatalities than any other venomous snake. It is irritable. When threatened, it coils tightly, hisses, and strikes with such speed that its victim has little chance of escaping. Its hemotoxic venom is a powerful coagulant, damaging tissue and blood cells.

Habitat: Variable, from farmlands to dense rain forests. It is commonly found around human settlements.

Length: Average 1 meter, maximum 1.5 meters

Distribution: Sri Lanka, south China, India, Malaysian Peninsula, Java, Sumatra, Borneo, and surrounding islands.

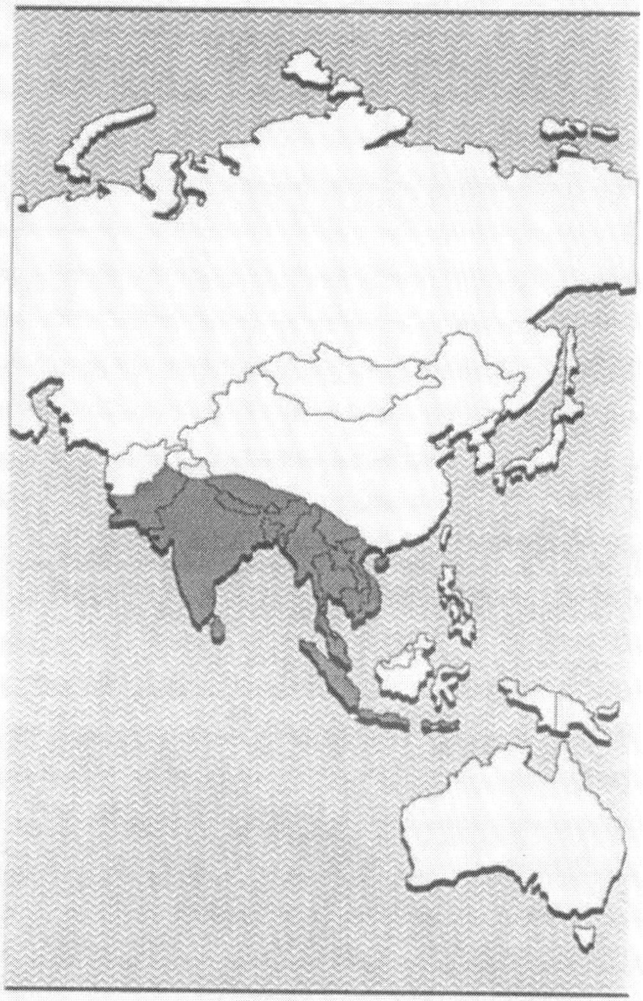

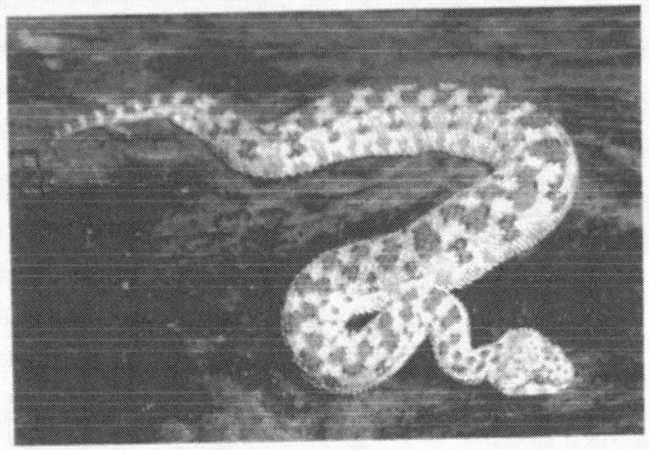

Sand viper
Cerastes vipera

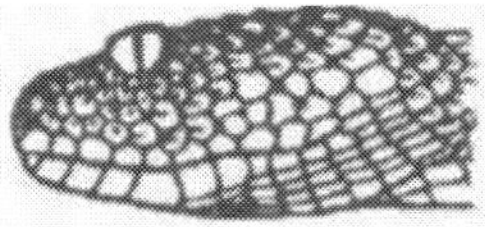

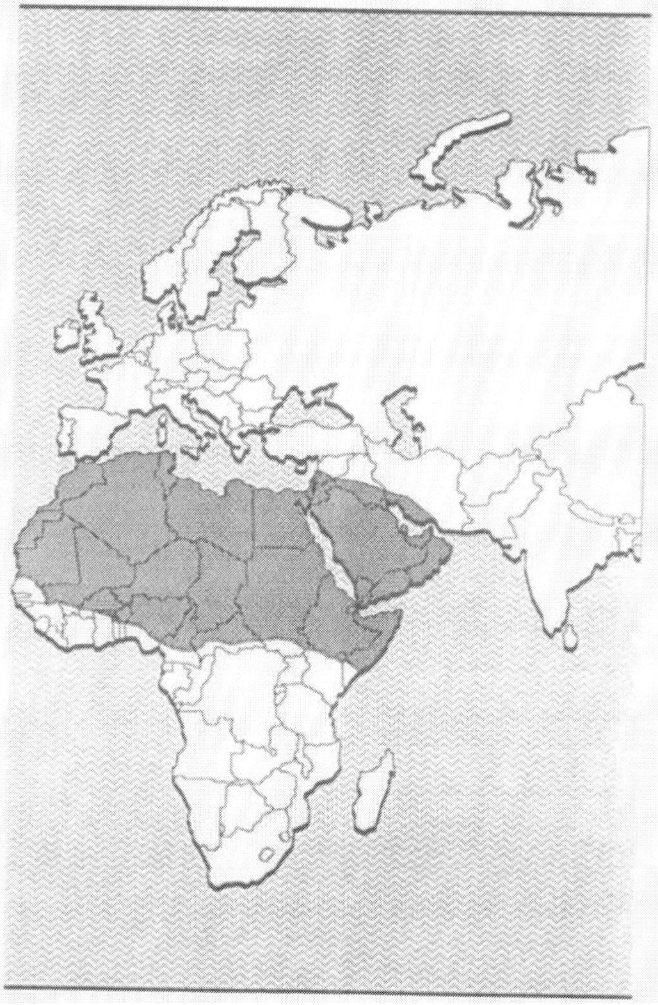

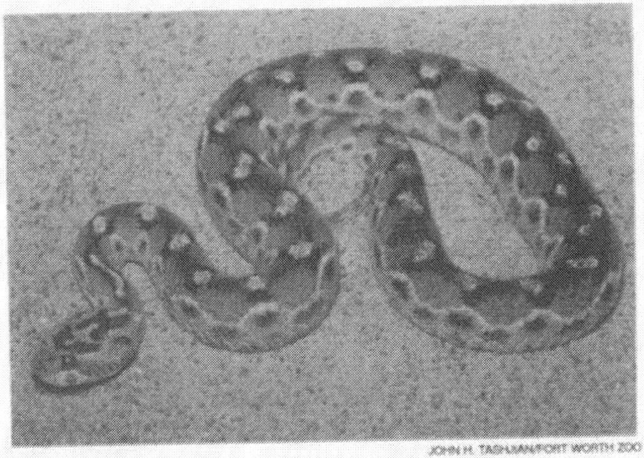

JOHN H. TASHJIAN/FORT WORTH ZOO

Saw-scaled viper

Echis carinatus

Description: Color is light buff with shades of brown, dull red, or gray. Its sides have a white or light-colored pattern. Its head usually has two dark stripes that start behind the eye and extend to the rear.

Characteristics: A small but extremely dangerous viper. It gets the name saw-scaled from rubbing the sides of its body together, producing a rasping sound. This ill-tempered snake will attack any intruder. Its venom is highly hemotoxic and quite potent. Many deaths are attributed to this species.

Habitat: Found in a variety of environments. It is common in rural settlements, cultivated fields, arid regions, barns, and rock walls.

Length: Average 45 centimeters, maximum 60 centimeters.

Distribution: Asia, Syria, India, Africa, Iraq, Iran, Saudi Arabia, Pakistan, Jordan, Lebanon, Sri Lanka, Algeria, Egypt, and Israel.

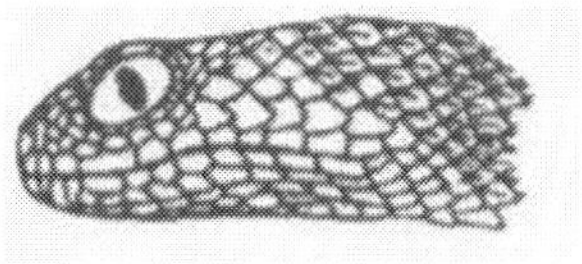

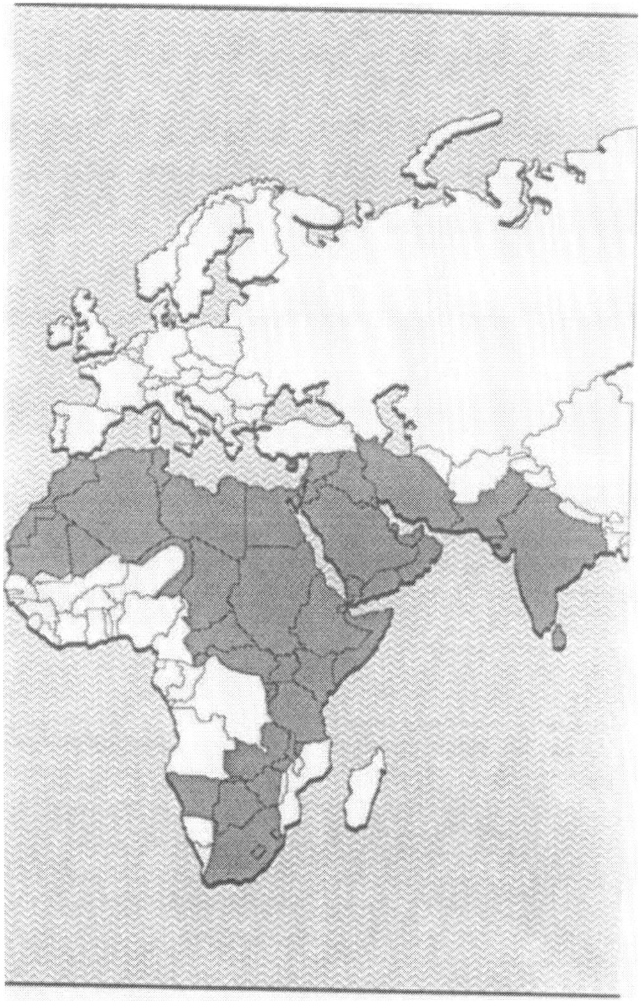

Wagler's pit viper or temple viper
Trimeresurus wagleri

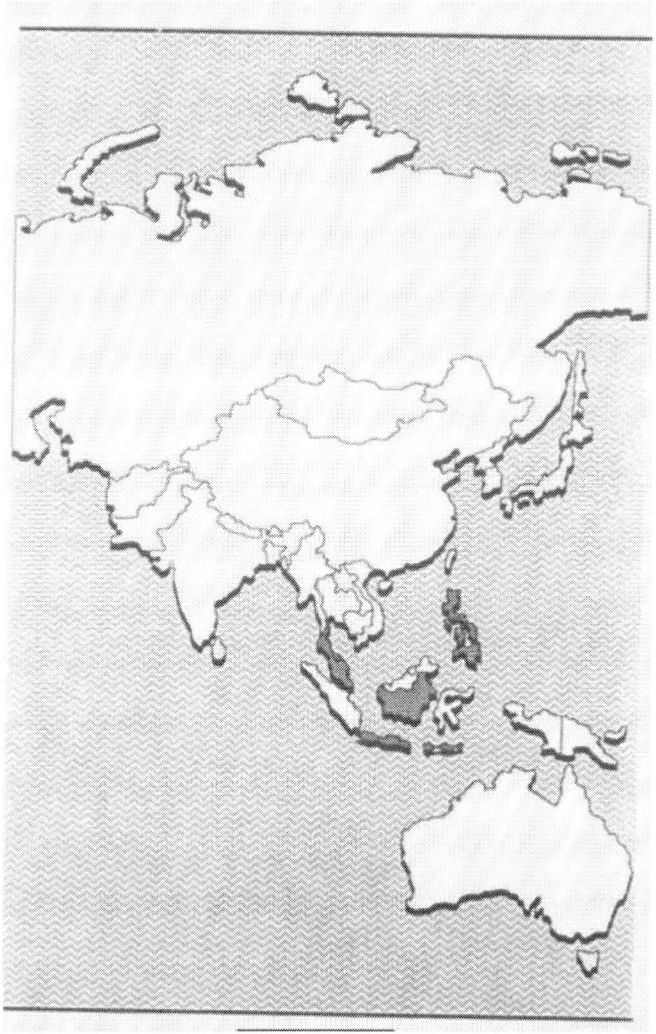

Australian copperhead

Denisonia superba

Description: Coloration is reddish brown to dark brown. A few from Queensland are black.

Characteristics: Rather sluggish disposition but will bite if stepped on. When angry, rears its head a few inches from the ground with its neck slightly arched. Its venom is neurotoxic.

Habitat: Swamps.

Length: Average 1.2 meters, maximum 1.8 meters.

Distribution: Tasmania, South Australia, Queensland, and Kangaroo Island.

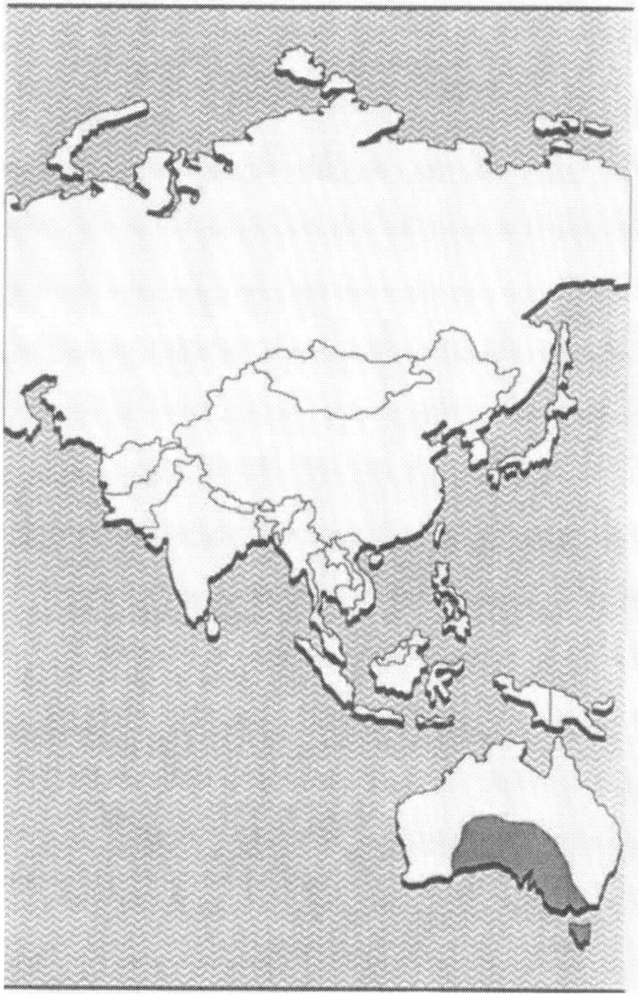

Death adder
Acanthophis antarcticus

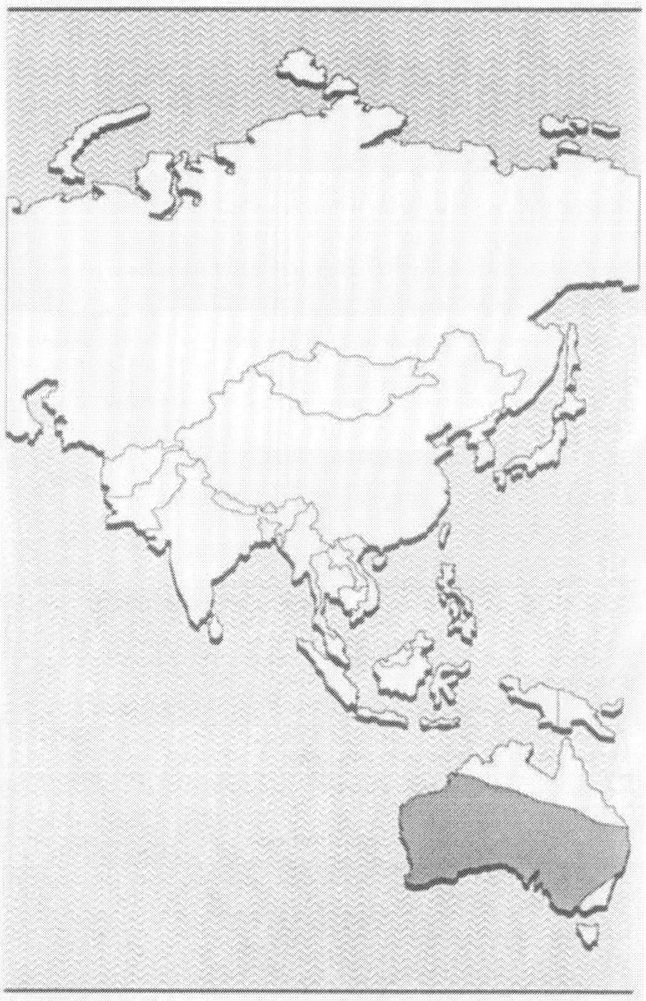

Taipan
Oxyuranus scutellatus

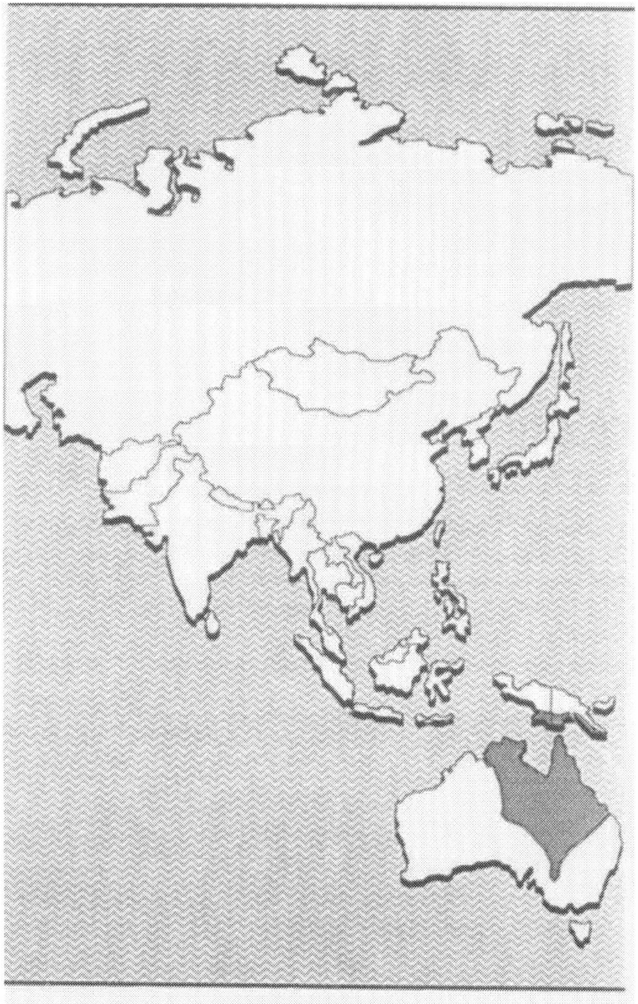

Tiger snake
Notechis scutatus

Description: Olive to dark brown above with yellowish or olive belly and crossbands The subspecies in Tasmania and Victoria is uniformly black.

Characteristics: It is the most dangerous snake in Australia. It is very common and bites many humans. It has a very potent neurotoxic venom that attacks the nervous system. When aroused, it is aggressive and attacks any intruder. It flattens its neck making a narrow band.

Habitat: Found in many habitats from arid regions to human settlements along waterways to grasslands.

Length: Average 1.2 meters, maximum 1.8 meters.

Distribution: Australia, Tasmania, Bass Strait islands, and New Guinea.

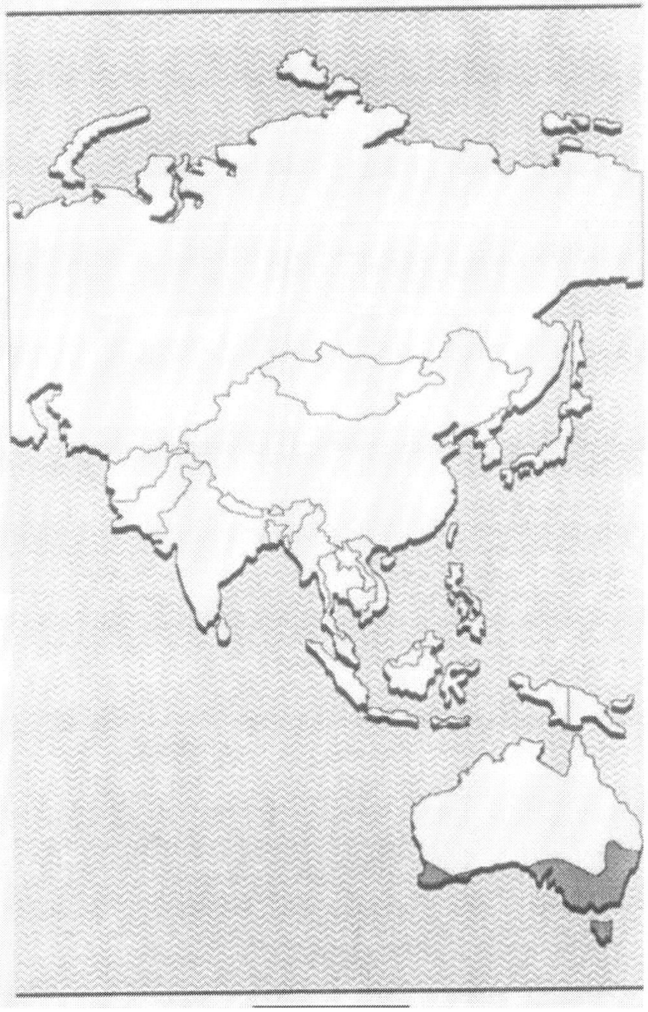

POISONOUS SEA SNAKES

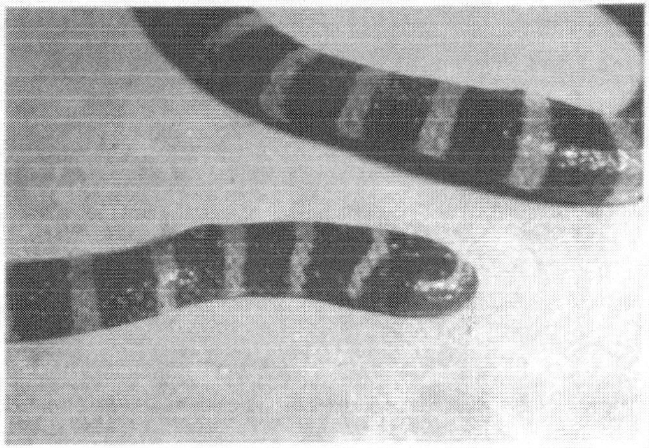

Banded sea snake
Laticauda colubrina

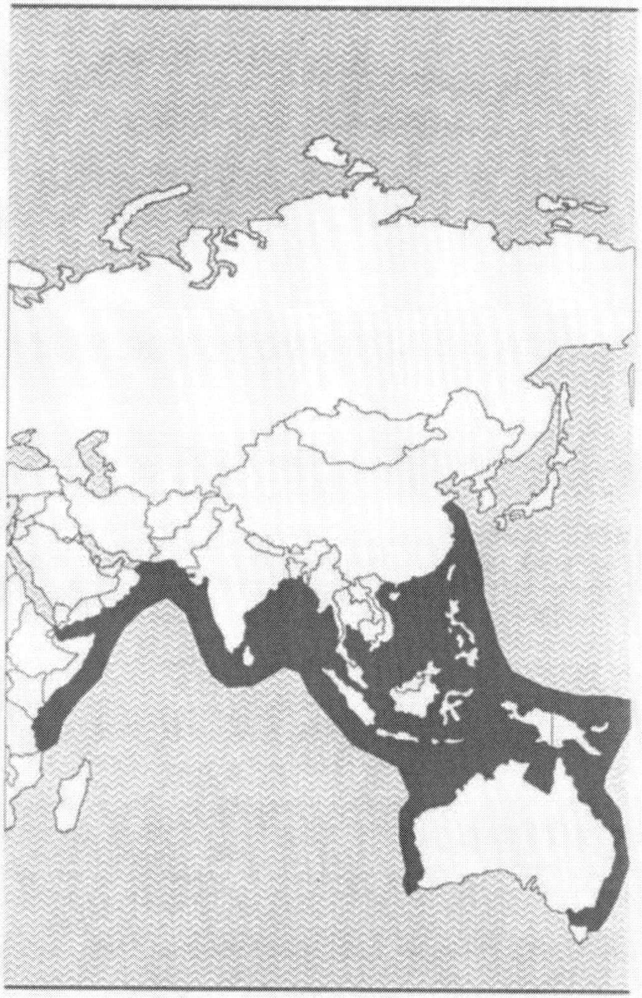

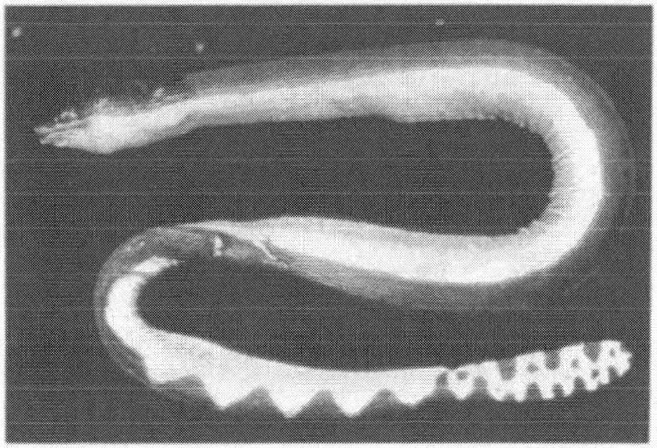

Yellow-bellied sea snake
Pelamis platurus

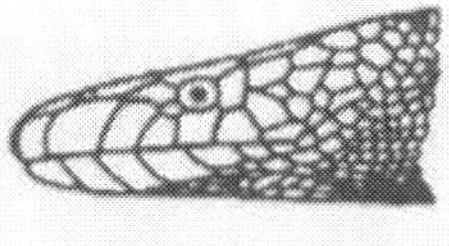

POISONOUS LIZARDS

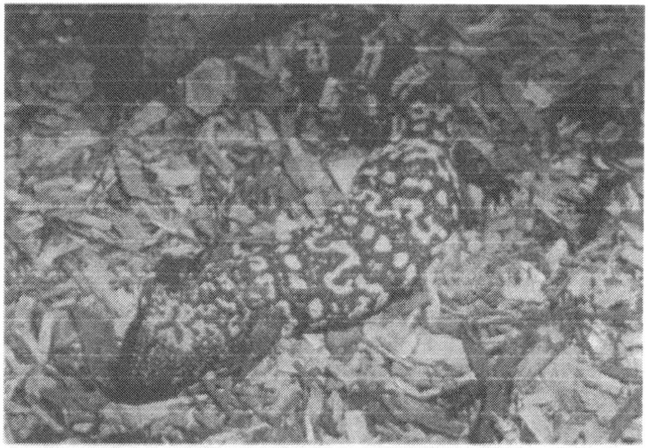

Gila monster
Heloderma suspectum

Description: Robust, with a large head and a heavy tail. Its body is covered with bead-like scales. It is capable of storing fat against lean times when food is scarce. Its color is striking in rich blacks laced with yellow or pinkish scales.

Characteristics: Not an aggressive lizard, but it is ready to defend itself when provoked. If approached too closely, it will turn toward the intruder with its mouth open. If it bites, it hangs on tenaciously and must be pried off. Its venom glands and grooved teeth are on its bottom jaw.

Habitat: Found in arid areas, coming out at night or early morning hours in search of small rodents and bird eggs. During the heat of the day it stays under brush or rocks.

Length: Average 30 centimeters, maximum 50 centimeters.

Distribution: Arizona, New Mexico, Utah, Nevada, northern Mexico, and extreme corner of southeast California.

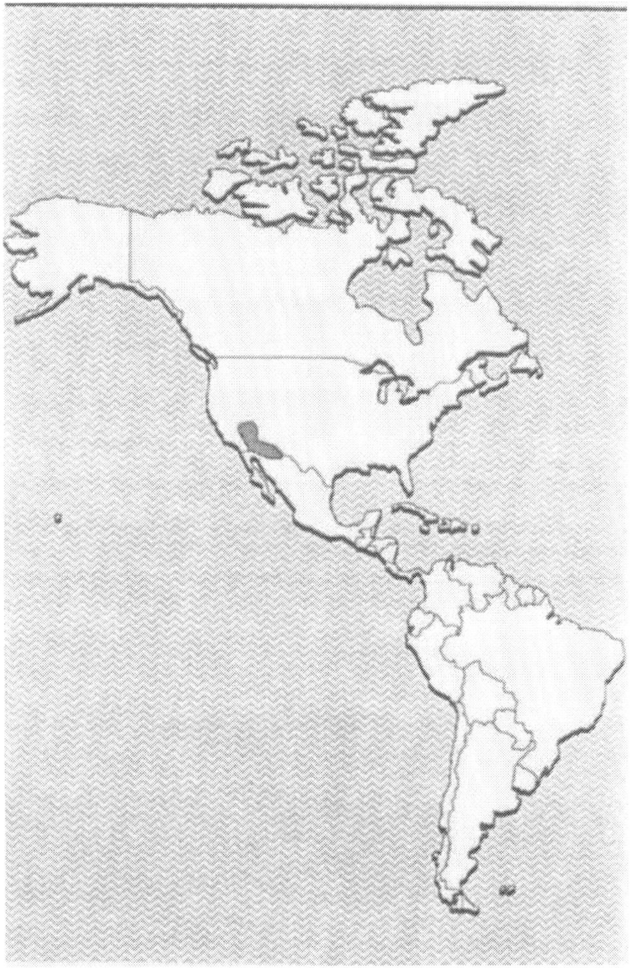

JOHN H. TASHJIAN/FORT WORTH ZOO

Mexican beaded lizard
Heloderma horridum

APPENDIX F

DANGEROUS FISH AND MOLLUSKS

Since fish and mollusks may be one of your major sources of food, it is wise to know which ones are dangerous to you should you catch them. Know which ones are dangerous, what the dangers of the various fish are, what precautions to take, and what to do if you are injured by one of these fish.

Fish and mollusks will present a danger in one of three ways: by attacking and biting you, by injecting toxic venom into you through its venomous spines or tentacles, and through eating fish or mollusks whose flesh is toxic.

The danger of actually encountering one of these dangerous fish is relatively small, but it is still significant. Any one of these fish can kill you. Avoid them if at all possible.

FISH THAT ATTACK MAN

The shark is usually the first fish that comes to mind when considering fish that attack man. Other fish also fall in this category, such as the barracuda, the moray eel, and the piranha.

Sharks

Sharks are potentially the most dangerous fish that attack people. The obvious danger of sharks is that they are capable of seriously maiming or killing you with their bite. Of the many shark species, only a relative few are dangerous. Of these, four species are responsible for most cases of shark attacks on humans. These are the white, tiger, hammerhead, and blue sharks. There are also records of attacks by ground, gray nurse, and mako sharks. See Figure F-1 for illustrations of sharks.

Avoid sharks if at all possible. Follow the procedures discussed **in** Chapter 16 to defend yourself against a shark attack.

Sharks vary in size, but there is no relationship between the size of the shark and likelihood of attack. Even the smaller sharks can be dangerous, especially when they are traveling in schools.

If bitten by a shark, the most important measure for you to take is to stop the bleeding quickly. Blood in the water attracts sharks. Get yourself or the victim into a raft or to shore as soon as possible. If in the water, form a circle around the victim (if not alone), and stop the bleeding with a tourniquet.

Other Ferocious Fish

In salt water, other ferocious fish include the barracuda, sea bass, and moray eel (Figure F-2). The sea bass is usually an open water fish. It is dangerous due to its large size. It can remove large pieces of flesh **from** a human. Barracudas and moray eels have been **Known** to attack man and inflict vicious bites. Be careful of these two species when near reefs and in shallow water. Moray eels are very aggressive when disturbed.

In fresh water, piranha are the only significantly dangerous fish. They are inhabitants of the tropics and are restricted to northern South

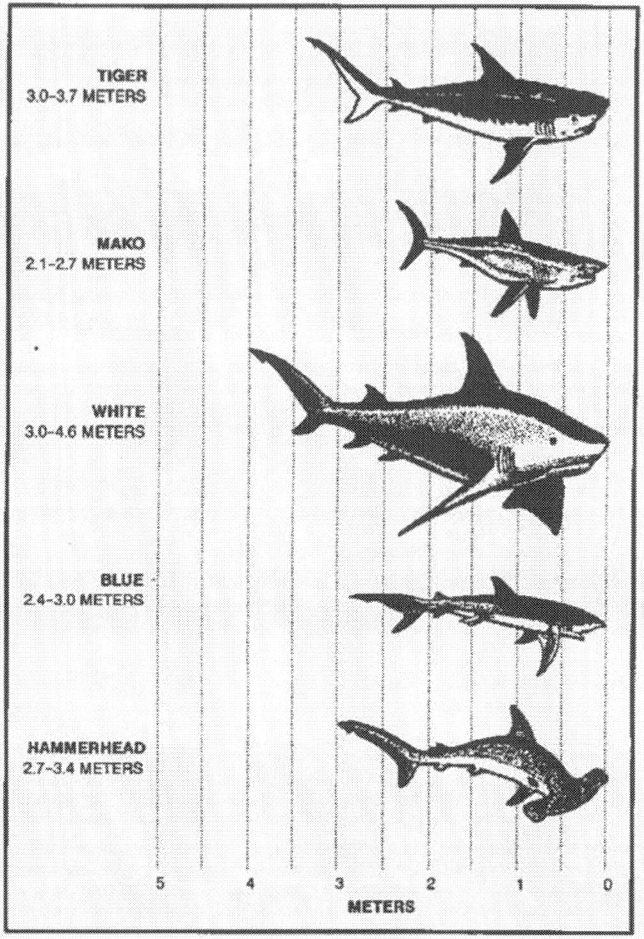

Figure F-1. Sharks.

America. These fish are fairly small, about 5 to 7.5 centimeters, but they have very large teeth and travel in large schools. They can devour a 135-kilogram nog in minutes.

VENOMOUS FISH AND INVERTEBRATES

There are several species of venomous fish and invertebrates, all of which live in salt water. All of these are capable of injecting poisonous venom through spines located in their fins, tentacles, or bites. Their venoms cause intense pain and are potentially fatal. If injured by one of these fish or invertebrates, treat the injury as for snakebite.

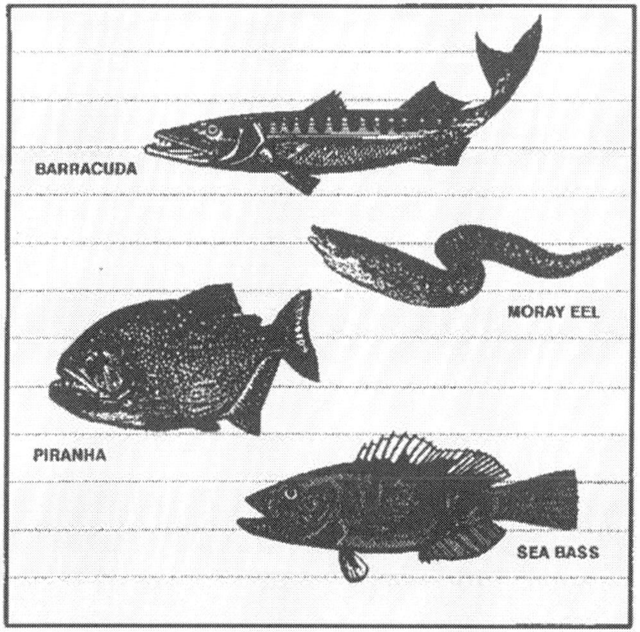

Figure F-2. Ferocious fish.

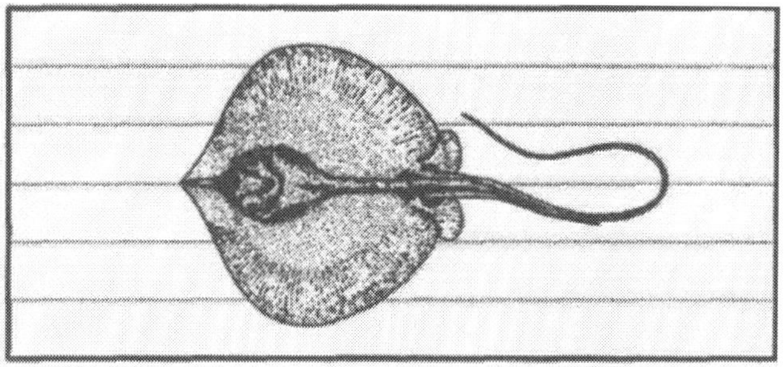

Stingrays
Dasyatidae species

Stingrays inhabit shallow water, especially in the tropics and in temperate regions as well. All have a distinctive ray shape but coloration may make them hard to spot unless they are swimming. The venomous, barbed spines in their tails can cause severe or fatal injury.

Rabbitfish
Siganidae species

Rabbitfish are found predominantly on reels in the Pacific and Indian oceans. They average about 30 centimeters long and have very sharp spines in their fins. The spines are venomous and can inflict intense pain.

Scorpion fish or zebra fish
Scorpaenidae species

Scorpion fish or zebra fish live mainly in the reefs in the Pacific and Indian oceans. They vary from 30 to 90 centimeters long, are usually reddish in coloration, and have long, wavy fins and spines. They inflict an intensely painful sting.

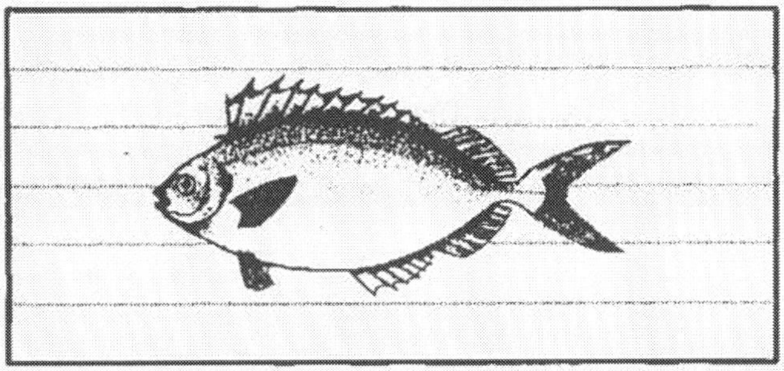

Siganus fish

The siganus fish is small, about 10 to 15 centimeters long, and looks much like a small tuna. It has venomous spines in its dorsal and ventral fins. These spines can inflict painful stings.

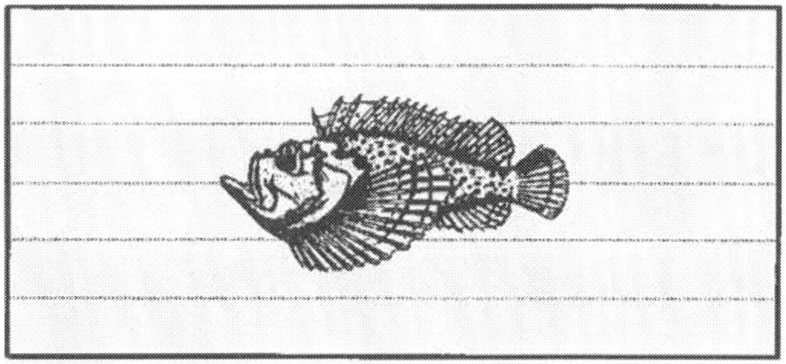

Stonefish
Synanceja species

Stonefish are found in the tropical waters of the Pacific and Indian oceans. Averaging about 30 centimeters in length, their subdued colors and lumpy shape provide them with exceptional camouflage. When stepped on, the fins in the dorsal spine inflict an extremely painful and sometimes fatal wound.

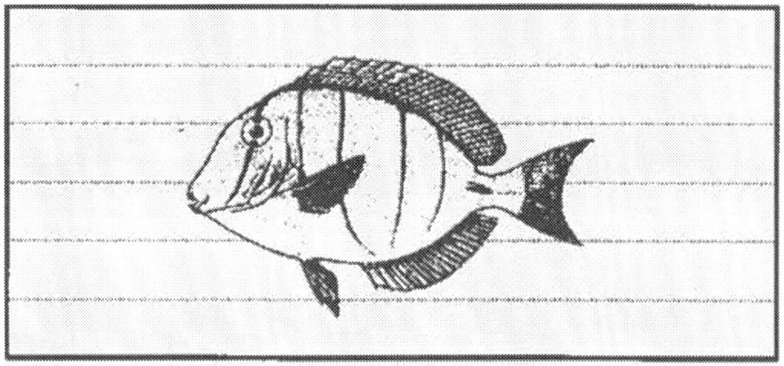

Tang or surgeonfish
Acanthuridae species

Tang or surgeonfish average 20 to 25 centimeters in length, with a deep body, small mouth, and bright coloration. They have needlelike spines on the side of the tail that cause extremely painful wounds. This fish is found in all tropical waters.

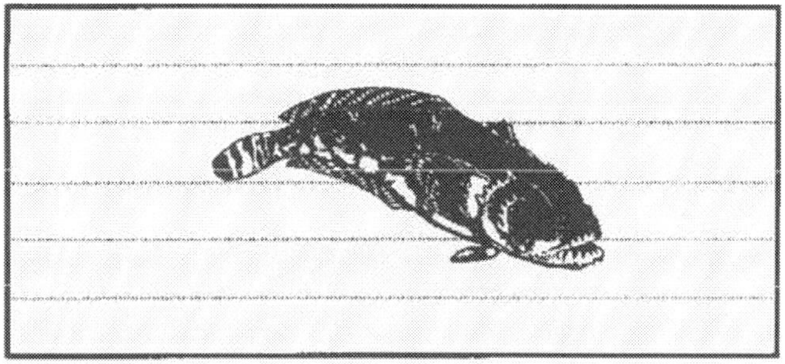

Toadfish
Batrachoididae species

Toadfish are found in the tropical waters off the coasts of South and Central America. They are between 17.5 and 25 centimeters long and have a dull color and large mouths. Tney bury ihemseives in the sand and may be easiiy stepped on. Tney have very sharp, extremely poisonous spines on the dorsal fin (back).

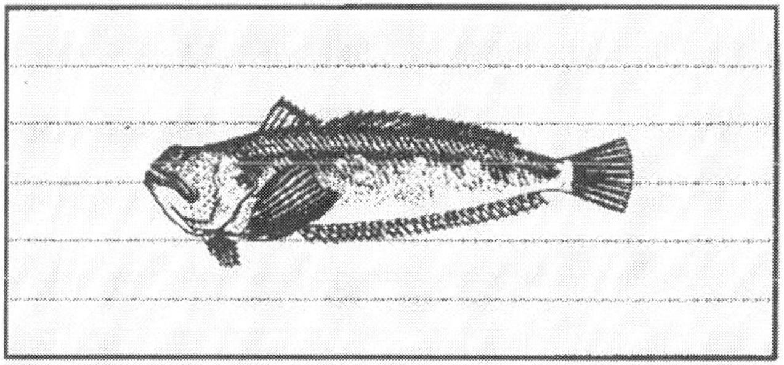

Weever fish
Trachinidae species

The weever fish is a tropical fish that is fairly slim and about 30 centimeters long. All its fins have venomous spines that cause a painful wound.

Blue-ringed octopus
Hapalochlaena lunulata

This small octopus is usually found on the Great Barrier Reef off eastern Australia, it is grayish-white with iridescent blue ringlike markings. This octopus usually will not bite unless stepped on or handled. Its bite is extremely poisonous and frequently lethal.

Portuguese man-of-war
Physalis species

Although it resembles a jellyfish, the Portuguese man-of-war is actually a colony of sea animals. Mainly found in tropical regions, the Gulf Stream current can carry it as far as Europe. It is also found as far south as Australia. The floating portion of the man-of-war may be as small as 15 centimeters, but the tentacles can reach 12 meters in length. These tentacles inflict a painful and incapacitating sting, but the sting is rarely fatal.

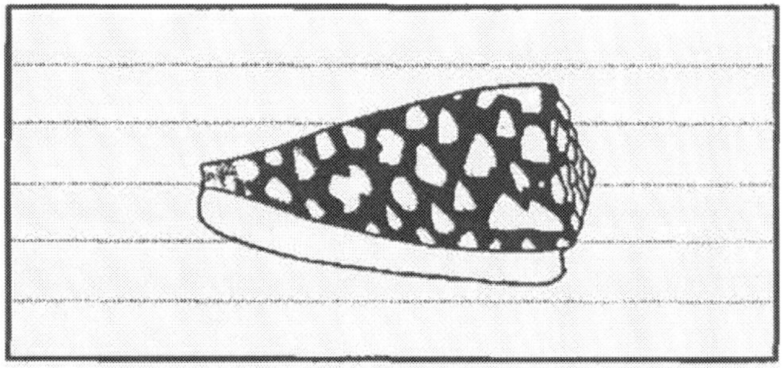

Cone shells
Conidae species

These cone-shaped shells have smooth, colorful mottling and long, narrow openings in the base of the shell. They live under rocks, in crevices or coral reefs, and along rocky shores of protected bays in tropical areas. Aii have tiny teeth thai are similar to hypodermic needles. They can inject an extremely poisonous venom that acts very swiftly, causing acute pain, swelling, paralysis, blindness, and possible death within hours. Avoid handling all cone shells.

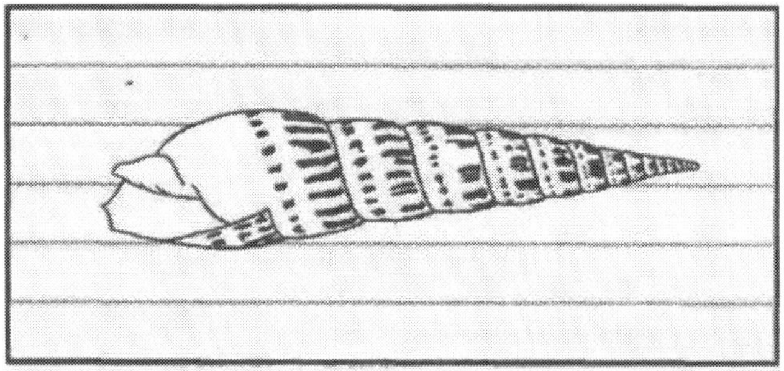

Terebra shells
Terebridae species

These shells are found in both temperate and tropical waters. They are similar to cone shells but much thinner and longer. They poison in the same way as cone shells, but the venom is not as poisonous.

FISH WITH TOXIC FLESH

There are no simple rules to tell edible fish from those with poisonous flesh. The most common toxic fish are shown in Figure 8-2. All of these fish contain various types of poisonous substances or toxins in their flesh and are dangerous to eat. They have the following common characteristics:

- Most live in shallow water around reefs or lagoons.
- Many have boxy or round bodies with hard shell-like skins covered with bony plates or spines. They have small parrotlike mouths, small gills, and small or absent belly fins. Their names suggest their shape.

In addition to the above fish and their characteristics, barracuda and red snapper fish may carry ciguatera, a toxin that accumulates in the systems of fish that feed on tropical marine reefs.

Without specific local information, take the following precautions:

- Be very careful with fish taken from normally shallow lagoons with sandy or broken coral bottoms. Reef-feeding species predominate and some may be poisonous.
- Avoid poisonous fish on the leeward side of an island. This area of shallow water consists of patches of living corals mixed with open spaces and may extend seaward for some distance. Many different types of fish inhabit these shallow waters, some of which are poisonous.
- Do not eat fish caught in any area where the water is unnaturally discolored. This may be indicative of plankton that cause various types of toxicity in plankton-feeding fish.
- Try fishing on the windward side or in deep passages leading from the open sea to the lagoon, but be careful of currents and waves. Live coral reefs drop off sharply into deep water and form a dividing line between the *suspected fish of the shadows* and the *desirable deep-water species.* Deepwater fish are usually not poisonous. You can catch the various toxic fish even in deep water. *Discard all suspected reef fish,* whether caught on the ocean or the reef side.

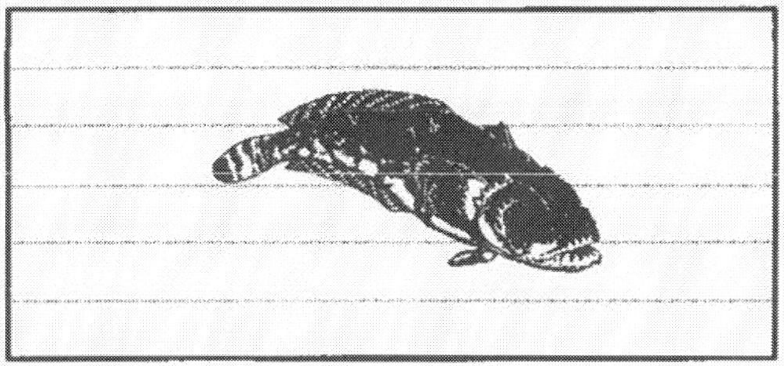

Toadfish
Batrachoididae species

Toadfish are found in the tropical waters off the coasts of South and Central America. They are between 17.5 and 25 centimeters long and have a dull color and large mouths. They bury themselves in the sand and may be easily stepped on. They have very sharp, extremely poisonous spines on the dorsal fin {back).

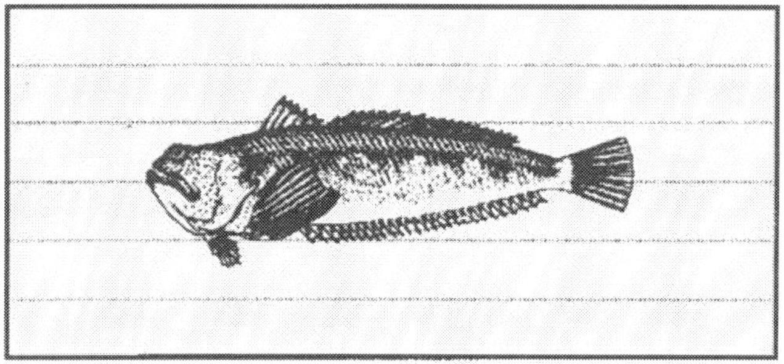

Weever fish
Trachinidae species

The weever fish is a tropical fish that is fairly slim and about 30 centimeters long. All its fins have venomous spines that cause a painful wound.

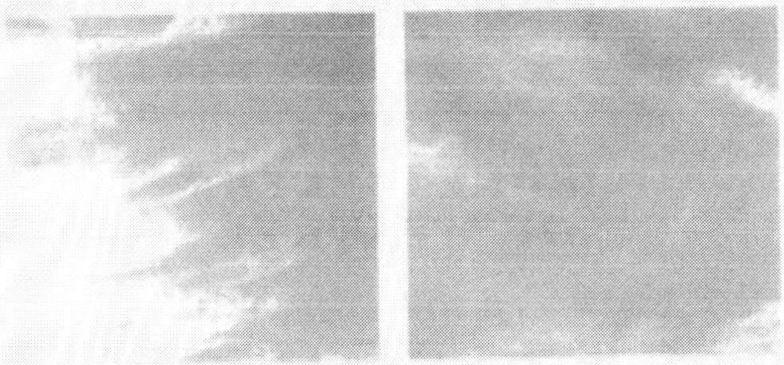

NATIONAL OCEANIC AND ATMOSPHERIC ADMINISTRATION

Cirrus clouds

Cirrus clouds are the very high clouds that look like thin streaks or curls. They are usually 6 kilometers or more above the earth and are usually a sign of fair weather. In cold climates, however, cirrus clouds that begin to multiply and are accompanied by increasing winds blowing steadily from a northerly direction indicate an oncoming blizzard.

NATIONAL OCEANIC AND ATMOSPHERIC ADMINISTRATION

Cumulus clouds

Cumulus clouds are fluffy, white, heaped-up clouds. These clouds, which are much lower than cirrus clouds, are often fair weather clouds. They are apt to appear around midday on a sunny day, looking like large cotton balls with flat bottoms. As the day advances, they may become bigger and push higher into the atmosphere. Piling up to appear like a mountain of clouds. These can turn into storm clouds.

NATIONAL OCEANIC AND ATMOSPHERIC ADMINISTRATION

NATIONAL OCEANIC AND ATMOSPHERIC ADMINISTRATION

Stratus clouds

Stratus clouds are very low, gray clouds, often making an even gray layer over the whole sky. These clouds generally mean rain.

NATIONAL OCEANIC AND ATMOSPHERIC ADMINISTRATION

NATIONAL OCEANIC AND ATMOSPHERIC ADMINISTRATION

Nimbus clouds

Nimbus clouds are ram clouds of uniform grayness that extend over the entire sky

NATIONAL OCEANIC AND ATMOSPHERIC ADMINISTRATION

NATIONAL OCEANIC AND ATMOSPHERIC ADMINISTRATION

Cumulonimbus clouds

Cumulonimbus is the cloud formation resulting from a cumulus cloud building up, extending to great heights, and forming in the shape of an anvil. You can expect a thunderstorm if this cloud is moving in your direction.

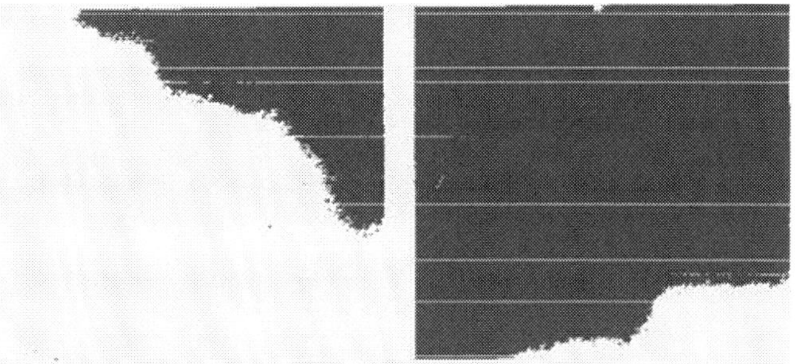

NATIONAL OCEANIC AND ATMOSPHERIC ADMINISTRATION

Cirrostratus clouds

Cirrostratus is a fairly uniform layer of high stratus clouds that are darker than cirrus clouds. Cirrostratus clouds indicate good weather.

NATIONAL OCEANIC AND ATMOSPHERIC ADMINISTRATION

Cirrocumulus clouds

Cirrocumulus is a small, white, round cloud at a high altitude. Cirrocumulus clouds indicate good weather.

NATIONAL OCEANIC AND ATMOSPHERIC ADMINISTRATION

Scuds

A loose, vapory cloud (scud) driven before the wind is a sign of continuing bad weather.

APPENDIX H

CONTINGENCY PLAN OF ACTION FORMAT

Properly planning for the possible contingencies that may occur during a mission is a positive step toward being able to cope successfully with the changes in situation. The contingency plan of action is a critical document to an individual soldier, or to a unit, faced with evading enemy forces. First, it is a plan that will provide evaders a starting point to begin operating effectively once evasion has begun. Second, it gives recovery forces the ability to know what the evaders are planning to do, thus making recovery, operations easier. A well-thought-out contingency plan of action that everyone can understand is an important document to the evader.

Note: Upon deployment, you may carry with you the information compiled in A through E of the SITUATION paragraph only.

I. SITUATION

A. Country Climatic Zones

1. Tropical Rainy Climate
2. Dry Climate
3. Temperate Climate
4. Cold Climate *(wet/dry)*
5. Polar

B. Climatic Land Zones *(whatever is applicable)*

1. Coasts - Seasons
   a. Temperature
   b. Precipitation
   c. General wind direction
   d. Cloud cover
2. Plains *(refer to coasts)*
3. Deserts *(refer to coasts)*
4. Plateaus *(refer to coasts)*
5. Mountains *(refer to coasts)*
6. Swamps *(refer to coasts)*

c. Light Data *(BMNT,* EEJVT, *Uoonrise, Moonset, Percent of Illumination)*

D. Terrain

1. Neighboring Borders
2. General Terrain Zones
   a. Coasts
      (1) General description and size
      (2) Vegetation
         (a) Natural
            1. Tundra
            1 Coniferous forest
            3; Deciduous forest
            4 Temperate grassland
            1 Marshland swamp
            4. Desert
            Zi Pastoral and arable land
            8; Tropical forest
            ⊲L Savanna
         (b) Cultivated
         (c) Concealment *(density)*
         (d) Growing seasons
         (e) Edible
            L Food value
            2. Procurement *(young or mature)*
            3. Preparation
            4 cooking
         (f) Poisonous
         (g) Medical use
         (h) Other uses
      (3) Animals and fish
         (a) Domestic
            L Food values
            2 Procurement
            3. Preparation
            4 Cooking
            5t Medical use
            (L Dangerous
            7. Poisonous
            8. Other uses
         (b) Wildlife *(animals, fish, insects, and reptiles) (see domestic)*
      (4) Water sources
      (a) Procurement
      (b) Potability
      (c) Preparation

b. Plains *(refer to coasts)*

c. Deserts *(refer to coasts)*

d. Plateaus *(refer to coasts)*

e. Mountains *(refer to coasts)*

f. Swamps *(refer to coasts)*

g. Rivers and lakes *(refer to coasts)*

3. Natural Land Barriers

a. Mountain ranges

b. Large rivers

E. Civilian Population

1. Numbers of Population

a. Totals and density *(by areas)*

b. Divisions of urban, suburban, rural, and nomads

2. Dress and Customs

3. Internal Security Forces

4. Controls and Restrictions *(explain)*

5. Border Area Security

F. Friendly Forces

1. FEBA/FLOT

2. Closest Units

3. Location of Friendly or Neutral Embassies, Liaisons, Consulates, etc.

4. Recovery Sites *(explain)*, LZs En Route.

G. Enemy Forces

1. Doctrine

2. Tactics

3. Intelligence Reports

a. Identification

b. Location

c. Activity

d. Strength

e. Night sighting devices

II. MISSION— Conduct Avoidance of Capture on Order From-To

III. EXECUTION

A. Overall Plan

1. When Do You Initiate Movement?

2. Location of Initial Movement Point

3. Actions at Initial Movement Point

4. Location of Hide Areas

5. Movement to Hide Areas

6. Actions Around the Hide Sites

7. Movement to Hide Sites

8. Actions at Hide Sites

a. Construction

b. Occupation

c. Movement out of hide site

9. Location of Hole-up Areas

10. Actions at Hole-up Areas

11. Location of Recovery Site(s)

B. Other Missions

1. Movement

a. Formation

b. Individual positions

c. Navigation

d. Stealth/listening

e. Security
(1) Noise
(2) Light
(3) All around security
f. Cover, concealment, and camouflage
g. Actions at breaks
(1) Listening *(5-10 minutes)*
(2) Long
h. Actions at danger areas *(enemy observation or fire)*
i. Actions for enemy sighting/contact
j. Rally points/rendezvous points
(1) Locations
(2) Actions

2. Actions in the Care of Sick or Injured
a. Initial movement point
b. Along the movement route
3. Actions for Crossing Borders
4. Actions at Recovery Site(s)
5. Other Actions
6. Training and Rehearsals
7. Inspections before starting movement

IV. SERVICE AND SUPPORT

A. Survival Aids

1. Health
a. First aid
b. Disease

2. Water
a. Procurement
b. Purification
c. Carrying

3. Food
a. Procurement
b. Preparation
c. cooking
d. Carrying

4. Shelter and Comfort/Warmth
5. Fire Starting
6. Recovery
7. Travel

B. Survival Kit(s)

C. Special Equipment

D. Inspections

1. Responsibilities
2. Equipment, Survival Items, and Kit(s)

V. COMMAND AND SIGNAL

A. Chain of Command
1. Senior Person
2. Team Leader

B. Signals To Be Used by Movement Teams
1. Along the Route
2. Rally/Rendezvous Points

C. Communications to Higher Headquarters *(radio)*

GLOSSARY

c	Celsius
cgy	centigray
cm	centimeter
CNS	central nervous system
CPA	contingency plan of action
CPR	cardiopulmonary resuscitation
F	Fahrenheit
FEBA	forward edge of the battle area
PLOT	forward line of own troops
HELP	heat escaping lessening posture
IV	intravenous
kg	kilogram
kph	kilometers per hour
m	meter
mg	milligram
MRE	meal, ready-to-eat
NBC	nuclear, biological, and chemical
RDF	radio direction finding
SERE	survival, evasion, resistance, and escape
SMCT	Soldier's Manual of Common Tasks
SOP	standing operating procedure

REFERENCES

RELATED PUBLICATIONS

The publications listed below are sources of additional information. They are not required in order to understand this publication.

U.S. Army Publications

AR 70-38 Research, Development, Test, and Evaluation of Materiel for Extreme Climatic Conditions, August 1979

FM 1-400 Aviator's Handbook, May 1983

FM 21-11 First Aid for Soldiers, October 1988

FM 31-70 Basic Cold Weather Manual, April 1968

FM 31-71 Northern Operations, June 1971

FM 90-3 Desert Operations, August 1977

FM 90-5 (HTF) Jungle Operations (How to Fight), August 1982

FM 90-6 Mountain Operations, June 1980

TC 21-3 Soldier's Handbook for Individual Operations and Survival in Cold Weather Areas, March 1986

GTA 21-7-1 Study Card Set, Survival Plants, Southeast Asia

Man and Materiel in the Cold Regions (Part I). US Army Cold Regions Test Center, Fort Greely, AK.

U.S. Air Force Publications

Air Force Manual 64-4 Survival Training, July 1985

Air Force Manual 64-5 Aircrew Survival, September 1985

Afoot in the Desert. Environmental Information Div, Air Training Command, Air University Library, Maxwell AFB, AL: Oct 80.

Arctic Survival Principles, Procedures, and Techniques. 3636th Combat Crew Training Wing (ATC), Fairchild AFB, WA: Sep 78.

Basic *Survival Medicine.* Environmental Information Div, Air Training Command Air University Library, Maxwell AFB, AL: Jan 81.

Cold Sea Survival. DTIC Technical Report AD 716389, AMRL-TR-70-72, Aerospace Medical Research Laboratory, Wright Patterson AFB, OH:

Shark. Information Bulletin No. 1,3636th Combat Crew Training Wing, ATC, Fairchild AFB, WA.

The Physiology of Cold Weather Survival. DTIC Technical Report AD 784268, Advisory Group for Aerospace Research and Development Report No. 620, Aerospace Medical Research Laboratory, Wright Patterson AFG, OH: Apr 73.

Toxic Fish and Mollush. Information Bulletin No. 12, Environmental Information Division, Air Training Command Air University Library, Maxwell AFB, AL: Apr 75.

U.S. Navy Publications

"Cold Water Survival, Hypothermia and Cold Water Immersion, Cold Weather Survival," SERE Newsletter, Vol. 1, NO. 7, FASOTRAGRUPAC, Jan 83.

"Deep Water Survival," SERE Newsletter, Vol. 7, No. 8, FASOTRAGRUPAC, Jan 83.

SERE Guide Soviet Far East, Fleet Intelligence Center Pacific, Box 500, FPO San Francisco, CA 96610, Mar 77.

Following are the stock numbers for deck of recognition cards, which were prepared oy the Naval Training Equipment Center, Orlando, FL.

National Stock No.

20-6910-00-820-6702 Device 9H5, Survival Plants, Pacific

20-6910-00-004-9435 Device 9H18 Study Card Set, Northeast Africa/ Mideast (Deck 1, Recognition Wildlife; Deck 2, Recognition Plantlife)

6910-00-106-4337/1 Device 9H15/1, Aviation Survival Equipment

National Stock No.

6919-00-106-4338/2 Device 9H15/2, Aviation Land Survival Techniques

6910-00-106-4352/3 Device 9H15/3, Aviation Sea Survival Techniques

6910-00-820-6702 Device 9H9A Study Cards, Survival Plant Recognition

OTHER PUBLICATIONS

Angier, Bradford. *Feasting Free on Wild Edibles.* Harrisburg, PA Stackpole Co., 1972.

Angier, Bradford. *Field Guide to Edible Wild Plants.* Harrisburg, PA Stackpole Co., 1974.

Angier, Bradford. *How to Stay Alive in the Woods.* Harrisburg, PA Stackpole Co., 1983.

Arnold, Harry L. *Poisonous Plants of Hawaii.* Rutland, VT: Tuttle & Co., 1968.

Buchman, Dian. Herbal Medicine: *The Natural Way to Get Well & Stay Well.* New York: David McKay Co., 1979.

Craighead, Frank C. Jr. and John J. Craghead. *How to Survive on Land and Sea.* Annapolis, MD: Naval Institute Press, 1984.

Ditmars, Raymond L. *Snakes of the World.* New York: MacmiUan Co., 1960.

Embertson, Tame. Pods: *Wildflowers and Weeds in Their Final Beauty.* New York: Charles Scribners Sons, 1979.

Cloudsley-Thompson, John. *Spiders, Scorpions, Centipedes, and Mites.* Oxford, Eng: Pergamon Press, 1958.

Gibbons, Euell. *Stalking the Wild Asparagus.New* York: David McKay Co., 1970.

Grimm, William C. *Recognizing Flowering Plants.* Harrisburg, PA Stackpole Co.

Grimm, William C. *Recognizing Native Shrubs.* Harrisburg, PA Stackpole Co.

Grimm, William C. *The Illustrated Book of Trees.* Harrisburg, PA: Stackpole Co., 1983.

Hall, Alan. *The Wild Food Trail Guide.* New York: Hole, Tinehard & Winston.

Medsger, Oliver P. *Edible Wild Plants.* New York: Macmillan Co., 1972.

Merlin, Mark D. *Hawaiian Forest Plants.* Honolulu: Orientala Publishing Co., 1978.

Minton, Sherman A. and Madge R. Minton. *Venomous Reptiles.* New York: Charles Scribners Sons, 1980.

Moore, Michael. *Medicinal Plants of the Mountain West.* Museum of New Mexico Press, 1979.

Parrish, Henry M. *Poisonous Snakebite in the United States.* New York: Vantage Press.

Russell, Findlay E. *Snake Venom Poisoning.* Philadelphia: J.P. Lippincott Company, 1983.

Squier, Thomas L. *U v i m Off The Land.* Rutland, VT: Academy Press, 1989.

Tomikel, John. *Edible Wild Plants of Pennsylvania and New York.* Pittsburgh, PA: Allegheny Press, 1973.

The Encyclopedia of Organic Gardening. Emmaus, PA: Rodale Press, 1978.

Wild Edible and Poisonous Plants of Alaska. Cooperative Extension Service, University of Alaska and U.S.D.A. Cooperating, Publication No. 28,1981.

INDEX

Made in United States
Orlando, FL
04 November 2026

90202051R00179